Lecture Notes in Electrical Engineering

Volume 689

The book series *Lecture Notes in Electrical Engineering* (LNEE) publishes the latest developments in Electrical Engineering - quickly, informally and in high quality. While original research reported in proceedings and monographs has traditionally formed the core of LNEE, we also encourage authors to submit books devoted to supporting student education and professional training in the various fields and applications areas of electrical engineering. The series cover classical and emerging topics concerning:

- Communication Engineering, Information Theory and Networks
- Electronics Engineering and Microelectronics
- Signal, Image and Speech Processing
- Wireless and Mobile Communication
- Circuits and Systems
- Energy Systems, Power Electronics and Electrical Machines
- Electro-optical Engineering
- Instrumentation Engineering
- Avionics Engineering
- Control Systems
- Internet-of-Things and Cybersecurity
- Biomedical Devices, MEMS and NEMS

For general information about this book series, comments or suggestions, please contact leontina.dicecco@springer.com.

To submit a proposal or request further information, please contact the Publishing Editor in your country:

China

Jasmine Dou, Associate Editor (jasmine.dou@springer.com)

India, Japan, Rest of Asia

Swati Meherishi, Executive Editor (Swati.Meherishi@springer.com)

Southeast Asia, Australia, New Zealand

Ramesh Nath Premnath, Editor (ramesh.premnath@springernature.com)

USA, Canada:

Michael Luby, Senior Editor (michael.luby@springer.com)

All other Countries:

Leontina Di Cecco, Senior Editor (leontina.dicecco@springer.com)

** Indexing: Indexed by Scopus. **

More information about this series at http://www.springer.com/series/7818

Mikhail V. Nesterenko ·
Victor A. Katrich · Yuriy M. Penkin ·
Sergey L. Berdnik · Oleksandr M. Dumin

Combined Vibrator-Slot Structures: Theory and Applications

Theoretical Aspects and Applications

Mikhail V. Nesterenko
V. N. Karazin Kharkiv National University
Kharkiv, Ukraine

Yuriy M. Penkin
V. N. Karazin Kharkiv National University
Kharkiv, Ukraine

Oleksandr M. Dumin
V. N. Karazin Kharkiv National University
Kharkiv, Ukraine

Victor A. Katrich
V. N. Karazin Kharkiv National University
Kharkiv, Ukraine

Sergey L. Berdnik
V. N. Karazin Kharkiv National University
Kharkiv, Ukraine

ISSN 1876-1100   ISSN 1876-1119   (electronic)
Lecture Notes in Electrical Engineering
ISBN 978-3-030-60179-9   ISBN 978-3-030-60177-5   (eBook)
https://doi.org/10.1007/978-3-030-60177-5

This Springer imprint is published by the registered company Springer Nature Switzerland AG
The registered company address is: Gewerbestrasse 11, 6330 Cham, Switzerland

It is impossible to solve the problem at the same level at which it arose. You need to get higher than this problem by rising to the next level.

Albert Einstein

Preface

At present, linear vibrator and slot radiators and coupling slots between electrodynamic volumes are widely used in variety of radio-electronic devices which are structural elements of antennas and antenna-feeder devices. Their multifunctional and extensive applications have become an objective prerequisite for developing theoretical methods aimed at studying electrodynamic characteristics of such devices. A special attention is paid to combined vibrator–slot structures significantly expanding possibilities for formatting electromagnetic fields with predefined spatial-frequency characteristics. Over the past decades, many authors, particularly authors of this book, have developed a modern theory of thin vibrator, slot, and combined vibrator–slot radiators. This theory combines fundamental asymptotic methods and the numerical–analytical approaches based on them which are intended for determining characteristics of stand along vibrator and slot elements. However, the electrodynamics theory of linear vibrator and slot radiators is far from being completed. First of all, this is related to further developing modern technology of transceiver antennas and antenna-feeder devices characterized by the following features: multifunctionality, multi-element applications, integration and modification of structural components to minimize weight and size parameters, electromagnetic compatibility of devices, usage of composites and metamaterials, and formation of required spatial-energy, spatial-polarization, and frequency–energy distributions of electromagnetic fields, including fields in dissipative media.

Multiparameter optimization of complex electrodynamic problems requires effective computing and software resources for mathematical modeling. Therefore, despite the rapid growth of the computing potential of modern computer technology, there is a need to develop new, more efficient methods of electrodynamic problem solutions concerning the design of antenna-feeder systems based on linear vibrator and slot structures with arbitrary geometric and electrophysical parameters. The effectiveness of mathematical modeling is determined by rigorous formulation and solution of corresponding boundary-value problems, the runtime of algorithms under condition that the minimum possible amount of RAM is used. The solution effectiveness directly depends on the degree of analytical developmental work spent during its derivation. That is, the more significant is the analytical component of the method, the higher is the overall effectiveness of its implementation. *In this regard, the development of rigorous numerical–analytical methods of electrodynamic analysis expanding the possibilities of physically correct mathematical modeling to new classes of boundary-value problems has been and remains an important problem for researchers in the field of theoretical and applied radio physics.*

Known numerical–analytical methods for analyzing single and multi-element structures of linear radiators and antenna-feeder devices created on their basis can be divided into two types. The first type includes analytical solution of integral equations for electric and magnetic currents in separate structural elements. The second type refers to the method of moments in which the expressions for matrix elements in the final system of linear algebraic equations (SLAE) can be solved analytically. Of course, methods of the first type are preferable due to their effectiveness and easiness of physical realization. However, their implementation is limited by the complexity of constructing asymptotic solutions of boundary-value problems for systems with large number of different constructive elements.

Characteristic property of electrodynamic problems concerning the linear vibrator and slot radiators is longitudinal dimensions of their elements which are comparable with the operating wavelength in the medium. Therefore, asymptotic long-wave or short-wave (quasi-optical) approximations cannot be applied. In resonant regions, asymptotic solutions of boundary-value problems for perfectly conducting rectilinear vibrators in free space were previously obtained by various methods, namely by the successive iteration method, weight function expansion in a series in powers of a small parameter, variational method, and key equation method. Similarly, the equivalent magnetic current for a slot cut in a thin perfectly conducting infinite screen was determined by using the small parameter method, variational method, and method of successive iterations. The electrodynamic characteristics of multi-element vibrator–slot systems were obtained by the methods of induced electromotive forces (EMF) and magnetomotive forces (MMF). However, within the framework of numerical–analytical methods of boundary-value problem, solutions for a class of linear radiators promising for practical applications were not proposed. This primarily concerns impedance vibrator radiators (scatterers) with irregular geometric and electrophysical parameters located in free space and other electrodynamic volumes, slot radiators cut in walls of various waveguide and resonator sections with perfectly conducting and impedance boundaries, and combined multi-element vibrator–slot structures. Beside the principles of asymptotic solutions of equation for electric and magnetic currents by the methods of induced EMF, MMF or the hybrid method of electro-magneto-motive forces (EMMF) for analyzing combined vibrator–slot structures have not been generalized.

The problem solutions concerning the electrodynamic characteristics of impedance vibrator and slot radiators with arbitrary geometric and electrophysical parameters were obtained within a framework of unified methodological approach to constructing asymptotic solutions of integral equations. This approach allowed us to investigate several combined vibrator–slot structures. The research results have revealed the possibilities of using these structures as basic elements of modern antenna-waveguide devices with advanced technical characteristics and function-alities operating in the ranges from meter to millimeter wavelengths.

Investigating transient near fields of electrically short radiators is usually carried out by the method of vector potential in time domain based on the small parameter approach. The features of electromagnetic fields in the near zone are applied to the study of interaction between elements of ultrawideband vibrator–slot structures similar to the Clavin element.

The monograph consists of nine chapters and five appendixes. For the conve-nience of working with the material of the book, a list of authors' literary sources is highlighted in a separate list (Authors' Publications List (APL)).

Chapter 1 is introductory in nature. Basic equations of macroscopic electrody-namics and methods for constructing Green's functions are presented. Besides issues related to surface impedances of various metal–dielectric structures are discussed. Asymptotic formulas for determining complex surface impedances of vibrators are presented in Subsection 1.3.2.6. These formulas allow us to calculate impedances for following metal–dielectric structures: a solid lossy cylindrical conductor, periodically corrugated or ribbed cylindrical conductor, cylindrical conductor coated by a magnetodielectric, cylindrical conductor with periodic dielectric transverse inserts, and dielectric cylinder. Formulas defining relationships between vibrator structural parameters and its surface impedance were used in mathematical models for various radiators. The information presented in this sec-tion allows the reader to perceive the information in further sections of the book without searching for special hard-to-reach literature.

In Chap. 2, a problem of the excitation (scattering or radiation) of electromag-netic fields by material bodies of finite dimensions in presence of coupling holes between two electrodynamic volumes is formulated in most general form. Initial systems of integral equations relative to densities of surface electric currents on vibrator and equivalent magnetic currents on coupling holes were derived based on the impedance boundary conditions on the vibrator surfaces and the continuity conditions of tangential components of magnetic fields at coupling holes. It was shown that the problem solution should be better based on integral equations with Green's function kernels for Hertz vector potentials. The asymptotic solutions of the equations in various electrodynamic volumes were obtained by using a general approach consisting in isolation of irregular part of the Green's presented by the Green's function of the free space. A physically correct transition from the derived integral equations to one-dimensional equations is made for currents on thin linear impedance vibrators with irregular geometric and electrophysical parameters along their length, in narrow slots arbitrary located relative to boundaries of cor-responding electrodynamic volumes, and in combined vibrator–slot structures. The fundamentals of the method of moments are presented, the advantages and disad-vantages of various approximate analytical methods for solving the integral equa-tions for currents are analyzed, and the application of the generalized method of induced EMF for studying multi-element vibrator–slot structures is considered. New aspects of thin impedance vibrator theory proposed by the authors are considered.

In Chap. 3, solutions of integral equations for currents in waveguide structures with single vibrator and slot elements are obtained by the averaging method and generalized methods of induced EMF and MMF. The choice of approximating functions for currents is justified by the results of comparative analysis of calculated and experimental data. A problem of electromagnetic wave scattering by a vibrator system in rectangular waveguides is also solved to demonstrate how transition from single to multi-element structures can be made.

Chapter 4, entitled "Combined Radiating Vibrator-Slot Structures in Rectangular Waveguides", is devoted to developing mathematical models and studying electrodynamic characteristics of structures consisting of transverse slot in a wide wall of a rectangular waveguide with impedance vibrators. Structures with or without vibrator–slot field interactions were considered; the second variant corresponds to a case when vibrator–slot field interactions are absent due to polarization decoupling. The possibility of controlling over a wide range reflection and transmission coefficients in waveguides with vibrator–slot structures by using passive vibrators of constant length with various distribution function of surface impedance is confirmed. It is also shown that the slot radiation coefficient in such structures can be close to unity.

In Chap. 5, T-shaped E-plane junctions consisting of infinite and semi-infinite waveguides coupling through a transverse slot cut in a wide wall of the infinite waveguide are considered. One or two impedance monopoles are located near the slot inside the main infinite waveguide. A characteristic feature of the T-junction structure model is simultaneous application of monopoles with variable surface impedance and the end wall of the semi-infinite waveguide with constant impedance distributed over its surface. Multiparameter studies of energy characteristics of T-junctions were carried out in single-mode regime of the coupling waveguides.

The main motive for the complicating the T-junction design is defined by the authors' desire to investigate conditions allowing its two-resonant operating mode by using two impedance vibrators inside the main waveguide. This possibility has been successfully confirmed by numerical modeling. Conditions necessary for transmitting the main part of input power from the T-junction main arm to the side waveguide at two resonant frequencies were determined. It was shown that impedance coatings of vibrators and end wall of side waveguide can be used as effective control elements for dividing the power between the T-junction output arms. It is also established that such a coating with frequency-dependent heterogeneous impedance can provide a three-resonant mode of power transmission from the main to side arm of T-junction.

In Chap. 6, a problem concerning of electromagnetic wave radiated by a Clavin-type element into a half space above an infinite perfectly conducting space is solved. A Clavin element consists of a longitudinal or transverse slot cut in a waveguide broad wall and two passive impedance vibrators located on both sides of slot longitudinal axis. The influence of vibrator length and distances between vibrators on the directional characteristics of the Clavin-type elements is analyzed under conditions that the relative level of E-plane lateral radiation and difference of the RP main lobe widths in the main polarization planes at the level of -3 dB are taken into account. It is shown that the radiation coefficient, directivity, and gain of the Clavin element can be controlled in a wide range by variating electric length of the vibrator and the distance between the vibrators ensuring a low level of radiation in the slot plane.

In Chap. 7 entitled "Combined Vibrator-Slot Structures Located on a Perfectly Conducting Sphere", two key problems are preliminarily solved: (1) a problem of electromagnetic waves radiated by a slot cut in the perfectly conducting sphere excited by the semi-infinite rectangular waveguide with/without impedance coating of the waveguide end wall; (2) a problem of electromagnetic wave excitation by a current distribution on the radial impedance vibrator located on the sphere. Based on the solution of these key problems, a combined vibrator–slot structure known as the Clavin-type element on a sphere is studied. To form narrow frequency–energy and spatial characteristics of spherical antennas, the configuration with a single or two-slot resonant diaphragm in the main waveguide forming a reentrant resonator with the waveguide end wall is considered.

In Chap. 8, the application of combined two-frequency antenna arrays with diode switching of the slot and vibrator activities is substantiated based on the Clavin-type vibrator–slot structures. It is proposed to use the Clavin-type elements with passive or active monopoles as the array radiators at the main or alternative frequencies. The features of radiation field formed by the combined antenna array at the fundamental and alternative frequencies are analyzed. It was shown that the alternative wavelength should not exceed the main wavelength at more than 25%. The electrodynamically rigorous mathematical model of a vibrator–slot structure, consisting of a narrow radiating slot in end wall of a rectangular waveguide and several thin impedance vibrators placed over the infinite plane, is also presented in this chapter. Simulation results of internal and external electrodynamic characteristics of antennas with optimized structural parameters confirmed the possibility of constructing combined radiating structures of Yagi–Uda type operating in microwave and extremely high frequencies. It was shown that the radiating structures with optimized parameters in narrow frequency bands can provide maximum directional coefficients with satisfactory matching of waveguide transmission line. Note that combined structures with vibrators characterized by various types of impedances can be studied within the framework of the developed model.

Chapter 9 is devoted to study of transient field of short vibrators and vibrator–slot structures similar to the Clavin element. On the basis of vector potential method in time domain, the analytical solution for the field of Hertzian dipole is obtained by taking into account additional terms in vector potential expansion. It helps us to increase the precision of the solution in the vicinity of the radiator in comparison with the classical expressions. Additionally, the new notation describes more clearly the process of electromagnetic wave formation near the source of given electrical current. The patterns at different distances and the wave zone boundary are analyzed for different time dependences of excitation current. The radiation in the case of more complicated impulse current distributions is studied for the case of a rail launcher. The research of ultrawideband vibrator–slot structure similar to

Clavin element is grounded on previous near-field research of Hertzian dipole. Its pattern, bandwidth, and time forms of radiated waves are presented.

Expressions for the electric and magnetic Green's functions of vector potentials for electrodynamic volumes considered in the monograph can be found in Appendix A. Representations of Green's functions for infinite medium in various systems of orthogonal coordinates are presented in Appendix B. To obtain formulas for the surface impedance calculation, solution of equations defining the electromagnetic fields for the magnetodielectric layer with linear variation of material parameters placed on the conducting plane is represented in Appendix C. Appendix D is devoted to the proof of theorems concerning thin vibrators considered in Sect. 2. The relationships between dimensions of physical quantities in the CGS and SI unit systems are given in Appendix E.

The monograph is written in an academic style; hence, it can be useful for teachers and students of technical universities. The results of numerical modeling presented in the book and asymptotic formulas obtained can be useful to engineers engaged in developing antenna and waveguide systems.

The authors would like to express sincere gratitude to Anatoliy M. Naboka for editing the English text.

Kharkiv, Ukraine

Mikhail V. Nesterenko
Victor A. Katrich
Yuriy M. Penkin
Sergey L. Berdnik
Oleksandr M. Dumin

Contents

1 Excitation of Electromagnetic Waves in Volumes with Coordinate
   Boundaries 1
   1.1 Helmholtz Equations in Electrodynamics 1
   1.2 Exact Boundary Conditions for Electromagnetic Fields 4
   1.3 Approximate Boundary Conditions for Electromagnetic Fields ... 5
        1.3.1 Impedance Boundary Conditions and the Limits
               of Their Correct Application 5
        1.3.2 Surface Impedance of Metal-Dielectric Structures 8
   1.4 Uniqueness Theorem and Duality Principle for Areas
        with Impedance Boundaries 35
        1.4.1 Features of the Application of the Uniqueness Theorem... 36
        1.4.2 Features of the Application of the Principle of Duality.... 38
   1.5 Tensor Green's Functions of the Vector Helmholtz Equation
        for Hertz Potentials................................. 39
        1.5.1 Properties of the Tensor Green's Function............. 39
        1.5.2 Construction of the Green's Tensor in Orthogonal
               Curvilinear Coordinate Systems 42
        1.5.3 Green's Tensor for Areas with Cylindrical Boundaries.... 44
   References 46

2 General Issues of the Theory of Thin Impedance Vibrators
   and Narrow Slots in a Spatial-Frequency Representation 51
   2.1 Problem Formulation and Initial Integral Equations 51
        2.1.1 Green's Function as Integral Equation Kernel 55
        2.1.2 Integral Equations for Electric and Magnetic Currents
               in Thin Vibrators and Narrow Slots 59
   2.2 Methods for Solving Integral Equations for Currents 63
        2.2.1 Basics of the Method of Moments 63
        2.2.2 Approximate Analytical Methods for Solving Current
               Integral Equations 67

2.3   New Aspects in the Development of the Theory of Thin
       Impedance Vibrators   78
       2.3.1   Presentation of Two Fundamental Theorems
               of the Vibrator Theory   78
       2.3.2   Alternative Green's Function of the Electric Field
               on the Vibrator   81
       2.3.3   Radiation of a Vibrator Located Above a Plane   84
       References   91

3   **Solution of Current Equations for Isolated Vibrator and Slot
     Scatterers**   95
     3.1   Vibrator with Variable Surface Impedance in Free Space   95
            3.1.1   Solving the Current Equation by the Averaging
                    Method   95
            3.1.2   Solving the Current Equation by the Generalized Method
                    of Induced EMF   100
            3.1.3   Justification of the Choice of Approximating Functions
                    for Current   101
     3.2   Vibrator with Variable Surface Impedance in a Rectangular
            Waveguide   102
            3.2.1   Solving the Current Equation by the Averaging
                    Method   102
            3.2.2   Solving the Current Equation by the Generalized Method
                    of Induced EMF   105
            3.2.3   Numerical and Experimental Results   108
     3.3   System of Two Impedance Monopoles in a Rectangular
            Waveguide   110
            3.3.1   Problem Formulation and Solution the System
                    of Equations for Currents   110
            3.3.2   Numerical and Experimental Results   113
     3.4   Narrow Slots in Rectangular Waveguide Walls   118
            3.4.1   Solving the Current Equation by the Averaging
                    Method   118
            3.4.2   Symmetric Transverse Slot in the Broad Waveguide
                    Wall   120
            3.4.3   Longitudinal Slot in the Waveguide Broad Wall   123
     References   126

4   **Combined Radiating Vibrator-Slot Structures in Rectangular
     Waveguide**   129
     4.1   Two-Element Vibrator-Slot Structure   129
            4.1.1   Problem Formulation and Solution the Equations
                    for Currents   129
            4.1.2   Numerical and Experimental Results   134

4.2   Three-Element Vibrator-Slot Structure Without Interaction
        Between the Vibrators and Slot   142
        4.2.1   Problem Formulation and Solution   142
        4.2.2   Numerical and Experimental Results   146
4.3   Three-Element Vibrator-Slot Structures with Interaction
        Between the Vibrators and Slot   151
        4.3.1   Problem Formulation and Solution   151
        4.3.2   Numerical and Experimental Results   153
References   156

**5   T-Junctions of Rectangular Waveguides with Vibrator-Slot
        Structures in Coupling Areas**   157
5.1   *E*-Plane T-Junction of Equal-Size Waveguides with the
        Two-Element Vibrator-Slot Structure   157
        5.1.1   Problem Formulation and Solution   157
        5.1.2   Numerical and Experimental Results   162
5.2   *E*-Plane T-Junction with Two Waveguides of Different
        Cross-Sections and Two-Element Vibrator Slot-Structure   169
        5.2.1   Features of the Problem Formulation   169
        5.2.2   Numerical and Experimental Results   170
5.3   Energy Characteristics of the T-Junction with the Three-Element
        Vibrator-Slot Structure   174
        5.3.1   Problem Formulation and Solution   176
        5.3.2   Numerical and Experimental Results   180
References   188

6   Waveguide Radiation of the Combined Vibrator-Slot Structures . . .   191
6.1   Problem Formulation and Initial Equations in the General Case
        for Longitudinal Slot Element   191
6.2   Clavin Element for the General Case   195
6.3   Vibrator-Slot Radiator Based on a Hollow Rectangular
        Waveguide   200
        6.3.1   Numerical and Experimental Results   204
6.4   Vibrator-Slot Radiator Based on a Rectangular Waveguide
        with the Tuning Vibrator (Numerical Results)   213
6.5   Problem Formulation and Initial Equations in the General Case
        for Transverse Slot Element   214
        6.5.1   Resonant Radiation Conditions for Structure with Clavin
                    Element and Transverse Slot   216
References   220

**7   Combined Vibrator-Slot Structures Located on a Perfectly
        Conducting Sphere**   223
7.1   Resonant Slot Radiator on a Sphere   223
        7.1.1   Problem Formulation and Solution   225

         7.1.2   Radiation Fields of the Slotted Spherical Antenna 232
         7.1.3   Numerical and Experimental Results 234
   7.2   Radial Impedance Vibrator Located on a Sphere 241
         7.2.1   Problem Formulation and Solution 241
         7.2.2   Numerical Results 246
   7.3   Combined Vibrator-Slot Structure Located on a Sphere 246
         7.3.1   Solution of the External Electrodynamic Problem 249
         7.3.2   Solution of the Equation System for Currents 250
         7.3.3   Fields Radiated by the Vibration-Slot Structure 251
         7.3.4   Numerical Results 252
   References 254

8   Combined Vibrator-Slot Radiators in Antenna Arrays 257
   8.1   Commutation Modes of Clavin Radiators in Antenna Arrays 259
         8.1.1   Radiation Fields of the Combined Antenna Array 261
         8.1.2   Numerical Results 264
   8.2   Yagi-Uda Combined Radiating Structures 267
         8.2.1   Formulation and Solution of the Diffraction Problem 267
         8.2.2   Numerical Results 271
   References 274

9   Ultrawideband Vibrator-Slot Structures 277
   9.1   Transient Near Field of Herzian Dipole 278
         9.1.1   Statement and Initial Expressions of the Problem 278
         9.1.2   Components of Radiated Field 281
         9.1.3   Numerical Simulations 282
         9.1.4   Free Field Formation 286
   9.2   Impulse Field of the System of Short Radiators 290
         9.2.1   Field of Two Parallel Dipoles 291
         9.2.2   Radiation of the System of Four and Six Dipoles 292
   9.3   Ultrawideband Combined Vibrator-Slot Radiator 294
         9.3.1   Statement of the Problem for Combined Radiator 295
         9.3.2   Results of Numerical Simulation 296
         9.3.3   Characteristics of Radiation of the Ultrawideband
                 Analogue of Clavin Radiator 297
   References 303

**Appendix A: Green's Functions of the Considered Electrodynamic
              Volumes** 305

**Appendix B: Representations of the Green's Function of Unlimited
              Space in Orthogonal Coordinate Systems** 315

**Appendix C: Strict Solution of Field Equations for a
              Magnetodielectric Layer on a Conducting Plane
              with a Linear Law of Change Permittivity** 319

Appendix D: Proofs of the Theorems for Thin Vibrator Radiators 321

**Appendix E: Electromagnetic Values in CGS and SI Systems
of Units** 329

Bibliography 333

Chapter 1
Excitation of Electromagnetic Waves in Volumes with Coordinate Boundaries

This chapter briefly presents basic equations, boundary conditions and tensor Green's functions for studying electromagnetic fields, which will be used throughout the book. This chapter with comments on some aspects of electrodynamics theory, will allow a reader to read the book without searching for additional references.

1.1 Helmholtz Equations in Electrodynamics

We will consider only harmonic oscillations, bearing in mind that non-harmonic oscillations can be decomposed into Fourier series or integrals. We also assume that all relevant quantities depend on time as $e^{i\omega t}$, where ω is the circular frequency of the monochromatic process. Then the relationships between vectors of electric and magnetic fields strength, $\vec{E}(t)$ and $\vec{H}(t)$, and their complex amplitudes, $\vec{E}$ and $\vec{H}$, can be written as

$$\vec{E}(t) = \mathrm{Re}\left\{ \vec{E}\, e^{i\omega t} \right\}, \qquad \vec{H}(t) = \mathrm{Re}\left\{ \vec{H}\, e^{i\omega t} \right\}, \tag{1.1}$$

where "Re" is the real part of the complex vector standing in the curly bracket, $\vec{E}(t)$ and $\vec{H}(t)$ are the intensity vectors of the electric and magnetic fields, $\vec{E}$ and $\vec{H}$ are the corresponding complex vector amplitudes. The complex amplitude of all time-dependent physical quantities in electrodynamics equations are defined in a similar way.

Let us introduce, along with real extraneous electric current $\vec{j}^E$, virtual extraneous magnetic current with densities $\vec{j}^M$, respectively Then for a stationary, homogeneous, and isotropic medium, the Maxwell equations, symmetric with respect to electric and magnetic quantities, can be written as

© The Author(s), under exclusive license to Springer Nature Switzerland AG 2020
M. V. Nesterenko et al., *Combined Vibrator-Slot Structures: Theory and Applications*, Lecture Notes in Electrical Engineering 689,
https://doi.org/10.1007/978-3-030-60177-5_1

$$\text{rot}\,\vec{E} = -ik\mu_1\vec{H} - \frac{4\pi}{c}\vec{j}^M, \qquad \text{rot}\,\vec{H} = ik\varepsilon_1\vec{E} + \frac{4\pi}{c}\vec{j}^E, \qquad (1.2)$$

where ε_1 and μ_1 are the complex permittivity and permeability of the medium in which the electromagnetic fields are excited, $k = \omega/c$ is the wave number, and $c \approx 2.998 \times 10^{10}$ cm/s is the speed of light in vacuum.

Throughout the book, the absolute Gaussian system of units will be used as the most convenient in theoretical physics as compared with the SI unit system. The Maxwell equations (1.2) in the SI system include four vectors: $\vec{E}$, $\vec{D}$, $\vec{H}$ and $\vec{B}$ defining the field at each point in space have different dimensions. Here $\vec{D}$ and $\vec{B}$ are the vector of electric and magnetic inductions. In the vacuum, the four vectors are connected in pairs by the relations: $\vec{D} = \varepsilon_0\vec{E}$ and $\vec{B} = \mu_0\vec{H}$, where the coefficients ε_0 and μ_0 play the role of the permittivity and permeability characterizing the electromagnetic properties of vacuum. These coefficients, devoid of real physical meaning, are included in all relationships for the fields, which is difficult to reconcile. However, if necessary, a mutual transition the CGS and SI systems is always possible (see Appendix E).

The charges are usually not included in Eq. (1.2), since, by virtue of the continuity equation the density of electric ρ^E and magnetic ρ^M charges are uniquely determined by currents as

$$\text{div}\,\vec{j}^E + i\omega\rho^E = 0, \quad \text{div}\,\vec{j}^M + i\omega\rho^M = 0. \qquad (1.3)$$

Applying to formula (1.3) the divergence operations and the identity div rot $\equiv 0$ to Eq. (1.2), we obtain

$$\text{div}\left(\varepsilon_1\vec{E}\right) = 4\pi\rho^E, \quad \text{div}\left(\mu_1\vec{H}\right) = 4\pi\rho^M. \qquad (1.4)$$

Thus, the charges and corresponding field equations may not be considered. Thus, both the charges themselves, and corresponding field equations, have no independent meaning and, therefore, are usually not considered. For the same reason, the wave equations for scalar potentials related to the vector potentials by the Lorentz or gradient invariance conditions will not be considered.

As can be seen, Eq. (1.2) are coupled with respect to complex vector amplitudes $\vec{E}$ and $\vec{H}$. Therefore, it seems convenient to convert the equations so that they include only one vector

$$\text{rot rot}\,\vec{E} = k^2\varepsilon_1\mu_1\vec{E} - \frac{4\pi}{c}\left(ik\mu_1\vec{j}^E + \text{rot}\,\vec{j}^M\right),$$

$$\text{rot rot}\,\vec{H} = k^2\varepsilon_1\mu_1\vec{H} - \frac{4\pi}{c}\left(ik\varepsilon_1\vec{j}^M - \text{rot}\,\vec{j}^E\right), \qquad (1.5)$$

Equation (1.5) can be easily transformed to inhomogeneous Helmholtz vector equations relative only $\vec{E}$ or $\vec{H}$ in the form

$$\Delta \vec{E} + k^2 \varepsilon_1 \mu_1 \vec{E} = \frac{4\pi}{c} \left(ik\mu_1 \vec{j}^E - \frac{1}{ik\varepsilon_1} \text{graddiv} \, \vec{j}^E + \text{rot} \vec{j}^M \right),$$

$$\Delta \vec{H} + k^2 \varepsilon_1 \mu_1 \vec{H} = \frac{4\pi}{c} \left(ik\varepsilon_1 \vec{j}^M - \frac{1}{ik\mu_1} \text{graddiv} \, \vec{j}^M - \text{rot} \vec{j}^E \right), \qquad (1.6)$$

where $\Delta \equiv \text{graddiv} - \text{rot rot}$ is the Laplace operator.

Equation (1.6) in many cases can be inconvenient in view of their right-hand side complexities. Therefore, it is useful to introduce electrodynamic vector potentials of the electric $\vec{A}^E$ and magnetic $\vec{A}^M$ types, or the Hertz polarization vector potentials, $\vec{\pi}^E$ and $\vec{\pi}^M$ which are interconnected, under the accepted assumptions, by the relations

$$\vec{A}^E = ik\varepsilon_1\mu_1 \vec{\pi}^E, \quad \vec{A}^M = ik\varepsilon_1\mu_1 \vec{\pi}^M. \qquad (1.7)$$

The formulas (1.7) emphasize the equivalence of both types of vector potentials. Therefore, the choice of one or another presentation, in essence, is a matter of habit or tradition. Further, we will use the equations for the Hertz vector potentials allowing to express the vectors $\vec{E}$ and $\vec{H}$ by the following relation

$$\vec{E} = \text{grad div} \, \vec{\pi}^E + k^2 \varepsilon_1\mu_1 \vec{\pi}^E - ik\mu_1 \text{rot} \, \vec{\pi}^M,$$

$$\vec{H} = \text{grad div} \, \vec{\pi}^M + k^2 \varepsilon_1\mu_1 \vec{\pi}^M + ik\varepsilon_1 \text{rot} \, \vec{\pi}^E. \qquad (1.8)$$

Substituting (1.8) into Eq. (1.6), we obtain the following inhomogeneous Helmholtz equations for Hertz vector potentials:

$$\Delta \vec{\pi}^E + k^2 \varepsilon_1\mu_1 \vec{\pi}^E = -\frac{4\pi}{i\omega\varepsilon_1} \vec{j}^E,$$

$$\Delta \vec{\pi}^M + k^2 \varepsilon_1\mu_1 \vec{\pi}^M = -\frac{4\pi}{i\omega\mu_1} \vec{j}^M. \qquad (1.9)$$

As can be seen, the right-hand part of Eq. (1.9) includes the densities of the excitation currents, therefore these equations are more suitable for solving electrodynamic problems rather than Eq. (1.6).

Further we will use the following notations: the vector $\vec{F}^{E(M)}$ denotes the vector either $\vec{F}^E$ or $\vec{F}^M$, and $\varepsilon_1(\mu_1)$ denotes either ε_1 or μ_1. Then, for example, Eq. (1.9) can written as

$$\Delta \vec{\pi}^{E(M)} + k^2 \varepsilon_1\mu_1 \vec{\pi}^{E(M)} = -\frac{4\pi}{i\omega\varepsilon_1(\mu_1)} \vec{j}^{E(M)}. \qquad (1.10)$$

The above equations are valid in any orthogonal coordinate system, allowing application of differential operators.

1.2   Exact Boundary Conditions for Electromagnetic Fields

The media parameters ε_1 and μ_1 involved in the equations of Sect. 1.1 as constant, can vary continuously or discreetly in space. For example, the parameters can have discontinuities at an interface between two media. The vectors of the electromagnetic fields can also have discontinuities at the interface, therefore, Eq. (1.2) must be supplemented with boundary conditions. As is known from classical electrodynamics, the boundary conditions at a smooth interface between two media have the following form

$$\mu_1\left(\overrightarrow{H}_1, \vec{n}\right) - \mu_2\left(\overrightarrow{H}_2, \vec{n}\right) = 4\pi\rho_S^M, \tag{1.11}$$

$$\varepsilon_1\left(\overrightarrow{E}_1, \vec{n}\right) - \varepsilon_2\left(\overrightarrow{E}_2, \vec{n}\right) = 4\pi\rho_S^E, \tag{1.12}$$

$$\left[\vec{n}, \overrightarrow{E}_1\right] - \left[\vec{n}, \overrightarrow{E}_2\right] = \frac{4\pi}{c}\vec{j}_S^M, \tag{1.13}$$

$$\left[\vec{n}, \overrightarrow{H}_1\right] - \left[\vec{n}, \overrightarrow{H}_2\right] = -\frac{4\pi}{c}\vec{j}_S^E, \tag{1.14}$$

where indices 1 and 2 correspond to the first and second media, $\rho_S^{E(M)}$ are the surface charge densities, $\vec{j}_S^{E(M)}$ are the surface current densities, and $\vec{n}$ is the vector of unit normal to the interface which is directed to the second medium. Equations (1.11) and (1.12) define variation of the normal components of the electromagnetic field at the interface between the two media. Note that conditions (1.13) and (1.14) follow directly from Maxwell equations (1.2).

If the extraneous electric currents on the interface surface S are absent, the condition (1.14) simplifies to

$$\left[\vec{n}, \overrightarrow{H}_1\right] = \left[\vec{n}, \overrightarrow{H}_2\right], \tag{1.15}$$

which defines the continuity of the tangential components of the magnetic field on the interface surface. Similarly, if $\vec{j}_S^M = 0$ the condition (1.13) defines the continuity of the tangential components of the electric fields at the interface.

If the conductivity of the second medium has infinite, i.e., the medium is perfectly conductive), the fields $\overrightarrow{E}_2$ and $\overrightarrow{H}_2$ are identically zero and the boundary condition can be written as

$$\left[\vec{n}, \vec{E}_1\right]\Big|_S = 0. \tag{1.16}$$

If the body conductivity is great, but finite, the exact boundary conditions (1.13) and (1.14) should be satisfied on the body surface. Hence, to determine electromagnetic fields outside a well-conducting body, the Maxwell equations must be solved both outside and inside the body using the boundary conditions at the interface between the two media. Such a problem is more complicated than that of determining the electromagnetic field only in one media and the boundary condition on the surface. Therefore, it is desirable to replace the boundary conditions (1.13) and (1.14) by the boundary condition connecting the values of the field vectors on the surface only in one media. Shchukin-Leontovich approximate boundary conditions or impedance boundary conditions [1, 2] satisfy this requirement.

1.3 Approximate Boundary Conditions for Electromagnetic Fields

1.3.1 Impedance Boundary Conditions and the Limits of Their Correct Application

The one-sided impedance boundary conditions allow to reduce the number of interfacing electrodynamic volumes which should be taken in the problem solution. Eliminating the need to determine fields inside the adjacent metal-dielectric elements at the problem formulation level is the main advantage of the impedance approach. The Shchukin-Leontovich impedance condition on the boundary surface S can be written in the following form [2, 3]

$$[\vec{n}, \vec{E}]|_S = \overline{Z}_S[\vec{n}, [\vec{n}, \vec{H}]]|_S, \tag{1.17}$$

where $\vec{E}$ and $\vec{H}$ are the vectors the electric and magnetic harmonic fields, $\vec{n}$ is the impedance surface normal, directed inside the impedance region, $\overline{Z}_S = Z_S/Z_0$ is the normalized surface impedance, and $Z_0 = 120\pi$ Ohm is the resistance of free space.

If the $\overline{Z}_S = 0$, i.e., interface surface is perfectly conducting, the formula (1.17) is reduces to (1.16). It should be noted that the boundary condition (1.17) is approximate, since, in this case, the solution of the electrodynamic problem represents the first term of the asymptotic expansion of the exact solution [4] in powers of the small parameter

$$|\overline{Z}_S| \ll 1. \tag{1.18}$$

Since only the tangential components are included in the boundary condition (1.17), there exists some restrictions on the surface S geometry. It is evident that condition (1.17) holds if the surface curvature radius is much greater than the length of the incident wave. The conditions which take into account the interface curvature can be read as [2, 5]:

$$E_{\tau 1} = \overline{Z}_S \left(1 + \frac{\chi_1 - \chi_2}{2ik\sqrt{\varepsilon_1 \mu_1}} \right) H_{\tau 2} \Big|_S,$$

$$E_{\tau 2} = -\overline{Z}_S \left(1 + \frac{\chi_2 - \chi_1}{2ik\sqrt{\varepsilon_1 \mu_1}} \right) H_{\tau 1} \Big|_S, \qquad (1.19)$$

where χ_1 and χ_2 are the main Gaussian curvatures of the surface, E_τ and H_τ are the tangential components of the electromagnetic fields on the interface surface.

The surface impedance of an electromagnetic field is usually interpreted as relationships determining links between the tangential components of the complex amplitudes on the surface S [6]. If the impedance value does not depend on the incidence angle and incident wave polarization, it is known as extraneous impedance [3]. If the impedance value does not depend on the wave incidence angle, but depends on the wave polarization and spatial orientation of the surface S, the surface impedance is a two-dimensional second-rank tensor which components are extraneous impedances. In the general case, a concept of anisotropic surface impedance is introduced as matrix

$$\widehat{\overline{Z}}_S = \left\| \begin{matrix} \overline{Z}_{S11} & \overline{Z}_{S12} \\ \overline{Z}_{S21} & \overline{Z}_{S22} \end{matrix} \right\|, \quad \overline{Z}_{Sjk} = \overline{R}_{Sjk} + i\overline{X}_{Sjk}, \, j, k \in \{1, 2\}, \qquad (1.20)$$

under conditions that inequalities

$$\overline{R}_{S11} \geq 0, \quad \overline{R}_{S22} \geq 0, \quad 4\overline{R}_{S11}\overline{R}_{S22} \geq \left| \overline{Z}_{S12} + \overline{Z}_{S21}^* \right|^2 \qquad (1.21)$$

hold. In (1.20) and (1.21) $\overline{Z}_{S21}^*$ is complex conjugate of $\overline{Z}_{S21}$. The inequalities (1.21) ensure that additional energy sources on the surface S and energy flows through this surface are absent. Of course, the impedance $\overline{Z}_S$ in (1.17) and (1.20) must be replaced by the tensor $\widehat{\overline{Z}}_S$. It should be emphasized that the surface S, on which the impedance boundary condition should be satisfied, does not have to coincide with the real impedance boundary surface and can be considered as a conditional boundary surface. A spectral analysis of complex structures and media should require introduction of a partial impedance, the value of which in the general case depends on the frequency and the number of spatial harmonics in the electromagnetic field representation. Such impedance problems are beyond the scope of this book.

First, let us consider possible formulations of the impedance conditions and the solution accuracies they can provide. According to results obtained in [4], the boundary condition (1.17) is applicable when the following requirements are met: a penetration depth of electric fields into an impedance material and a field wavelength should be small compared to an incident wave wavelength, a distance from a field source, and curvature radii of a boundary surface S. In addition, variations of the material parameters of the impedance layer at distances comparably with the field wavelength or penetration depth should be small. In the general case, the accuracy of the formula (1.17) was estimated to be proportional to $\sim \left|\overline{Z}_S\right|^2$, since only the first term of the solution obtained as asymptotic series with respect to the normalized impedance $\overline{Z}_S$ was used. Leontovich obtained a similar estimate by comparing the plane wave reflection obtained in the impedance approximation with that of the exact Fresnel solution [1]. However, for a certain class of propagation models, corrections to (1.17) are proportional to cubic but not quadratic terms in the small parameter $\left|\overline{Z}_S\right|$ [4].

Even though the accuracy estimates of the condition (1.17) were obtained on the basis of the skin-effect theory for of conducting body surfaces [1, 4], they can be uniquely extended to the more general case of impedance domains [3]. All the above requirements can be integrated into one of a purely physical nature: the field at the impedance surface must be a plane wave propagating in a direction normal to a boundary S. This condition is always fulfilled for electrically thin impedance structures, including film coatings.

However, the presented requirement cannot be strictly fulfilled if the impedance surface is excited by waves at small incident or Brewster angles. In the first case, the reflected and refracted rays are gliding near the surface, while in the second case they must be mutually perpendicular. In both cases, the directions of the refracted rays do not coincide with the direction of the boundary surface normal. Therefore, it is customary to distinguish between three separate cases of the impedance conditions depending upon the wave incidence angle: (1) normal incidence, when the Shchukin-Leontovich condition are valid, (2) Brewster angle incidence, and (3) tangential incidence. Apparently for the first time, the condition (1.17) was corrected and used in [7] where problem of the reflection of electromagnetic waves from the surface of real soil at angles close to the sliding incidence angle was solved. Subsequently, similar cases were studied for a number of other media including inhomogeneous plasma. The analysis of various variants of the approximate impedance boundary conditions are presented in [8–12].

The accuracy justification of the impedance boundary condition (1.17) cannot give a complete answer to the question: with what accuracy the specific characteristics of the wave fields can be calculated for arbitrary angles of a plane wave incidence on the media interface. General conclusions concerning for the smallest errors relative to exact values for the whole range of incidence angles were formulated in [13] for perpendicular and parallel polarizations relative to the interface surface. For the perpendicular polarization, the reflection coefficients should be calculated based on the Shchukin-Leontovich approximate boundary conditions. For a wave of parallel polarization, the formulas valid for the Brewster angle are preferable. These conclusions were made based on the exact formulas obtained in [14].

The above analysis of the boundary condition (1.17) accuracy was performed under condition that surface impedance $\overline{Z}_S$ presented by a power series only terms proportional to the first degree of a small parameter were taken into account. However, this simplification permits only the small $\overline{Z}_S$, and secondly, and does not provide the necessary accuracy for solving the diffraction problem when the wave is incident at the Brewster angle or tangential to the surface interface. These shortcomings can be eliminated by the method proposed in [15], where a generalized impedance approximation was formulated as:

$$\vec{E}_\tau + \overline{Z}_S\left[\vec{H}_\tau, \vec{n}\right] + \frac{1}{2}\overline{Z}_S^3 n^2\left(N_1 + \sum_{s=1}^{\infty}\left(\overline{Z}_S n\right)^{2s}\frac{(2s-1)!!}{2^s(s+1)!}N_{2s+1}\right)\vec{H}_\tau = 0,$$

(1.22)

where the matrices $N_m (m = 2s + 1)$ are defined as

$$N_m = \left\|\begin{array}{cc} 0 & 1 \\ m & 0 \end{array}\right\|,$$

(1.23)

$n = k/k_0$ is a dimensionless refraction parameter, k and k_0 are wave numbers in the impedance medium and external space, respectively. In the first approximation, Eq. (1.22) coincides with the boundary condition (1.17), but the vectors $\vec{E}_\tau$ and $\vec{H}_\tau$ are related through the refraction factor n, which in a reflection problem is uniquely connected to a wave incidence angle.

Thus, the approximate Shchukin-Leontovich condition (1.17) valid for a small surface impedance $\overline{Z}_S$, is generalized for arbitrary $\overline{Z}_S$ as the series expansion (1.22), which expand the applicability of the impedance approach. Since the exact boundary condition (1.22) is decomposed in a series in odd powers of the parameter $\overline{Z}_S$, the Shchukin-Leontovich condition linear in $\overline{Z}_S$ differs from the exact only by terms proportional to $\sim \left|\overline{Z}_S\right|^3$. That is, the fields obtained using condition (1.17) are correct up to $\sim \left|\overline{Z}_S\right|^2$. Thus, the accuracy of condition (1.17) application turns out to be higher than it could be supposed based on the results obtained in [4].

1.3.2   Surface Impedance of Metal-Dielectric Structures

The key stage of the impedance approach application is the problem of determining the surface impedance for a specific spatial structure [16, 17]. In this subsection, we analyze metal-dielectric structures, which theoretical estimates of surface impedance are well-known. Let us first consider a problem of a plane electromagnetic wave incidence on a flat interface between two media [5, 7, 14], to demonstrate a general approach for obtaining surface impedance formulas.

Let a plane in a rectangular coordinate system XOY be the interface between two media with parameters $(\varepsilon_1, \mu_1, \sigma_1)$ and $(\varepsilon_2, \mu_2, \sigma_2)$, and a conductivity of the second medium $\sigma_2 \gg 1$. Consider reduced electromagnetic fields in the two media:

$$\vec{E}_1 = \sqrt{\varepsilon_1'}\,\vec{\tilde{E}}_1, \ \vec{H}_1 = \sqrt{\mu_1'}\,\vec{\tilde{H}}_1, \ \vec{E}_2 = \sqrt{\varepsilon_2'}\,\vec{\tilde{E}}_2, \ \vec{H}_2 = \sqrt{\mu_2'}\,\vec{\tilde{H}}_2, \qquad (1.24)$$

where $\{\vec{\tilde{E}}_1, \vec{\tilde{H}}_1\}$ and $\{\vec{\tilde{E}}_2, \vec{\tilde{H}}_2\}$ are the true fields, $\varepsilon_1 = \varepsilon_1' + 4\pi i\sigma_1/\omega$ and $\varepsilon_2 = \varepsilon_2' + 4\pi i\sigma_2/\omega$ are complex permittivity's of the media. Here, the periodic dependence of the fields on time t is preserved, as in [5], in the form $e^{-i\omega t}$. A plane wave incident at an angle θ_1 measured from the normal to the interface can be represented as $\vec{\tilde{E}}_1 = (\tilde{E}_1, 0, 0)$, where $\tilde{E}_1 = E_0 e^{ik_1(y \sin \theta_1 - z \cos \theta_1)}$. Therefore, we can write

$$\vec{H}_1 = \frac{1}{ik_1}\text{rot}\,\vec{E}_1 = \frac{1}{k_1}\left[\vec{k}_1, \vec{E}_1\right], \qquad (1.25)$$

where $\vec{k}_1 = (0, \sin \theta_1, -\cos \theta_1)$, $k_1 = \frac{\omega}{c}\sqrt{\varepsilon_1'\mu_1'}$, $\left|\vec{E}_1\right| = \left|\vec{H}_1\right|$.

Since a plane wave of the same polarization is excited in the second medium, and the density of the surface current is zero,

$$\vec{E}_2 = (\tilde{E}_2, 0, 0), \ \vec{H}_2 = \frac{1}{k_2}\left[\vec{k}_2, \vec{E}_2\right], \qquad (1.26)$$

where $\vec{k}_2 = (0, \sin \theta_2, -\cos \theta_2)$, $k_2 = \frac{\omega}{c}\sqrt{\varepsilon_2'\mu_2'}$, $\left|\vec{E}_2\right| = \left|\vec{H}_2\right|$. The tangential components of the electromagnetic field in the second medium are equal to $\tilde{E}_{2\tau} = \tilde{E}_{2x} = \tilde{E}_2$ and $\tilde{H}_{2\tau} = \tilde{H}_{2y} = \tilde{H}_2 \cos \theta_2$, where θ_2 is the wave propagation angle in this medium. Then, the ratio of the tangential components of the electromagnetic field in the second medium can be immediately determined as

$$\frac{\tilde{E}_{2x}}{\tilde{H}_{2y}} = \frac{\tilde{E}_2}{\tilde{H}_2 \cos \theta_2} = \frac{1}{\cos \theta_2}. \qquad (1.27)$$

The cosine of the angle θ_2 can be easily fined by using the Snell law

$$\frac{\sin \theta_1}{\sin \theta_2} = n = \sqrt{\frac{\varepsilon_2\mu_2}{\varepsilon_1\mu_1}}$$

where n is the refractive index between the two media. Since $\sigma_2 \gg 1$, $|\varepsilon_2'| \gg 1$ and $n \gg 1$. Then

$$\cos\theta_2 = \sqrt{1 - \sin^2\theta_2} = \sqrt{1 - \frac{1}{n^2}\sin^2\theta_1}. \tag{1.28}$$

If the inequality $\left|\frac{\sin\theta_1}{n}\right| \ll 1$ holds, the $\cos\theta_2 \approx 1$, and we obtain the Shchukin-Leontovich condition, i.e., according to (1.27), $\frac{\widetilde{E}_{2x}}{\widetilde{H}_{2y}} \approx 1$. For the true fields we obtain

$$\frac{\widetilde{E}_{2x}}{\widetilde{H}_{2y}} = \sqrt{\frac{\mu_2}{\varepsilon_2}}. \tag{1.29}$$

Further, surface currents are absent, we will use the continuity of the tangential components of the electric and magnetic fields and take into account the boundary condition (1.17). The relation for the fields only in the first medium can be written as

$$E_{1x} = \overline{Z}_S H_{1y}, \tag{1.30}$$

therefore, the surface impedance of the second medium is equal to

$$\overline{Z}_S = \sqrt{\mu_2/\varepsilon_2}. \tag{1.31}$$

Since the value of the surface impedance $\overline{Z}_S$ is determined as the square root of the complex value, the branch of the root, for which the imaginary part $\operatorname{Im}\overline{Z}_S < 0$ should be selected. In this case $\operatorname{Im}\sqrt{\varepsilon_2'} > 0$, and the waves propagating in the second media are damping. For another polarization of the incident wave, we obtain the expression

$$E_{1y} = -\overline{Z}_S H_{1x}. \tag{1.32}$$

If the angle $\theta_1 \approx \pi/2$, we obtain using the expression (1.28) that $\cos\theta_2 = \sqrt{1 - \frac{1}{n^2}\sin^2\theta_1} = \sqrt{1 - \frac{1}{n^2}}$ valid for the arbitrary refractive index n. The surface impedance for the sliding waves can be easily obtained based on this result and expression (1.27). Analogously, the impedance for the wave incident at the Brewster angle can be obtained based on expression (1.28). Below in this section, the impedances of various structures are considered only for the normal incidence of the excitation wave on the flat interface between impedance surface and free half-space. The formula (1.32) can also be used to determine the impedance value if the material parameters of a medium filling the second region have been previously determined.

1.3.2.1 Real Metals

As is known, electromagnetic waves penetrate into metals at a depth small compared to a free space wavelength, λ. For superconductors and normal metals, the penetration depth at high and microwave frequencies is equal to about $10^{-2} \div 10\,\mu$m. Due to the small penetration depth, the fields components normal to the surface are much greater than their tangential components. The penetration depth Δ^0 can be determined using the expression [18]

$$\Delta^0 = \omega/k\sqrt{2\psi\omega\mu}. \tag{1.33}$$

Since the phenomenon of the electromagnetic field concentration near the surface of the body is related to the skin effect, it is argued that the impedance boundary conditions (1.17) occur when strong skin-effect is present, i.e., the skin layer thickness Δ^0 is small compared to all values with length dimensions that characterize the electrodynamic structure. First of all, the inequalities $\Delta^0 \ll \lambda/(2\pi)$ and $\Delta^0 \ll R$ hold, where R is the distance from the impedance surface to the source. The skin layer thickness should be small as compared with to the body dimensions, $\Delta^0 \ll l$, in all directions, and to the curvature radii of the body surface, $\Delta^0 \leq D$.

In the general case, for the time dependence of the fields $e^{i\omega t}$, the complex depth of field penetration into the metal is introduced [8, 19]:

$$\delta = \frac{1}{H_{\tau 1}|_{z=0}} \int_0^\infty H_{\tau 1}(z)\mathrm{d}z = \delta_1 - i\delta_2, \tag{1.34}$$

where $\{0, z\}$ is the axis directed inside the metal along the surface normal. The symbols δ_1 and δ_2 denote the resistive and inductive skin layer depth. In this case, the surface impedance can be written as

$$\overline{Z}_S = \overline{R}_S + i\overline{X}_S = k\delta_2 + ik\delta_1, \tag{1.35}$$

where $\overline{R}_S$ and $\overline{X}_S$ are the normalized surface resistance and reactance.

Let us analyze the case when the metal is located in an electromagnetic field at room temperature. A current in any point inside the metal is defined by two factors: first electrons are accelerated under the action of the electric field $\overrightarrow{E}$, and, second, the path between two successive collisions with the lattice is limited by the free path l of the electrons. When the current is forming, the fields existing on the length l should be taken into account. Since the free path l of electrons in metals at room temperature is much less than the depth of the skin layer, the field $\overrightarrow{E}$ in the process of current formation can be considered to be constant. Hence, in this case, the current density $\overrightarrow{j}$ is determined only by the magnitude of the field at that point. Under these conditions, the skin-effect is called by the classic skin-effect. To find a local relationship between the quantities $\overrightarrow{j}$ and $\overrightarrow{E}$, a simple model of free electrons can be used to obtain [19]:

$$\vec{j} = \sigma_2 \vec{E} /(1 + i\omega\tau), \tag{1.36}$$

where $\tau = l/v_F$, v_F is the Fermi velocity.

When relaxation effects can be neglected, i.e., when the condition $\omega\tau \ll 1$ is fulfilled, the formula (1.36) transfers into the traditional Ohm's law, $\vec{j} = \sigma_2 \vec{E}$. Then, for an isotropic homogeneous metal, the formulas defining the normalized surface impedance and penetration depth can be determined as [20]

$$\overline{Z}_S^{cl} = (1 + i)\sqrt{k/2\sigma_2 Z_0}, \tag{1.37}$$

$$\delta_{cl} = \sqrt{2/(kZ_0\sigma_2)} = 2\delta_1. \tag{1.38}$$

As can be seen, the essential feature of the formulas is equality of surface resistance and reactance, $\overline{R}_S^{cl} = \overline{X}_S^{cl} = \sqrt{k/2\sigma Z_0}$.

If the electron mean free path l is comparable to or greater than the penetration depth, then the formation of a current in the vicinity any point of the metal is be determined by collision processes in an area where the electric field differs markedly from the field at that point. In this case, the current density $\vec{j}$ depends on the fields defined in the vicinity of this point with the radius l. If $l \approx \delta_{cl}$ or even $l \gg \delta_{cl}$, the effect becomes typical for pure metals at low temperatures and is known as the *anomalous skin-effect* Really, when the temperature is decreasing, the average free-path length l increases as σ_2, while decreases δ_{cl} as $\sigma_2^{-1/2}$. Of course, with such free path lengths, the theory of the classical skin-effect is no longer applicable and a more general consideration is required. For example, for the pure copper $l \sim 5 \times 10^{-2}$ μm, the ratio $l/\delta_{cl} = 3 \times 10^{-2}$ at $300^0 K$ and 10 GHz. The pure copper conductivity σ_2 at helium temperatures can increase by a factor of 10^5 [21], while the ratio l/δ_{cl} can be about 10^6. Of course, in this case the formula for the classical skin-effect is no longer applicable and a more general consideration is required.

A rigorous theory of the anomalous skin-effect based on the free electron model was developed by Reuter and Sondheimer [22]. The theory assumes that if external perturbations are absent, the electrons at some point in the metal are distributed in the momentum space sphere of radius mv_F, where m is the electron mass. If, for some reason, the sphere is deformed, the total electron momentum arises appearing, which determines the current in the metal. When the anomalous skin-effect is present, the free path length l is comparable to or greater than the field penetration depth, the fields at an arbitrary point will be defined by the fields of other region where electrons were located before entering the considered place. To take into account this effect, the effective field $\vec{E}_{eff}$ in the metal is included in the expression for current density, similar to (1.36). This approach resembles accounting for the secondary fields due to induced currents on the scatterer the diffraction problems.

However, the question arises: how to correctly define the current $\vec{j}$ at any point, if it lies at a distance less than l from the metal boundary? To so the boundary conditions for the reflection of electrons on the metal surface should be taken into account. One of the possible assumptions consists in that the electrons colliding with the interface completely lose information about the field in which they were before the collision, and are reflected equiprobable in all directions, i.e., the reflection is diffuse. Moreover, in the absence of external influences outside the metal, the field $\vec{E} = 0$. Another assumption consists in that the electrons collides with a surface, the reflection can be specular. In this case, an electron moving to a flat boundary and reflected back to the observation point after the collision with the boundary to can be considered as moving from free space in the field which is the mirror-symmetric with respect to the interface. That is, the field outside the metal surface is assumed to be a mirror-symmetric field inside the metal. In the intermediate case of the two regimes, when a part of electrons p is mirror reflected and the remaining part $(1 - p)$ is diffuse reflected. The specularity coefficient p is equal to zero or one for diffuse or mirror reflections.

As a result of a rather complicated solution of the general problem the correct expressions for impedances in specular $\overline{Z}_S^{mir}$ and $\overline{Z}_S^{dif}$ mode of electron reflections can be found in [19]

$$\overline{Z}_S^{mir} = \frac{2ikl}{\pi} \int\limits_0^\infty \frac{d\tau}{\tau^2 + i\alpha\, k(\tau)}, \tag{1.39}$$

$$\overline{Z}_S^{dif} = ikl\pi \left/ \int\limits_0^\infty \ln(1 + i\alpha\, k(\tau)/\tau^2)d\tau, \right. \tag{1.40}$$

where

$$k(\tau) - \frac{2}{\tau^3}[(1 + \tau^2)\text{arctg } \tau - \tau], \alpha = \frac{3}{4}kZ_0\left(\frac{l}{\sigma_2}\right)^2, \sigma_2^{mir} = \frac{3}{2}\left(\frac{l}{\delta_{cl}}\right)^2. \tag{1.41}$$

The plots of expressions (1.39) and (1.40) as functions of σ_2 are similar. For small α, i.e., when mean free path of electrons is small, the formulas $\overline{Z}_S \sim \sigma_2^{-1/2}$, $\overline{R}_S = \overline{X}_S$ are valid for the diffuse and specular reflection, and are consistent with the classical skin-effect (1.37). For the large α the impedances $\overline{Z}_S^{mir(dif)}$ tend to the limits

$$\overline{Z}_S^{mir(dif)}\bigg|_\infty = \tilde{k}_{mir(dif)} \left(\frac{\sqrt{3}}{4\pi}\right)^{1/3} \left(\frac{l}{\sigma_2}\right)^{1/3} \left(\frac{kZ_0}{2}\right)^{2/3} Z_0(1 + i\sqrt{3}), \tag{1.42}$$

where the coefficients $\tilde{k}_{mir} = 8/9$ and $\tilde{k}_{dif} = 1$. As can be seen, the limiting values for diffuse and specular reflections differ only by the coefficients. In this case, the surface resistance and reactance are related as

$$X_S^\infty = \sqrt{3}R_S^\infty. \tag{1.43}$$

Table 1.1 The values of the
coefficients in the formulas
(1.44)

p	F_R	F_X	G
0	1.157	0.473	0.2757
1	1.376	0.416	0.3592

Of course, values of the impedances can only be calculated by numerical integration of formulas (1.39) and (1.40) valid for anomalous skin-effect. However, in, Chambers [23] have obtained simple interpolation formulas that allow to quickly calculate the values for the intermediate region between the classical and anomalous limits

$$R_S = R_S^\infty(1 + F_R\alpha^{-G}); \quad X_S = X_S^\infty(1 + F_X\alpha^{-G}), \tag{1.44}$$

where the values of the constant $F_{R(X)}$ and G for are given in Table 1.1.

The resistance R_S and reactance X_S calculated by the interpolation formulas (1.44) and by expressions (1.39) and (1.40) with an accuracy of 0.1% [19]. For the arbitrary p, Hartman and Luttinger in [24] have obtained the highly accurate solution for the extremely anomalous region in the form

$$\overline{Z}_S^\infty(p) = 2\overline{Z}_S^{dif\infty}\left[1 - \cos\left(\frac{2}{3}\arccos p\right)\right]/(1 - p). \tag{1.45}$$

The surface impedance of superconductors is of a separate fundamental interest for the researchers. As known that the electrical resistance of many pure metals, alloys and compounds at the DC disappears sharply at a critical temperature T_{cr}, which for all known superconductors are in the region of low temperatures. The highest critical temperature for pure metals, 9.3 K, has niobium, while for compound Nb_3Ge it equals 22.3 K. Perfect conductivity, $\sigma \to \infty$, i.e., total absence of resistance at the DC is considered the only fundamental property of superconductors. Meissner and Ochsenfeld in [25] have found that the magnetic flux is pushed out of the conductor when it goes into the superconducting state. This effect cannot be explain by the perfect conductivity directly from the ideal conductivity and is another important fundamental property of superconductors. It should also be noted that the surface resistance of superconductors in alternating fields, in contrast to the DC resistance, is not zero, since transitions between adjacent quasiparticle excitations that occur in a superconductor at temperatures $T > 0$ can be induced.

Phenomenological theories of superconductivity (see a review in [19]) describe a number of superconductor properties, but in many cases, they give only an approximate description and often they cannot adequately describe the specific superconductor parameter. Therefore, the microscopic theory of superconductivity, the BCS theory [26] is of great importance. This theory is based on a fact established by Cooper in [27] that an arbitrarily weak attraction between two electrons can lead to the formation of a bound state, which energy is less than the sum of the energies of the individual electrons.

The attraction can be explained by the polarization of the ion lattice by one electron, which leads to the attraction of the second electron. If attraction exceeds the repulsive Coulomb interaction, the bound pairs of electrons, the Cooper pairs are formed. Electron pairs do not exist independently of each other, but form a condensate that provides a single quantum state of the superconductor. If the Cooper pairs are exposed to external forces, for example, created by an electric field, their momentum is increased due to accelerating acquire a pulse, the same for all pairs. Thus a continuous electric current is arising. By acquiring momentum the kinetic energy of the couples is increased, but when it excides a binding energy, the Cooper pairs are destroyed.

The electrodynamics of superconductors, based on the microscopic theory [26], makes it possible to obtain expressions for the normalized impedances specular reflection and diffuse modes in the temperature range $0 < T < T_{cr}$

$$\widetilde{Z}_S^{mir} = \frac{2ik}{\pi} \int\limits_0^\infty \frac{d\tau}{\tau^2 + kZ_0\widetilde{Q}(\tau, \omega)/\omega}, \tag{1.46}$$

$$\widetilde{Z}_S^{dif} = ik\pi \left/ \int\limits_0^\infty \ln(1 + kZ_0\widetilde{Q}(\tau, \omega)/\omega\tau^2)d\tau. \right. \tag{1.47}$$

The most difficult problem related to formulas (1.46) and (1.47) consists in determining the integral kernel $\widetilde{Q}(\tau, \omega)$. Halbritter had carried out a rigorous theoretical study of the kernel $\widetilde{Q}(\tau, \omega)$ and had obtained the relations for the real $\widetilde{R}_S$ and imaginary $\widetilde{X}_S$ parts of the surface impedance based on the expressions (1.46) and (1.47) [28]. The formulas thus obtained are rather cumbersome and are not presented here. However, it should be noted that the integral kernel can be represented as $\widetilde{Q}(\tau, \omega) = \widetilde{Q}_A + \widetilde{Q}_P + \widetilde{Q}_C$, where the first two terms are real and only the third term $\widetilde{Q}_C$ have both real and imaginary part, which determine the losses. Since the real parts of the impedances $\widetilde{Z}_S^{mir}$ and $\widetilde{Z}_S^{dif}$ in most external problems are small, they can be neglected, and the more convenient formulas for the surface reactance can be used for calculations [19].

1.3.2.2 Rough and Corrugated Metal Screens

Above, we considered the surface impedance of real metals and superconductors with absolutely smooth surfaces on which the Schukin-Leontovich boundary condition (1.17) are valid. However, surface roughness, i.e. imperfections associated with the surface deviation from geometrically perfect form is unavoidably existing on solid material surfaces. The surface imperfections can be caused by a corpuscular structure of matter, technological defects as result of the surface treatment, ets.

There exist many approaches to study the effect of roughness upon the surface impedance. In several publications (for example, [29]), increase of the surface resistance and reactance was associated with a proportional increase of the actual area of a rough surface as compared to a flat one. However, this technique is valid only if characteristic dimensions noticeably exceed the penetration depth.

The most common techniques to study rough surfaces is statistical approach (for example, [30]), when a real surface is described by random function. Small deviations of the boundary shape from the plane are described by a set of random functions describing the boundary deviation from the plane $z = 0$ at the point ρ ($\overrightarrow{\rho}$ is a two-dimensional vector in the plane $z = 0$). This approach is often used for studying electromagnetic waves scattering by rough surfaces, when the effective surface scattering impedance is introduced [30]. In the common sense, the impedance is a characteristic of metal surfaces, determined by energy accumulated in the metal and its losses [19], while the scattering impedance describes the scattering properties of the surface and characterizes the energy loss by the coherent field component due to its transformation into a scattered component. In this case, the scattering processes are determined by diffraction effects which depend on the ratio between a wavelength and irregularity dimension. This effect can be significant, for example, in guide systems of large length, when, due to scattering, the energy of the fundamental mode is transformed into non-fundamental modes.

Naturally, the impedance boundary conditions are not satisfied at each point of the surface, and only the effective surface impedance $Z_S^{eff} = R_S^{eff} + i X_S^{eff}$ can be defined. It is usually assumed that characteristic dimensions of the surface irregularities (average height h and average horizontal dimension d) are much smaller than the free space wavelength and distance at which the incident wave field varies considerably. In other words, it is assumed that the impedance boundary conditions are valid on some plane surface that corresponds to a rough surface and is determined by the shape of the object at a given place. If the surface roughness is isotropic and the average dimensions and radii of curvature of the surface elements are much larger than the penetration depth, the impedance can be defined in the following form [19]:

$$R_S^{eff} = \tilde{\tilde{k}} R_S, \quad X_S^{eff} = \tilde{\tilde{k}} X_S, \tag{1.48}$$

where $\tilde{\tilde{k}} = \frac{1}{\Delta S_0 |H_{0\tau}|^2} \int_{\Delta S} |H_\tau|^2 ds$, ΔS_0 is the flat surface part corresponding to the area of the rough surface ΔS. The parameter $\tilde{\tilde{k}}$ is known as the roughness coefficient [31], which is a ratio between the surface resistance or reactance of the rough surface and that for the flat surface. The general approach for determining the parameter $\tilde{\tilde{k}}$ can be found in the monograph [19], where explicit formulas were obtained for some types of rough surfaces.

In the case of rectangular, periodically located notches (corrugations) in a conducting screen, the effective surface impedance can be determined in a different way, based on electrodynamic methods of diffraction grating analysis. If the corrugations are small, the equivalent boundary conditions can be written in the following form [32]: $E_z = Z_S^{eff} H_x$, $E_x = 0$ or $E_z = 0$, $E_x = Z_S^{eff} H_z$ for transverse or longitudinal notches with respect to the axis $\{0, z\}$. The axis $\{0, y\}$ is supposed to be directed vertically to the corrugated surface. Then according to [32]

$$\overline{Z}_S^{eff} = i(2g/L)\text{tg}kc, \tag{1.49}$$

where g is the notch width, c is the notch height, and L is the period of the corrugations.

1.3.2.3 Layered Dielectric Structures

Among of layered magnetodielectric structures which material parameters of the medium are piecewise constant functions of one coordinate, dielectric materials are used most frequently, since the layers are nonmagnetic materials in most applications. Electromagnetic waves propagation outside such structures (for example, above the underlying surface), can be analyzed by introducing the surface impedance, as for the air-dielectric interface over the half-space (formula (1.30)). Initially, such a formulation of the problem was caused by the interest in simulating of radio waves propagation over real layered soils. In some cases, it turns out to be methodically expedient to classified the layered dielectrics into natural and artificial structures.

Now, the method of radioimpedance sounding allowing to define the physical properties of inhomogeneous structures by application of interpretation models for experimental frequency dependences of surface impedance. This method applies to studies of the surface layer structure of the earth's crust (for example, [33]), and to a bio-impedance analysis of human body composition (for example, [34]). However, these questions are beyond the scope of this book.

As for as the artificial structures are concerned, the most interesting from the point of view of practical applications are multilayer plane-parallel systems. Multilayer interference structures (MIS) are a set of various dielectric layers of small thicknesses which is of order or less than an operating wavelength. The MIS operation is based on interference effects occurring inside the system with multiple wave reflections at the interfaces between layers with different wave parameters. The material of the individual layers, layer numbers, sequence order and thickness are chosen depending on the spectral characteristics of the system as a whole.

The MIS are widely used in optics; however, these structures are increasingly being used in microwave techniques for creating matching devices, filters, wave energy absorbers, rejection elements and other waveguide sections. An important advantage of the MIS waveguide elements is absence of the wave mode conversion at the flat interface between media with different permeability. This is especially important in the millimeter wavelength range, since diffraction losses increase when the frequency is increasing and the waveguide dimension is decreasing. This effect can be explained by the wave front distortion at inhomogeneities of the waveguide path and the main mode transformation into rapidly damping higher modes, which are the source of the main wave energy losses.

The MIS are effective phase shifters: phases of the waves reflected by and transmitted through the structure can be easily controlled by variation of the structure parameters. Therefore, layered dielectric structures can be used as mirrors, band-pass and single-frequency filters, impedance matchers, absorbing materials, power dividers, and high Q resonant elements. As is known, it is very difficult to manufacture conventional diaphragm-pin impedance transformers for the millimeter and submillimeter wavelengths, while production of the high quality layered dielectric structures is quite simple.

Let an area occupied by the layered structure coincides with the half-space $-\infty < z < 0$, $(-\infty < x, y < +\infty)$ in the Cartesian coordinate system and the properties of the structure itself can vary only along the coordinate z. Then the electromagnetic field as function of $\vec{R}(x, y, z)$ can be represented as the discrete or continuum superposition of spatial inhomogeneous plane waves

$$\vec{E}(\vec{R}) = \vec{E}(\vec{k}, z)e^{-i\vec{k}\vec{r}}, \quad \vec{H}(\vec{R}) = \vec{H}(\vec{k}, z)e^{-i\vec{k}\vec{r}}, \tag{1.50}$$

where $\vec{k} = (k_x, k_y, 0)$ is a spectral parameter, $\vec{E}(\vec{k}, z)$ and $\vec{H}(\vec{k}, z)$ are the vector amplitude of electric and magnetic wave, and $\vec{r} = (x, y, 0)$. Then, the impedance boundary condition (1.17) can be written as

$$\vec{E}_\tau(\vec{k}, z) = \hat{\vec{Z}}_S(\vec{k})[\vec{z}_0, \vec{H}_\tau(\vec{k}, z)], \tag{1.51}$$

which establishes a relationship between the tangential components of the vector amplitudes on the plane $z = 0$. More precisely, limit values of the tangential components when $z \to -0$ or $z \to +0$ coincide in the formula (1.51) by virtue of the boundary condition defining continuity of the tangential field components. In the above formula, $\vec{z}_0$ is the unit vector of the axis $\{0, z\}$ considered to be directed upwards normal to the interface directed vertically upwards. The non-local impedance $\hat{\vec{Z}}_S(\vec{k})$ known also as frequency-dependent impedance is a dyadic function of the spectral parameter $\vec{k}$ that fully describes the interaction of the electromagnetic field with the medium in the lower half-space.

For an isotropic layered medium, the tensor $\hat{\vec{Z}}_S(\vec{k})$ is expressed in terms of two scalar values, which have the meaning of scalar impedance for waves of vertical and horizontal polarization, respectively. The scalar impedances can be calculated numerically using the Riccati equation. For a piecewise homogeneous medium, it can be determined analytically using recurrent formulas [14]. As an example, we present the result of problem solving for a layer of a magnetodielectric with material parameters (ε, μ) located on an perfectly conducting screen, when a layer is excited by a normally incident monochromatic plane wave

$$\overline{Z}_S = i\sqrt{\frac{\mu}{\varepsilon}}\text{tg}(kh_d\sqrt{\varepsilon\mu}), \tag{1.52}$$

where h_d the layer thickness. More detailed formulas for this case are presented in Sect. 1.3.2.7.

For an arbitrarily anisotropic layered medium, the dyad $\widehat{\vec{Z}}_S(\vec{k})$ is determined by four scalar quantities. In the general case, the dyad can be constructed only numerically by solving the matrix Riccati equation [35]. However, for an anisotropic medium consisting of homogeneous flat layers of uniaxial magnetodielectric, the impedance $\widehat{\vec{Z}}_S(\vec{k})$ can be obtained analytically [36]. The values of permeability, orientation of the optical axis in each layer, thickness of the layers, and total number of layers can be arbitrary. The method proposed in [36] consists in the sequential calculation for each layer of nonlocal impedance dyad of the upper boundary, starting from the lowest, by the known impedance dyad of its lower boundary.

Application of the constant surface impedance (1.52) is usually limited by a condition associated with the provision of a single-mode wave propagation (including surface wave propagation) in the layered dielectric structures. For more general conditions, the frequency-dependent impedance (1.51) should be used. However, in some cases, even under the multimode regime of the layered structure, a constant surface impedance could be used. For example, if a dielectric layer on a metal screen is excited by a vertically oriented electric dipole, then the amplitudes of the reflected field can be correctly determined by using the constant impedance [37]. This is explained by the fact that of the possible for propagation of two or three types of waves inside a layer, in the cases considered, only the highest one will be effectively excited, thus providing the conditional single-mode mode.

Let us now consider a dielectric structure in which one of the layers is the planar volume of a cold plasma [38]. The equivalent permeability ε_{eff} and conductivity σ_{eff} of such layer is determined by the relations $\varepsilon_{eff} = \varepsilon_0\left[1 - \omega_p^2/(\omega^2 + v^2)\right]$, $\sigma_{eff} = \varepsilon_0 v \omega_p^2/(\omega^2 + v^2)$, where ω_p is the angular plasma frequency, $\omega_p^2 = n_e e^2/(\varepsilon_0 m_e)$, ε_0 is the vacuum permittivity, e is the electron charge, m_e is the electron mass, n_e is the electron concentration, and v is the frequency of electron-neutral collisions.

At present, interest in the studies of magnetoelectric materials and multiferroics, which are characterized by the interrelation of magnetic and electrical properties, has increased significantly. The magnetoelectric effect of such substances, i.e., to polarize in a magnetic field and to magnetized in an electric field, opens up a whole series of new areas for their practical use. The properties of magnetoelectric are described in the extensive review [39], where a special attention is paid to media that have magnetoelectric properties at room temperature, for example, materials based on bismuth ferrite, films of garnet ferrites, etc., since these materials are most promising for practical applications. However, it turns out that the greatest magnetoelectric effects can be observed for composite multilayer structures. For example, the coefficient of the magnetoelectric effect in composites based on a piezoelectric-ferromagnetic, reaches $\sim 10^{-1}$, which is almost three orders of magnitude grater as compared to the best samples of single-phase multifferoics. Of course, these multi-layered structures can be described, as mentioned above, using the impedance approach. It is important, that the effective surface impedance can be varied by external electric and magnetic fields.

1.3.2.4   Thin Dielectric Frequency-Selective and Chiral Layers

Flat metal-dielectric frequency-selective surface (FSS) are now actively used to create microwave circuits and antennas, high-quality filters and resonators, wave-guiding structures, etc. The electrodynamic theory of the frequency selective surfaces based on metal screens have already being formed. For example, in monographs [40, 41], the effective transmission coefficients ware determined for several FFS structures. The transmission coefficients can be used to defined the effective surface impedance of the FFS. For doubly periodic dielectric structures [42], new problems have arisen associated with applicability of various cell and array geometries, the cell material parameters and metallic inclusions.

The mechanism of frequency selectivity in dielectric frequency-selective structures differ from that in ordinary metal frequency-selective surfaces. Thin metal screens separated by a layer of a uniform dielectric with thickness of about a quarter wavelength can function as filters only for main waves. If a ratio of distances between layers and an operating frequency is large, higher modes can be excited in dielectric plates. The selectivity of the dielectric FSS is based on the behavior of higher modes excited in the plate, when these modes interfere with the main mode. At high frequencies, the characteristics of such structures strongly depend on a wave incidence angle. However, the dielectric structure at frequencies significantly lower than a cutoff frequency behaves as a homogeneous anisotropic material at a fundamental mode [43]. In this case, the properties of the artificial layer can be described by using the effective dielectric constant tensor, which, of course, allows us to introduce into consideration the effective surface impedance. Similar assertion can be made for the case of plane wave scattering at a two-periodic gyrotropic layer [44].

Among flat structures that are being actively investigated at the present time, thin chiral layers, structures with the property of chirality or enantiomorphism (from the Greek "χερ"—"hand"), should also be singled out. The chirality is a property of a living or non-living object that cannot be superposed on its image in a flat mirror for any movement and rotation. At microwave frequencies, the chiral properties can have only artificial media. The chiral inclusions at microwave frequencies are artificial conductive one-, two-, or three-dimensional microelements having a mirror-asymmetric shape, which dimensions are significantly smaller than the length of an excitation wave [45–50]. The chiral medium has a spatial dispersion; therefore, mirror-asymmetric microelements should be placed periodically at distances commensurate with a wavelength. The orientation of the geometrical axes of microelements must be chaotic, therefore, the chiral medium is biisotropic at the macroscopic level. If all the chiral microelements are oriented in one direction, then the structure becomes anisotropic.

Various elements can be used in the chiral structures: three-dimensional objects (right-and-left handed metallic spirals [51, 52], spherical particles with spiral conductivity [53], open rings with protruding ends [54]) and two-dimensional microscopic objects (S-strip elements and their mirror equivalents [55–57], flat multi- thread spirals, Möbius tapes [58], and others). The chiral layer is called planar if its elements are conductive microscopic strips of mirror-asymmetric shapes that are uniformly distributed on a dielectric or ferrite substrate. From the point of view of technical implementations of the chiral structure, the planar model is more preferable, however, its degree of chirality is less than that of the three-dimensional chiral structure. An increase in the chirality structure can be achieved by creating multilayer chiral meta-structures, at the layer interfaces on which the stripe chiral microelements are distributed. The electrodynamic properties of single-layer arrays based on chiral strip elements are detailly studied [55–57, 59, 60].

Microwave chiral media can be described with the help of three material parameters: the relative permittivity ε, permeability μ, physical chirality parameter χ. The material equations for the chiral medium can be presented in the CGS unit system as

$$\vec{D} = \varepsilon \vec{E} \mp i\chi \vec{H}, \; \vec{B} = \mu \vec{H} \pm i\chi \vec{E}, \tag{1.53}$$

where $\vec{D}$ and $\vec{B}$ are the electric and magnetic inductions. The upper plus or minus signs in the formulas (1.53) correspond to the physically chiral medium based on the right mirror forms of chiral elements, while the lower signs correspond to that based on the left mirror forms. Since the physically chiral medium is gyrotropic, the material equations include the tensor permittivity $\hat{\varepsilon}^{(\pm)}$ and permeability $\hat{\mu}^{(\pm)}$, [50]

$$\vec{D} = \hat{\varepsilon}^{(\pm)} \vec{E}, \quad \vec{B} = \hat{\mu}^{(\pm)} \vec{H},$$

$$\hat{\varepsilon}^{(\pm)} = \begin{bmatrix} \varepsilon & \pm i\chi_E & 0 \\ \mp i\chi_E & \varepsilon & 0 \\ 0 & 0 & \varepsilon \end{bmatrix}, \; \hat{\mu}^{(\pm)} = \begin{bmatrix} \mu & \mp i\chi_H & 0 \\ \pm i\chi_H & \mu & 0 \\ 0 & 0 & \mu \end{bmatrix}, \tag{1.54}$$

where $\chi_E = \chi/\eta$, $\chi_H = \chi\eta$, $\eta = \sqrt{\mu/\varepsilon}$. The upper plus or minus signs correspond to the chiral medium based on right spirals, while the lower signs correspond to the medium based on left spirals. The chirality parameter is included into the off-diagonal elements of tensors for the biotropic medium. The gyrotropic axis is directed along the axis $\{0z\}$. The determinants of matrices in (1.54) $\mathrm{Det}\hat{\varepsilon}^{(\pm)} = \varepsilon(\varepsilon^2 - \chi_E^2)$ and $\mathrm{Det}\hat{\mu}^{(\pm)} = \mu(\mu^2 - \chi_H^2)$ depend on the medium chirality parameters.

The parameter of physical chirality is determined by the dimension, shape and concentration of the microparticles in the medium, that is, it takes into account the resonant properties of the chiral element itself. Therefore, a physically chiral medium must be created for a specific, fairly narrow range of incident wave frequency, near the resonant frequency of the chiral element. This extremely complicates the theoretical and experimental study of the chirality parameter for practical media implementation. Strictly speaking, the description of wave processes in a bounded chiral medium is currently an intractable problem. Therefore, there is a need to develop approximate methods and algorithms for calculating the characteristics and parameters of limited chiral structures. We will not analyze geometric-chiral structures classified in a separate class, since they cannot be implemented in a thin-film format, consisting of a large number of layers, each of which is an ordered chiral composition.

If physically chiral layers are thin with respect to the wavelength, approximate impedance boundary conditions of two classes are used: bilateral with the chiral layer located between two arbitrary conducting media excluding the perfectly conducting one and one-sided when the chiral layer is located on a perfectly conducting screen. Until recently, only one-sided impedance type boundary conditions were known for a thin chiral layer with a flat boundary shape located on an perfectly conducting plane [49], however, these conditions did not take into account the phenomenon of cross-polarization. The similar approximate boundary conditions which consider the cross-polarization of the field is considered in [50]. The most general approach to obtaining impedance boundary conditions as quadrature formulas for thin chiral layers with a coordinate surface shape described in an arbitrary system of generalized orthogonal curvilinear coordinates is presented in [61]. The chirality parameter, obtained in this article for structures implemented in practice can be preliminarily determined experimentally [62]. A method for measuring the chirality parameter for a plane-parallel sample of an artificial medium using scanning microwave radiation (at predefined wavelength) for known values of relative permittivity ε and permeability μ is proposed in the patent [62], where the chirality parameter is calculated as the ratio of the polarization rotation angle to the absolute value of the wave number of the microwave radiation.

1.3.2.5 Layered Dielectrics with Metal Inclusions

As is known [63, 64] that the main problem in the implementation of wide-angle antenna scanning arrays in the centimeter and millimeter wavelength range is the creation of phase shifters, since their dimensions are limited by the size of the periodic array cell. In turn, the cell dimensions can be substantially less than the operating wavelength to ensure scanning mode without exciting additional interference maxima in the operating frequency band. One possible solution of this problem is to use reconfigurable radiators in reflective antenna arrays, where the phase modulation of the reflected wave is provided by geometry variation (reconfiguration) of the radiators [65]. The reconfiguration of the radiators leads to variation of the surface impedance of the antenna reflectors [66]. The reflector made using multilayer printing technology forms a coating with controlled surface impedance, which is referred to as smart coating technology or *Smart Skin* technology [67]. This technology, can be used to create airborne conformal antennas, scanning antennas of cylindrical or more complicated forms for mobile communication systems, aircraft radar covers, and much more.

The technology of intelligent coatings allows to integrate radiating and control element in one design. Among such integrated systems are reconfigurable antennas, in which radiation phase controlling the is incorporated in the design of their element Microwave switches produced by using a technology of micro- and nano-electromechanical systems can be used as actuating elements for reconfiguration of the radiators. However, the switches have also significant drawbacks, since, in practical designs, it is difficult to provide the necessary isolation between the control and power circuits, and the data channel at the UHF or EHF wavelength. These shortcomings can be eliminated by using optocouplers, semiconductor photoresistive microwave switches [68] based on photoconductivity (optocoupler microwave switches).

Structures with passive metal elements periodically distributed in the dielectric layer are known as the FSS. Usually these structures are a thin layer of periodically metal elements distributed on a dielectric substrate or apertures in a metal screen. Often these thin layers are stacked one on another separated by layers of a dielectric material. The results of multiyear studies of the FSS have been summarized in monographs [69, 70].

We will not present here a comparative analysis of approaches to modeling various layered structures, but we simply emphasize that the problem can be solved if amplitudes, energy coefficients of the reflected and transmitted waves, their phases and polarizations are known. In other words, the general solution of the problem, allows to find a nonlocal impedance $\widehat{\vec{Z}}_S(\vec{k})$ for the structure based on the relation (1.51). Of particular importance for antenna applications are thin single-layer film-type structures that require separate consideration.

In recent years, artificial spatially inhomogeneous structures with a periodically varying refractive index, called electromagnetic crystals also known in optics as photonic crystals, were studied quite intensively. By analogy with optics, the properties of both 3-D crystals and 2-D thin-film structures are investigated. In such materials there are transmission and non-transmission zones, i.e., frequency intervals in which waves (photons) can or cannot propagate. However, at the radio frequencies, the electromagnetic crystals with periodic structures should be distinguish from effective media (artificial dielectrics) or FSS. The key point that determines the difference between them is that the component of the wave vector of the electromagnetic field, parallel to the direction of periodicity, is comparable to the reciprocal of the crystal lattice period. In this case, the band structure is inevitable, therefore, the effective permittivity and hence the nonlocal surface impedance cannot complete describe the properties of the electromagnetic crystal.

1.3.2.6 Surface Impedance of Electrically Thin Vibrators

The surfaces of radiating or scattering vibrator antennas may have similar characteristics, with some of the impedance surfaces discussed above, i.e., they may differ from a perfectly conducting surface, hence the vibrators to be characterized as impedance ones.

Linear impedance vibrators are widely used in radio engineering and radio electronic complexes for various purposes, as stand along receiving and transmitting structures, elements of antenna systems, and devices of antenna-feeder paths. Wide application and multi-functional use of the impedance vibrators, including multi-element structures, is an objective prerequisite for theoretical studies of electrodynamic characteristics of such systems. Since longitudinal dimensions of the vibrators are comparable with the operating wavelength in the surrounding space, asymptotic long-wave or short-wave (quasi-optical) approximations for their analysis cannot be used. A correct mathematical modeling of real vibrator structures without increasing the complexity of formulating a corresponding electrodynamic problem, an impedance concept is successfully used (for example, the monograph [17] and references in it).

To determine the electrodynamic characteristics of electrically thin impedance vibrators, formulas for the numerical evaluation of the vibrator surface impedance should be obtained. With this purpose in mind, we consider an auxiliary problem of axially symmetric excitation of an infinite two-layer cylinder (external and internal radiuses are r and r_i) by a converging cylindrical wave. Let us introduce a cylindrical coordinate system ρ, φ, z so that the axis $\{0z\}$ is directed along the cylinder axis. Then, due to the problem symmetry, the electromagnetic field has only components E_z and H_φ, which depend only on the coordinate ρ. Permittivity and permeability of the media are ε, μ and ε_i, μ_i in the region $\rho \in [r_i, r]$ and $\rho \leq r_i$, respectively.

The normalized surface impedance $\overline{Z}_S = E_z/H_\varphi$ can be found according to (1.17), under the condition as a solution of Maxwell equations in terms of the Bessel $I_{0,1}$ and $N_{0,1}$ Neumann functions

$$
\begin{Bmatrix} i E_z \\ H_\varphi \end{Bmatrix} = \begin{Bmatrix} I_0(k\sqrt{\varepsilon\mu}r) + N_0(k\sqrt{\varepsilon\mu}r) \\ I_1(k\sqrt{\varepsilon\mu}r) + N_1(k\sqrt{\varepsilon\mu}r) \end{Bmatrix}
$$
$$
\times \sqrt{\frac{\mu}{\varepsilon}} \frac{\sqrt{\frac{\varepsilon}{\mu}} N_1(k\sqrt{\varepsilon\mu}r_i) I_0(k\sqrt{\varepsilon_i\mu_i}r_i) - \sqrt{\frac{\varepsilon_i}{\mu_i}} N_0(k\sqrt{\varepsilon\mu}r_i) I_1(k\sqrt{\varepsilon_i\mu_i}r_i)}{\sqrt{\frac{\varepsilon_i}{\mu_i}} I_0(k\sqrt{\varepsilon\mu}r_i) I_1(k\sqrt{\varepsilon_i\mu_i}r_i) - \sqrt{\frac{\varepsilon}{\mu}} I_1(k\sqrt{\varepsilon\mu}r_i) I_0(k\sqrt{\varepsilon_i\mu_i}r_i)}. \quad (1.55)
$$

Assuming that $r_i = 0$ and $|\varepsilon| \gg 1$ ($\varepsilon = \varepsilon' + 4\psi/i\omega$), we obtain the well-known formula the surface impedance of a cylindrical conductor which takes into account the skin-effect [71, 72]

$$
\overline{Z}_S = \frac{k'}{120\psi} \frac{I_0(k'r)}{I_1(k'r)}, \quad (1.56)
$$

where $k' = (1 - i)/\Delta^0$, $\Delta^0 = \omega/k\sqrt{2\psi\omega\mu}$ is the skin layer thickness, and σ is metal conductivity.

It is possible to average the impedances along the cell period for the corrugated ($L_1 \sim L_2$) or the ribbed ($L_1 \ll L_2$) conductors (here L_1 is the ridge thickness with $\overline{Z}_S = 0$, L_2 is the cavity width with $\overline{Z}_S \neq 0$) with the cells period ($L_1 + L_2) \ll \lambda/\sqrt{\varepsilon\mu}$ and at $|\varepsilon_i| \gg 1$. Then, taking into account (1.55), we have:

$$
\overline{Z}_S = -i \frac{L_2}{L_1 + L_2} \sqrt{\frac{\mu}{\varepsilon}} \frac{I_0(k\sqrt{\varepsilon\mu}r) N_0(k\sqrt{\varepsilon\mu}r_i) - I_0(k\sqrt{\varepsilon\mu}r_i) N_0(k\sqrt{\varepsilon\mu}r)}{I_1(k\sqrt{\varepsilon\mu}r) N_0(k\sqrt{\varepsilon\mu}r_i) - I_0(k\sqrt{\varepsilon\mu}r_i) N_1(k\sqrt{\varepsilon\mu}r)}, \quad (1.57)
$$

which is also true for conductors with an insulating coating of an electrically-magneto-electrician ($L_1 = 0$), as well as for metal cylinders ($r_i = 0$) with transverse dielectric inserts ($L_2 \gg L_1$).

Of particular practical interest is the case of thin vibrators, $\left|(k\sqrt{\varepsilon\mu}r)^2 \ln(k\sqrt{\varepsilon\mu}r_i)\right| \ll 1$, when the surface impedance does not depend on the excitation method of a conductor, and the corresponding boundary conditions become external [3], i.e. they coincide for exciting fields of any structure. Then, according to (1.55)–(1.57), we obtain expressions for the vibrator impedance in the thin wire approximation

$$\overline{Z}_S = \frac{1+i}{120\psi\Delta^0} \tag{1.58}$$

- the solid metallic cylinder of the $r \gg \Delta^0$ radius ($\overline{Z}_S = 0$ for the case of the perfect conductivity, when $\sigma \to \infty$).

Within the framework of above approximations, the expression (1.58) is consistent with the formula (1.37), which indirectly confirms the possibility of using the formulas for the surface impedances of flat surface and for vibrators. This assertion is valid for the following cases: the classical skin-effect (1.37), anomalous skin-effect (1.39)–(1.43), and skin-effect superconductor mode (1.46), (1.47).

$$\overline{Z}_S = \frac{1}{120\psi h_0 + ikr(\varepsilon - 1)/2} \tag{1.59}$$

- the dielectrical metalized cylinder with covering, made of the metal of the $h_0 \ll \Delta^0$ thickness, hence at $\varepsilon = 1$

$$\overline{Z}_S = \frac{1}{120\psi h_0} \tag{1.60}$$

- the tube metallic cylinder of the $r \gg \Delta^0$ radius (for the "nano-radius" vibrator [72] $h_0 = r, r \ll \Delta^0$), and at $h_0 = 0$ in (1.59).

$$\overline{Z}_S = -i\frac{2}{kr(\varepsilon - 1)} \tag{1.61}$$

- the dielectrical cylinder.

$$\overline{Z}_S = -i\frac{L_2}{L_1 + L_2}\frac{2}{kr\varepsilon} \tag{1.62}$$

- the metal-dielectrical cylinder.

$$\overline{Z}_S = \frac{1}{120\psi h_0 - i/kr\mu \ln(r/r_i)} \tag{1.63}$$

- the magnetodielectrical metalized cylinder with the inner conducting cylinder, hence (1.60) at $r = r_i$, and at $h_0 = 0$

$$\overline{Z}_S = ikr\mu \ln(r/r_i) \tag{1.64}$$

- the metallic cylinder with covering, made of magnetodielectric of the $r - r_i$ thickness or the ribbed cylinder.

The surface impedance for a vibrator with spiral conductivity, i.e., for the kiral objects [47], in a particular case of a monopillar metal helix can be found using the formula (1.47) as

$$\overline{Z}_S = (i/2)kr \, \mathrm{ctg}^2 \psi. \tag{1.65}$$

where r $(kr \ll 1)$ is helix radius and ψ is an angle of helix.

The formulas (1.58)–(1.65) obtained within the framework of a general impedance concept are valid for thin infinite and finite cylinders, located in free space. If the vibrator is located in a material medium with parameters ε_1 and μ_1, a multiplier $\sqrt{\mu_1/\varepsilon_1}$ should be included in the formulas. If the medium parameters ε and μ can be smoothly varied by a static electric or magnetic field, then the radiation characteristics of the system with fixed geometry can be controlled by these fields. As can be seen from formulas (1.64) and (1.65) the characteristics of the vibrators with the purely inductive surface impedance can be described by using a concept of effective vibrator length, defined by the formula

$$2L_{eff} = \left[1 + \frac{\mu \ln(r/r_i)}{2\ln(2L/r)}\right]2L, \quad 2L_{eff} = \left[1 + \frac{\mathrm{ctg}^2\psi}{4\ln(2L/r)}\right]2L. \tag{1.66}$$

Thus, the electrodynamic characteristics of the impedance vibrator with the length $2L$ is equivalent to a perfectly conducting vibrator with a length $2L_{eff}$, so that $2L_{eff} > 2L$.

Separately, it is necessary to emphasize the possibility of calculating the surface resistance of a carbon nanotube [72–74]. For example, if the nanotube is located entirely in a dielectric with permittivity ε and permeability $\mu = 1$, the surface resistance can be determined using the following relation

$$\rho_S = i\pi^2 a\hbar^2(\omega - iv)/(2e^2 v_F), \tag{1.67}$$

where a is nanotube radius, v_F is Fermi velocity, $v_F = 9.71 \times 10^5$ m/s, ω is cyclic frequency, v is relaxation frequency, $v = 3.33 \times 10^4$ Hz, e is electron charge, $\hbar$ is Planck constant.

1.3.2.7 Surface Impedance of a Magnetodielectric Layer on a Metal Substrate Surface

I. **Surface impedance of a uniform magnetodielectric layer on a perfectly conducting plane.** In this subsection, we present formulas determining the impedance $\overline{Z}_S$ for some specific examples of the physical implementation of impedance surfaces. First, we consider an auxiliary problem concerning a normal incidence of a plane electromagnetic wave on a dielectric layer that separates two half-spaces. The layer thickness is h_d, complex permittivity and permeability are ε_1, and μ_1, and the wave number $k_1 = k\sqrt{\varepsilon_1\mu_1}$. The incidence wave propagates in the free upper half-space ($\varepsilon = \mu = 1$), while the lower half space is characterized by material parameters ε_2, μ_2.

The boundary value problem solution can be easily obtained by taking into account the boundary conditions for the electric and magnetic fields on both surfaces of the dielectric layer. Comparing this solution with the requirements of the Shchukin-Leontovich impedance boundary condition (1.17) on the upper boundary of the dielectric layer, we obtain the rigorous expression for the distributed surface impedance

$$\overline{Z}_S = \overline{Z}_1 \frac{i\overline{Z}_1 \text{tg}(k_1 h_d) + \overline{Z}_2}{\overline{Z}_1 + i\overline{Z}_2 \text{tg}(k_1 h_d)}, \tag{1.68}$$

where $\overline{Z}_1 = \sqrt{\mu_1/\varepsilon_1}$ and $\overline{Z}_2 = \sqrt{\mu_2/\varepsilon_2}$.

If the layer of magnetodielectric is on the perfectly metal surface, the formula (1.68) after substitution $\overline{Z}_2 = 0$ is reduced to the relation similar to (1.52)

$$\overline{Z}_S = i\sqrt{\mu_1/\varepsilon_1}\,tg(k_1 h_d), \tag{1.69}$$

where h_d is magnetodielectric thickness. In the case of an arbitrary incident field, formula (1.69) is approximate and becomes more accurate if the inequality $|\varepsilon_1\mu_1| \gg 1$ better inequality holds (approximation of geometrical optics [3]). For example, if the incident field is the H_{10}-wave propagating in a rectangular waveguide, the formulas (1.68) and (1.69) can be written as:

$$\overline{Z}_S = \overline{Z}_1 \frac{i\overline{Z}_1 \text{tg}(k_g h_d) + \overline{Z}_2}{\overline{Z}_1 + i\overline{Z}_2 \text{tg}(k_g h_d)}, \tag{1.70}$$

$$\overline{Z}_S = i(k_1/k_g)\sqrt{\mu_1/\varepsilon_1}\,tg(k_g h_d), \quad k_g = \sqrt{k_1^2 - (\pi/a)^2}. \tag{1.71}$$

When the condition $|\varepsilon_1\mu_1| \gg 1$ is fulfilled, the Eqs. (1.70) and (1.71) are converted to (1.68) and (1.69), respectively.

As shown in Refs. [75–77], real parts of permittivity and/or permeability of magnetodielectric metamaterials can become negative. Therefore, the formulas (1.69) and (1.71) can be written as

$$\overline{Z}_S = \pm i\sqrt{\mu_1/\varepsilon_1}\,\text{tg}(k_1 h_d), \qquad (1.72\text{a})$$

$$\overline{Z}_S = \pm i\,(k_1/k_g)\sqrt{\mu_1/\varepsilon_1}\,\text{tg}(k_g h_d), \qquad (1.72\text{b})$$

where the sign plus or minus correspond to the cases $\text{Re}\varepsilon_1 > 0$ or $\text{Re}\varepsilon_1 < 0$. The frequency plots of the surface impedance for the metamaterial layer on a metal substrate are presented in Sect. 5.3.

If a layer is electrically thin ($|k_1 h_d| \ll 1$ or $|k_g h_d| \ll 1$), when the quasi-stationary approximation [3, 78]), the relations (1.69), (1.71) and (1.72) allow to obtain that $\overline{Z}_S \approx ik\mu_1 h_d$ [78], i.e. the normalized surface impedance does not depend on the material permittivity, as is the case for thin impedance vibrators (see formula (1.64).

If a thin conductive film with a thickness h_R $((h_R/\Delta^0 \ll 1))$, is deposited on a layer of a magnetodielectric metamaterial located on a metal plane, the surface impedance defined by the formula (1.68) is equal to

$$\overline{Z}_{SR} = \frac{\overline{R}_{SR}}{1 + \overline{R}_{SR}/\overline{Z}_S}, \qquad \overline{R}_{SR} = \frac{1}{Z_0\sigma_1 h_R}, \qquad (1.73)$$

where $\overline{Z}_S$ is defined by formulas (1.69), (1.71) and (1.72).

Further in the book, the impedance $\overline{Z}_S$ (1.71) and (1.72b) will be used. However, if other structures of impedance coatings are to be analyze, the impedance $\overline{Z}_S$ can be obtained in other ways, as it will be demonstrated below.

II.   **Surface impedance of a magnetodielectric layer with inhomogeneous dielectric permittivity on a perfectly conducting plane.** Modern technologies to produce thin-film coatings make it possible to apply both homogeneous and inhomogeneous structures [79–81]. In this subsection, approximate analytical expressions will be obtained for the distributed surface impedance of a magnetodielectric layer with an inhomogeneous permittivity located on a perfectly conducting plane. The formulas are valid if variations of the layer permittivity are small.

Consider a plane layer of a magnetodielectric $-\infty < x, y < \infty$, $-h_d \leq z \leq h_d$ specified in Cartesian coordinate system (x, y, z) with magnetic permeability μ_1 and dielectric permittivity ε_1, placed on a perfectly conducting plane at $z = h_d$. Let a plane monochromatic electromagnetic wave $E_x(z) = E_{0x}e^{-ikz}$ be incident on a layer of a magnetodielectric located from the free half-space $z = -\infty$. Then the distributed surface impedance for this layer determined by the expression [3] can be written as

$$\overline{Z}_S = E_{0x}(-h_d)/H_{0y}(-h_d), \tag{1.74}$$

where the fields $E_{0x}(-h_d)$ and $H_{0y}(-h_d)$, i.e. $E_x(z)$ and $H_y(z)$ at the plane $z = -h_d$ inside the magnetodielectric layer with material parameters $\mu_1 = const$, $\varepsilon_1 = \varepsilon_1(z)$ which can be found as solution of the following differential equations

$$\frac{d^2 E_x(z)}{dz^2} + k^2 \mu_1 \varepsilon_1(z) E_x(z) = 0, \tag{1.75a}$$

$$H_y(z) = \frac{i}{k\mu_1} \frac{dE_x(z)}{dz} \tag{1.75b}$$

with the boundary conditions on the surfaces $z = \pm h_d$. Equations (1.75) are valid for any functions $\varepsilon_1(z)$; the relation (1.74) for normal incidence of plane waves on the plane layer of the magnetodielectric is exact.

The solutions are quite complex, and in each case are expressed by special functions. Some example of these solution can be found in Appendix C. If the permittivity is a relatively slow varying function within the layer, an approximate solution in a class of elementary functions can be obtained for various functions $\varepsilon_1(z)$. Let us now several examples allowing to find analytical expressions for surface impedance.

(1) **Surface impedance of a magnetodielectric layer with permittivity varying as a power-law function**. Let us assume that $\varepsilon_1(z)$ is represented as

$$\varepsilon_1(z) = \varepsilon_1(0)[1 - \varepsilon_r f(z)], \tag{1.76}$$

where the function $\varepsilon_r = [\varepsilon_1(0) - \varepsilon(-h_d)]/\varepsilon_1(0)$ is the relative constant in the layer ($|\varepsilon_r| < <1$), $f(z) = (-z/h_d)^n$, ($n = 1, 2, 3 \ldots$) is a predefined function. Then Eq. (1.75a) is transformed into a non-homogeneous differential equation with constant coefficients:

$$\frac{d^2 E_x(z)}{dz^2} + k_1^2 E_x(z) = \varepsilon_r k_1^2 f(z) E_x(z), \tag{1.77}$$

where $k_1^2 = k^2 \mu_1 \varepsilon_1(0)$.

Suppose that the right-hand side of Eq. (1.77) is known, then applying the method of variation of arbitrary constants, we have

$$E_x(z) = C_1 e^{-ik_1 z} + C_2 e^{ik_1 z} + \varepsilon_r k_1 \int\limits_{-h_d}^{z} f(z') E_x(z') \sin k_1(z - z') dz'. \tag{1.78}$$

We will seek the solution of the integral Eq. (1.78) in power series

$$E_x(z) = E_{x0}(z) + \varepsilon_r E_{x1}(z) + \varepsilon_r^2 E_{x2}(z) + \ldots + \varepsilon_r^m E_{xm}(z) + \ldots \qquad (1.79)$$

After substituting (1.79) into Eq. (1.78) and keeping the zero- and first-order terms proportional to the small ε_r we obtain the following expressions for the electric and magnetic fields inside the layer

$$E_x(z) = C_1 e^{-ik_1 z}[1 + \varepsilon_r f_E(k_1 z)] + C_2 e^{ik_1 z}[1 + \varepsilon_r f_E^*(k_1 z)],$$
$$H_y(z) = (1/\overline{Z}_1)\{C_1 e^{-ik_1 z}[1 + f_H(k_1 z)] - C_2 e^{ik_1 z}[1 + f_H^*(k_1 z)]\}, \qquad (1.80)$$

where $*$ is the sign of complex conjugation, $\overline{Z}_1 = \sqrt{\mu_1/\varepsilon_1(0)}$, the functions $f_E(k_1 z)$ and $f_H(k_1 z)$ depend upon the function $f(z)$. Consider the two simplest cases, when the function $f(z)$ is linear or quadratic function. In the first case, the permittivity $\varepsilon_1(z) = \varepsilon_1(0)\left(1 + \varepsilon_r \frac{z}{h_d}\right)$, and the functions $f_E(k_1 z)$, $f_H(k_1 z)$ are equal to

$$f_E(k_1 z) = -\frac{1}{4k_1 h_d}[(k_1 z) + i(k_1 z)^2 - i/2],$$
$$f_H(k_1 z) = -\frac{1}{4k_1 h_d}[-(k_1 z) + i(k_1 z)^2 + i/2]. \qquad (1.81)$$

In the second case, the permittivity $\varepsilon_1(z) = \varepsilon_1(0)\left(1 - \varepsilon_r \frac{z^2}{h_d^2}\right)$, and the functions $f_E(k_1 z)$, $f_H(k_1 z)$ are equal to

$$f_E(k_1 z) = \frac{1}{(2k_1 h_d)^2}\left[(k_1 z)^2 - \frac{1}{2} + i\frac{2(k_1 z)^3}{3} - i(k_1 z)\right],$$
$$f_H(k_1 z) = \frac{1}{(2k_1 h_d)^2}\left[-(k_1 z)^2 + \frac{1}{2} + i\frac{2(k_1 z)^3}{3} + i(k_1 z)\right]. \qquad (1.82)$$

The unknown constants C_1 and C_2 can be determined based on the boundary conditions of the field components on the planes $z = \mp h_d$. The expression for the surface impedance can be obtained in the form determining in (1.80) from the boundary conditions of the continuity of the field components for and if the electric component is zero for, we obtain the desired

$$\overline{Z}_S = \overline{Z}_1 \frac{[1 + \varepsilon_r f_E(-k_1 h_d)][1 + \varepsilon_r f_E^*(k_1 h_d)]e^{i4k_1 h_d} - [1 + \varepsilon_r f_E^*(-k_1 h_d)][1 + \varepsilon_r f_E(k_1 h_d)]}{[1 + \varepsilon_r f_H^*(-k_1 h_d)][1 + \varepsilon_r f_E(k_1 h_d)] + [1 + \varepsilon_r f_H(-k_1 h_d)][1 + \varepsilon_r f_E^*(k_1 h_d)]e^{i4k_1 h_d}}. \qquad (1.83)$$

When $\varepsilon_r = 0$ and $2h_d \to h_d$, the formula (1.83) reduces to the relation (1.69) defining the impedance of the homogeneous magnetodielectric layer on the perfectly conducting surface.

Neglecting the terms proportional to ε_r^2 in the formula (1.83) after substitution $2h_d \to h_d$, we finally have

$$\overline{Z}_S = i\overline{Z}_1 \frac{\text{tg}(k_1 h_d)}{1 + \varepsilon_r f_{Lin}(k_1 h_d)\text{tg}(k_1 h_d)},$$

$$f_{Lin}(k_1 h_d) = \left(\frac{1}{2k_1 h_d} + \frac{i}{2}\right), \qquad (1.84)$$

for the linear function $\varepsilon_1(z)$, and

$$\overline{Z}_S = i\overline{Z}_1 \frac{\text{tg}(k_1 h_d) + \varepsilon_r f_{Sq1}(k_1 h_d)}{1 + \varepsilon_r f_{Sq2}(k_1 h_d)\text{tg}(k_1 h_d)},$$

$$f_{Sq1}(k_1 h_d) = \frac{1}{k_1 h_d} - \frac{\text{tg}(k_1 h_d)}{(k_1 h_d)^2}, \quad f_{Sq2}(k_1 h_d) = \frac{k_1 h_d}{6}, \qquad (1.85)$$

for the quadratic function $\varepsilon_1(z)$.

The expression (1.83) is valid not only for the power function $\varepsilon_1(z)$, since the representation of the permittivity (1.76) can be also applied for other functions $f(z)$: trigonometric $f(z) = \left(1 - \cos\frac{\pi z}{2h_d}\right)$ and $f(z) = -\sin\frac{\pi z}{2h_d}$; fractal functions $f(z) = 1 - F(z)$, where $F(z) = \frac{1}{C_N}\sum_{n=1}^{N}\eta^{(D-2)n}\cos\left(\eta^n\frac{\pi z}{2h_d}\right)$ is the Weyerstrasse functions [82], $C_N = \sum_{n=1}^{N}\eta^{(D-2)n}$ is the normalization factor, $1 < D < 2$ is the fractional fractal dimension, $\eta > 1$ is the function representation parameter; etc.

(2) **Surface impedance of a magnetodielectric layer when permittivity is an exponential function**. Let the permittivity a of magnetodielectric layer ($0 \leq z \leq h_d$) is an exponential function

$$\varepsilon_1(z) = \varepsilon_1(0)e^{kz}. \qquad (1.86)$$

In this case, the presentation (1.76) allowing to isolate a small parameter is no longer valid, but some other representations are possible, which make it possible to reduce Eq. (1.75a) to an integral equation with a small parameter and solve it by the above method. In our opinion, the WKB approximation [14, 83] is the most suitable for the equation solution (1.75a), which can be written as

$$E_x(z) = \frac{1}{\sqrt[4]{\varepsilon_1(z)}}\left\{C_1 e^{-ik\sqrt{\mu_1}\int\limits_0^z \sqrt{\varepsilon_1(z)}dz} + C_2 e^{ik\sqrt{\mu_1}\int\limits_0^z \sqrt{\varepsilon_1(z)}dz}\right\}. \qquad (1.87)$$

The expression (1.87) is valid if the function $\varepsilon_1(z)$ has no zeros or poles on the interval $[0, z]$. The criterion of the WKB approximation applicability is the following inequality [83]

$$\left|\frac{d\varepsilon_1(z)/dz}{k\sqrt{\mu_1}[\varepsilon_1(z)]^{3/2}}\right|_{max} \ll 1, \qquad (1.88)$$

which for the exponential function $\varepsilon_1(z)$ reduces to $\left|\frac{1}{\sqrt{\mu_1\varepsilon_1(0)}}\right| \ll 1$.

If $(h_d/\lambda)^2 \ll 1$, substitution of (1.86) into (1.87) allows us to obtain the following expressions:

$$E_x(z) = \frac{e^{-kz/4}}{\sqrt[4]{\varepsilon_1(0)}}\left(C_1 e^{-ik_1 z} + C_2 e^{ik_1 z}\right), \qquad (1.89a)$$

$$H_y(z) = \frac{1}{\overline{Z}_1}\frac{e^{-kz/4}}{\sqrt[4]{\varepsilon_1(0)}}\left\{C_1 e^{-ik_1 z} f_H[\varepsilon_1(0)\mu_1] - C_2 e^{ik_1 z} f_H^*[\varepsilon_1(0)\mu_1]\right\}, \qquad (1.89b)$$

where $f_H[\varepsilon_1(0)\mu_1] = 1 - \frac{i}{4\sqrt{\varepsilon_1(0)\mu_1}}$. When the arbitrary constants C_1 and C_2 are determined by applying the boundary conditions at the planes $z = 0$ and $z = h_d$, the normalized surface impedance is equal to

$$\overline{Z}_S = i\overline{Z}_1 \frac{\operatorname{tg}(k_1 h_d)}{1 + f_{Exp}[\varepsilon_1(0)\mu_1]\operatorname{tg}(k_1 h_d)},$$

$$f_{Exp}[\varepsilon_1(0)\mu_1] = \frac{1}{4\sqrt{\varepsilon_1(0)\mu_1}}, \qquad (1.90)$$

which transfers to $\overline{Z}_s \approx ik\mu_1 h_d$ if $|k_1 h_d| \ll 1$.

Thus, we can conclude that formulas (1.84), (1.85), and (1.90) defining the distributed surface impedance of a magnetodielectric layer on the perfectly conducting plane, valid for various distribution function of the dielectric layer permittivity obtained by different methods, have a similar structure and, in the limiting cases $\varepsilon_1 = const$ and $|k_1 h_d| \ll 1$ are reduced to the identical relations.

These results allow us to generalize the approach for the case of a thin metal film with a thickness less than the skin layer depth deposited on the magnetodielectric layer located on the metal plane. Then the surface impedance of this structure can be determined by the expression (1.73), while $\overline{Z}_S$ can be calculated by the formulas (1.84), (1.85), or (1.90).

(3) *Examples of calculating the surface impedance of a magnetodielectric layer on the perfectly conducting plane.* Let us present some results of numerical calculations of the distributed surface impedance, illustrating the applicability of the proposed method. Since the parameters of magnetodielectric depend on the frequency, we will consider only frequencies corresponding to single-mode regime of standard rectangular waveguides operating in the centimeter wavelength.

Figure 1.1 shows the plots of the real and imaginary parts of the complex surface impedance $\overline{Z}_S = \overline{R}_S + i\overline{X}_S$ as function the of the magnetodielectric layer of thickness. The material parameters of TDK IR-E110 at the frequency $f = 7 \div 12$ GHz corresponding to operating wavelength of the H_{10}-wave of the rectangular waveguide with cross-section 22.86×10.16 mm^2 are $\varepsilon_1 = 8.84 - i0.084$ and $\mu_1 = 2.42 - 0.0825 f - i0.994$ [84]. As can be seen, if the layer thickness is equal to

Fig. 1.1 The surface impedance of a magnetodielectric layer on a metal plane as function of the layer thickness, at wavelength $\lambda = 30.0$ mm for various functions $\varepsilon_1(z)$: 1—constant, 2—linear, 3—quadratic, 4—exponential

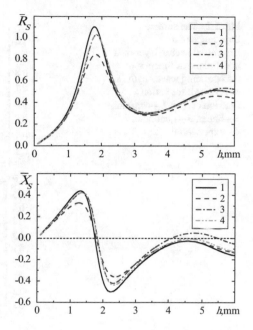

the quarter wavelength in the magnetodielectric ($h_d \approx 1.8$ mm at $\lambda = 30.0$ mm), the real impedance has a pronounced maximum for all considered functions $\varepsilon_1(z)$ (here and below $\varepsilon_r^2 = 0.04$ for linear and quadratic functions). When the thickness layer is further increased, the $\overline{R}_S$ tends to some constant value, which is equal to $\text{Re}\overline{Z}_1$ in the lossy magnetodielectric. The imaginary part of the layer impedance as function of the layer thickness is of an alternating character, and $\overline{X}_S = 0$ for given ε_1, μ_1, λ and $h_d \approx 1.8$ mm. The reflection coefficient of a plane wave from this structure is minimal, therefore the layer is known as antireflection coating [85]. Note that the greatest difference in magnitude from the case of a homogeneous magnetodielectric occurs with a linear distribution law.

The plots of the surface impedance as functions of the wavelength for the magnetodielectric TDK IR-E110 with various curves $\varepsilon_1(z)$ are shown in Fig. 1.2. The layer thickness $h_d \approx 1.6$ mm was selected [84], where the problem of radiation from a longitudinal slot cut in a wide wall of a rectangular waveguide ($a = 22.86$ мм, $b = 10.16$ мм) placed over an impedance plane was solved. As can be seen from the plots, the layer inhomogeneity leads to decrease of $\overline{R}_S$ in the wavelength of H_{10}-wave. The imaginary part of the impedance can be larger or smaller than the that for a uniform material. The surface impedance $\overline{Z}_S$ of the layer with the quadratic and exponential $\varepsilon_1(z)$ almost coincide.

The problem of an impedance slot diaphragm was solved by using the Shchukin-Leontovich approximate impedance boundary condition (1.17), which can be apply if the inequality $|\overline{Z}_S| \ll 1$ holds. This condition can be fulfilled for the thin layers as is shown in Fig. 1.3 (magnetodielectric TDK IR-E110, $h_d = 0.3$ mm). In this case, the real and imaginary parts of the impedance monotonously decrease with

Fig. 1.2 The surface impedance of a magnetodielectric layer on a metal plane as function of the operating wavelength (h_d = 1.6 mm): for various functions $\varepsilon_1(z)$: **1**—constant, **2**—linear, **3**—quadratic, **4**—exponential

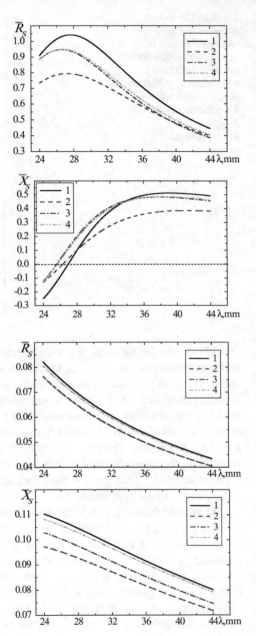

Fig. 1.3 The surface impedance of a magnetodielectric layer on a metal plane as function of the operating wavelength (h_d = 0.3 mm): for various functions $\varepsilon_1(z)$: **1**—constant, **2**—linear, **3**—quadratic, **4**—exponential

Fig. 1.4 The surface impedance of the magnetodielectric layer covered by the resistive film on the metal plane as function of the operating wavelength ($\overline{R}_{SR} = 0.3$, $h_d = 0.3$ mm for various functions $\varepsilon_1(z)$: **1**—constant, **2**—linear, **3**—quadratic, **4**—exponential

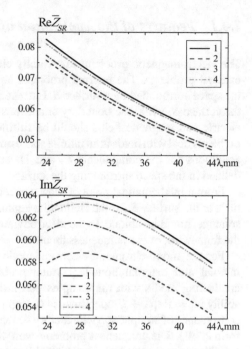

increasing wavelength for all type of the function $\varepsilon_1(z)$. The imaginary part of the impedance $\overline{X}_S$ is always positive, that is, it has an inductive character.

Figure 1.4 shows that if the magnetodielectric surface is coated by a resistive film with low conductivity (for example, nichrome), the real part of the surface impedance $\mathrm{Re}\overline{Z}_{SR}$ of the structure has the same magnitude and varies almost as in the previous case (see Fig. 1.3). The imaginary part of the impedance $\mathrm{Im}\overline{Z}_{SR}$ turn out to be much smaller and has a maximum in the short-wave part of the single-mode waveguide regime.

Thus, the analysis shows that the influence of the magnetodielectric inhomogeneity on the surface impedance is relatively small—not more than 25% compared to the homogeneous layer, and turns out to be the most significant for the linear function $\varepsilon_1(z)$ as compared with quadratic and exponential functions. However, the inhomogeneity of the layer can be as additional means to control electrodynamic parameters of the waveguide-slot microwave devices.

1.4 Uniqueness Theorem and Duality Principle for Areas with Impedance Boundaries

Since later in the monograph electrodynamic problems will be investigated using impedance boundary conditions, it is useful to discuss some aspects of using such boundary conditions.

1.4.1 Features of the Application of the Uniqueness Theorem

The electromagnetic problems are usually classified into two classes: internal and external problems. The internal problems concern with fields in a limited part of the space surrounded by a surface S. Extraneous currents $\vec{j}^{E(M)}$ are specified inside the surface S, while the boundary conditions for the tangential components of the electric and magnetic fields should be fulfilled on the surface itself. The external problems deal with fields in an infinite space outside the surface S, where the boundary conditions for the components E_τ and H_τ are specified, and external currents are defined in the space surrounding the surface.

From a mathematical point of view, the requirements to satisfy boundary conditions at the surface S for the Helmholtz equations (1.6) or (1.10) determine that the solution should be unique. In the electrodynamics theory this statement is proved in the framework of the uniqueness theorem.

For the mono-chromatic electromagnetic fields uniqueness of solution of the internal and external boundary value problems for the Maxwell equations in conducting media with the complex permittivity $\varepsilon_1 = \mathrm{Re}\varepsilon_1 + i\mathrm{Im}\varepsilon_1$ and permeability $\mu_1 = \mathrm{Re}\mu_1 + i\mathrm{Im}\mu_1$ ($\mathrm{Im}\varepsilon_1 > 0$, $\mathrm{Im}\mu_1 > 0$) are intensively studied (here, for convenience of comparison with literary sources, we choose a time dependence in the form $e^{-i\omega t}$). The uniqueness problems were thoroughly studied when the boundary conditions for electric and magnetic fields are given apart from each other [18]. Since the uniqueness theorem proof for boundary value problems with impedance surfaces satisfying the boundary conditions (1.17) are covered in literature contradictorily, additional consideration of this question can be found in [86].

A key point in the proof of the uniqueness theorem is a lemma asserting that the homogeneous Maxwell equations do not have solutions other than zero. The lemma is applied to difference electromagnetic fields $\left(\vec{E}, \vec{H} \right)$, and it is proved based on the complex Pointing theorem [18, 87] or the Lorentz lemma [8, 86].

If the tangential component of the electric and magnetic field are explicitly specified on the surface S: the electric field is defined on the surface S_1, the part of the surface S, and the tangential component of the magnetic field is defined on the surface S_2 ($S = S_1 + S_2$), the difference field $\left(\vec{E}, \vec{H} \right)$ should satisfy homogeneous Maxwell equations with uniform boundary conditions on the surface S $\vec{E}_\tau = 0$ on the surface S_1 and $\vec{H}_\tau = 0$ on the surface S_2.

If the Shchukin-Leontovich impedance boundary condition (1.17) are specified on the surface S, it is easy to verify that the difference field $\left(\vec{E}, \vec{H} \right)$ should also satisfy the homogeneous linear Maxwell equations with impedance boundary conditions (1.17) on the surface S. The distinction in the boundary conditions for difference fields dictates the need for the proof of the uniqueness theorem for the boundary value problems with impedance surfaces.

There exist several approaches to determining the surface impedance in the literature, different are used. In some works, the surface impedance defined as a linear operator connecting the tangential components $\vec{E}_\tau$ and $\left[\vec{H}_\tau, \vec{n}\right]$, and in other it is determined from the relation between $\vec{E}_\tau$ and $\left[\vec{H}_\tau, \vec{n}\right]$. Therefore, it is always necessary to keep a check on definitions of the surface impedance.

For the general case, we introduce definition of the anisotropic surface impedance in the form of a matrix

$$\widehat{Z}_S = \left\| \begin{matrix} Z_{11} & Z_{12} \\ Z_{21} & Z_{22} \end{matrix} \right\|, \tag{1.91}$$

where $Z_{jk} = R_{jk} + X_{jk}$, indices j, k are equal to 1,2. To ensure the absence of additional energy sources on the impedance surface, or more precisely, the energy fluxes through the surface the inequalities should hold:

$$R_{11} \geq 0, \quad R_{22} \geq 0, \quad 4R_{11}R_{22} \geq \left| Z_{12} + Z_{21}^* \right|^2, \tag{1.92}$$

where Z_{21}^* is the complex conjugate of Z_{21}. It should be emphasized that when the normal vector to the surface S is directed inside the impedance body, the physically correct using the impedance boundary condition (1.17) is possible only if the requirements $\mathrm{Re} Z_S \geq 0$ and (1.92) are met for the isotropic and anisotropic surface impedance Z_S.

The authors of this book has proved the uniqueness theorem applying the Lorentz lemma to the difference electromagnetic field $\left\{ \vec{E}_d, \vec{H}_d \right\} = \left\{ \vec{E}, \vec{H} \right\}$ $\left\{ \vec{E}^*, \vec{H}^* \right\}$ assuming that electrical and magnetic losses are nonzero at each point of the spatial region D_e, and applying the impedance boundary condition (1.20) on the given surface. The final expression for the case of isotropic surface impedance can be presented as

$$\oint_S \mathrm{Re} Z_S \left| \left[\vec{n}, \vec{H}_d \right] \right|^2 \mathrm{d}s + \omega \int_{D_e} \left(\mathrm{Im}\varepsilon_1 \left| \vec{E}_d \right|^2 + \mathrm{Im}\mu_1 \left| \vec{H}_d \right|^2 \right) \mathrm{d}v = 0, \tag{1.93}$$

where $\mathrm{d}v$ is the volume element. As can be seen, the expression (1.93) includes the integral terms taking into account the electric and magnetic losses in the medium and the surface integral, proportional to $\mathrm{Re} Z_S$. Obviously, to satisfy condition (1.21) with and, and also when, everywhere in the region there should be and. Obviously, on the assumption that the inequalities $\mathrm{Im}\varepsilon_1 > 0$, $\mathrm{Im}\mu_1 > 0$, and $\mathrm{Re} Z_S \geq 0$ hold, the condition (1.21) can be fulfilled if everywhere in the region D_e the equalities $\vec{E}_d = 0$ and $\vec{H}_d = 0$ are valid. The expression (1.93) turns out to be valid even if the time dependence of monochromatic wave processes is chosen in the form $e^{i\omega t}$. In this case, the signs before ω, $\mathrm{Im}\varepsilon_1$, and $\mathrm{Im}\mu_1$ in (1.93) should be opposite.

Thus, it can be stated that the solutions of internal and external boundary value problems of the Maxwell system of equations for areas bounded by impedance surfaces in conducting media where the imaginary parts of permittivity and permeability satisfy the relations $\mathrm{Im}\varepsilon_1 > 0$ and $\mathrm{Im}\mu_1 > 0$ are unique if the conditions $\mathrm{Re}Z_S \geq 0$ and (1.92) are met in the case of an isotropic and anisotropic impedance surfaces.

1.4.2   Features of the Application of the Principle of Duality

The duality principle commonly used in the formulation of Weinstein [18], establish a relation between two diffraction problems: (1) diffraction in free space on a flat infinitely thin perfectly conducting plate; (2) diffraction in free space on a flat, infinitely thin perfectly conducting screen with an aperture that coincides with the plate in the first problem. The duality theorem is analogous to the Babinet theorem which relates diffraction phenomena on mutually complementary screens within the framework of Huygens scalar principle. An analysis of the duality principle for areas with impedance surfaces was absent in references prior to publications [86, 88].

The duality principle is based on the property of permutation symmetry or permutation invariance of Maxwell equations with respect to external electric $\vec{j}^{E}$ and magnetic $\vec{j}^{M}$ currents. This property allows the following two-sided permutation

$$\vec{j}^E \Leftrightarrow -\vec{j}^M;\ \ \vec{H}_1 \Leftrightarrow \vec{E}_2;\ \ \vec{E}_1 \Leftrightarrow \vec{H}_2;\ \ \varepsilon \Leftrightarrow -\mu. \tag{1.94}$$

The choice of one or other pair of coupled electrodynamic problems with various geometry, differing by excitation conditions $\left(\vec{j}^E \Leftrightarrow -\vec{j}^M\right)$, which in fact determine the content of the reciprocity principle for each of the possible options can be justified on the basis of these permutations. Problems with only one type of the excitation currents should be solved by using the Maxwell equations, asymmetrical with respect to extraneous currents. In these cases, it is difficult to directly use the property of permutation invariance, and separate proofs of the duality principle are required, for example, as it was done in [18].

The bilateral permutations (1.94) are valid for solution of electrodynamic problems in infinite space if the field boundedness condition at infinity is satisfied. The boundary value problems for impedance surfaces have one significant feature based on the permutation principle (1.94)—the boundary conditions should be also transformed.

Since the impedance boundary condition (1.17) are asymmetric with respect to replacing only the fields $\vec{E}_1 \Leftrightarrow \vec{H}_2$ and $\vec{H}_1 \Leftrightarrow \vec{E}_2$, it is also necessary to make the replacement of the surface impedance. It is easy to verify that if the impedance is isotropic, the permutations (1.94) must be supplemented by

$$Z_S \Leftrightarrow -1/Z_S. \tag{1.95}$$

To verify the legitimacy of requirement (1.95), let us consider the problem of the normal incidence of a plane electromagnetic wave on a flat interface between the free half-space and a half-space with a homogeneous medium characterized by complex permittivity ε_1 and permeability μ_1. The formula of the wave impedance in the form $Z_0 = \sqrt{\mu_1/\varepsilon_1}$ was obtained as solution of this problem. In this case, the permutation (1.95) is directly resulted from permutations for the fields from (1.94) and $\varepsilon_1 \Leftrightarrow -\mu_1$.

Analyzing the permutation (1.95), the following conclusion can be made: in any electrodynamic volumes containing impedance boundaries, the duality principle can be realized only using permutation (1.95) in related problems. Since the relation $\mathrm{Re}\, Z_S > 0$ is a condition of physical feasibility for passive impedance structures, the of this permutation is possible only for surfaces characterized by the purely imaginary impedance. This conclusion remains valid if the surface impedance is represented in the matrix form (1.19). Since the algebra of rank-2 matrices coincides with the algebra of linear operators, the condition (1.95) can be written as

$$\widehat{Z}_S \Leftrightarrow -\widehat{Z}_S \Big/ \det\left[\widehat{Z}_S\right], \tag{1.96}$$

where $\det\left[\widehat{Z}_S\right]$ is the determinant of the impedance matrix $\widehat{Z}_S$. In this case, the permutation (1.96) is possible only when $Z_{12} = Z_{21}$, i.e. matrix $\widehat{Z}_S$ is symmetric.

1.5 Tensor Green's Functions of the Vector Helmholtz Equation for Hertz Potentials

1.5.1 Properties of the Tensor Green's Function

As mentioned earlier, the vector Helmholtz equations (Subsect. 1.1) and boundary conditions (Sects. 1.2 and 1.3) are a single boundary value problem. This boundary value problem can be analytically solved by the eigenfunctions and Green's function method.

The of eigenfunctions approach (see, for example, [18]), is based on solving differential equations by the separation of variables. The eigenfunctions are solutions of homogeneous ordinary differential equations which a separation constants satisfy the corresponding boundary conditions at the interval ends of the independent variables. The values of the separation constant, valid for specified boundary conditions, are called eigenvalues. This method will not be used in this book, therefore it is not discussed here in detail.

The second method is more physically obvious, since Green's function represent a field at the observation point generated by a point source. To obtain the field generated by sources distributed in space, one should find the corresponding volume integral over the entire region where the sources are given.

Let us consider the basic properties of the tensor Green's function of the vector Helmholtz equation [89]:

(a) in the system of orthogonal curvilinear coordinates, the tensor Green's function $\widehat{G}(q_1, q_2, q_3; q_1', q_2', q_3') = \widehat{G}(\vec{q}, \vec{q}')$ satisfies the inhomogeneous Helmholtz equation

$$\Delta \widehat{G}(\vec{q}, \vec{q}') + k^2 \widehat{G}(\vec{q}, \vec{q}') = -4\pi \hat{I} \frac{\delta(q_1 - q_1')\delta(q_2 - q_2')\delta(q_3 - q_3')}{h_1 h_2 h_3}, \qquad (1.97)$$

where $\hat{I}$ is the unit tensor, (q_1', q_2', q_3') are coordinates of the source, $\delta(q - q')$ is the Dirac delta function, h_n are Lame coefficients, and the Laplacian is applied to all components of the tensor;

(b) the Green's function is the symmetric tensor (affinor) of the second rank defined by nine components, since each of the three components of a vector source can create three field components. The tensor symmetry ensures the equality

$$\vec{F}(\vec{q})\,\vec{\vec{G}}(\vec{q}, \vec{q}') = \vec{\vec{G}}(\vec{q}, \vec{q}')\,\vec{F}(\vec{q}). \qquad (1.98)$$

holds;

(c) all components of the Green's tensor are invariant with respect to the mutual change of the of the source $(\vec{q}')$ and observation $(\vec{q})$ coordinates;

(d) when the source and observation points coincide, all components of the Green's tensor have an integrable singularity

$$1 \Big/ \sqrt{(q_1 - q_1')^2 + (q_2 - q_2')^2 + (q_3 - q_3')^2};$$

(e) the solution of the vector inhomogeneous Helmholtz equation can be obtained in integral form by using the Green's tensor.

The solution for vector Hertz potentials is presented in the following form [89]

$$\vec{\pi}^{E(M)}(\vec{q}) = \frac{1}{i\omega\varepsilon_1\mu_1} \int\limits_V \vec{j}^{E(M)}(\vec{q}')\widehat{G}^{E\,(M)}(\vec{q}, \vec{q}')dv$$

$$+ \oint\limits_S \left\{ \mathrm{div}\,\vec{\pi}^{E(M)}(\vec{q}')\widehat{G}^{E(M)}(\vec{q}, \vec{q}')\vec{n} - \mathrm{div}\widehat{G}^{E(M)}(\vec{q}, \vec{q}')\vec{\pi}^{E(M)}(\vec{q}')\vec{n} \right.$$

$$+ \left[\vec{n}, \widehat{G}^{E(M)}(\vec{q}, \vec{q}')\right] \mathrm{rot}\, \vec{\pi}^{E(M)}(\vec{q}') - \left[\vec{n},\, \vec{\pi}^{E(M)}(\vec{q}')\right] \mathrm{rot} \widehat{G}^{E(M)}(\vec{q}, \vec{q}') \Big\} \mathrm{d}s',$$

$$(1.99)$$

where $\vec{n}$ is the unit vector of the outward normal to the surface S, $\mathrm{d}v$ is the volume element ($\mathrm{d}s'$ is the area element in the dashed coordinates). The volume integral is taken over the entire volume bounded by the surface S, the surface integral is taken over the entire surface S, and differentiation is performed with respect to operations the dashed coordinates Note that the expression in braces is a vector.

Thus, the solution of the inhomogeneous Helmholtz equation can be represented as the sum of the volume and surface integrals. The integrands of surface integrals include boundary values of the unknown function and its derivatives, that reduces the range of solvable problems. If the boundary values of the unknown function are not known in advance, the equalities (1.99) are converted into integral equations whose solution is no less difficult than the initial differential equations.

The surface integrals can be excluded from (1.99) by constructing the Green's function in a special way. If components of the Green's function $\widehat{G}(\vec{q}, \vec{q}')$ satisfy the boundary conditions on the surface S coincide with components of the vector potentials $\vec{\pi}^{E(M)}(\vec{q})$, the surface integrals disappear, since the components of the integrand vector vanish. Hence, the solution for $\vec{\pi}^{E(M)}(\vec{q})$ will be represented only as the volume integral.

Substituting this solution into Eq. (1.8), we obtain the following expressions for the electromagnetic field

$$\vec{E}(\vec{q}) = \frac{1}{4\pi\, i\omega\varepsilon_1}\left[\mathrm{graddiv} + k_1^2\right]\int\limits_V \vec{j}^E(\vec{q}')\widehat{G}^E(\vec{q}, \vec{q}')\mathrm{d}v$$

$$- \frac{1}{4\pi}\mathrm{rot}\int\limits_V \vec{j}^M(\vec{q}')\widehat{G}^M(\vec{q}, \vec{q}')\mathrm{d}v,$$

$$\vec{H}(\vec{q}) = \frac{1}{4\pi\, i\omega\mu_1}\left[\mathrm{graddiv} + k_1^2\right]\int\limits_V \vec{j}^M(\vec{q}')\widehat{G}^M(\vec{q}, \vec{q}')\mathrm{d}v$$

$$+ \frac{1}{4\pi}\mathrm{rot}\int\limits_V \vec{j}^E(\vec{q}')\widehat{G}^E(\vec{q}, \vec{q}')\mathrm{d}v. \qquad (1.100)$$

Let us now represent of the electromagnetic field through the dyadic Green's functions [90]

$$\begin{bmatrix} \vec{E}(\vec{q}) \\ \vec{H}(\vec{q}) \end{bmatrix} = \int\limits_V \begin{bmatrix} \hat{g}^E(\vec{q}, \vec{q}') & \hat{g}^{EM}(\vec{q}, \vec{q}') \\ \hat{g}^{ME}(\vec{q}, \vec{q}') & \hat{g}^M(\vec{q}, \vec{q}') \end{bmatrix} \times \begin{bmatrix} \vec{j}^E(\vec{q}') \\ \vec{j}^M(\vec{q}') \end{bmatrix}\mathrm{d}v, \qquad (1.101)$$

where $\hat{g}^E(\vec{q}, \vec{q}')$, $\hat{g}^M(\vec{q}, \vec{q}')$, $\hat{g}^{EM}(\vec{q}, \vec{q}')$, and $\hat{g}^{ME}(\vec{q}, \vec{q}')$ are electric, magnetic, electro-magnetic, and magneto-electric functions. Then, comparing expressions

(1.100) and (1.101), we can define the dyadic Green's functions through the Green's tensors of the Hertz vector potentials as

$$\hat{g}^E(\vec{q},\vec{q}') = \frac{1}{4\pi\, i\omega\varepsilon_1}[\text{graddiv} + k_1^2]\widehat{G}^E(\vec{q},\vec{q}'),$$

$$\hat{g}^{EM}(\vec{q},\vec{q}') = -\frac{1}{4\pi}\text{rot}\widehat{G}^M(\vec{q},\vec{q}'),$$

$$\hat{g}^{ME}(\vec{q},\vec{q}') = \frac{1}{4\pi}\text{rot}\widehat{G}^E(\vec{q},\vec{q}'),$$

$$\hat{g}^M(\vec{q},\vec{q}') = \frac{1}{4\pi\, i\omega\mu_1}[\text{graddiv} + k_1^2]\widehat{G}^M(\vec{q},\vec{q}'). \qquad (1.102)$$

As indicated above, the Green's functions for vector potentials $\widehat{G}^{E(M)}(\vec{q},\vec{q}')$ in the source region have an integrable singularity. In the case of the field Green's functions (1.102) the singularity is non-integrable. Therefore, the apparatus of generalized functions should be used to regularize the singularity which significantly complicates the Green's tensors constructing.

In this connection, the excitation problems of various electrodynamic volumes by external currents will be solved by using the tensor Green's function of the Helmholtz equations for Hertz vector potentials $\overrightarrow{\pi}^{E(M)}(\vec{q})$. When the potentials $\overrightarrow{\pi}^{E(M)}(\vec{q})$ is determined, the electromagnetic fields $\vec{E}(\vec{q})$ and $\vec{H}(\vec{q})$, can be found from (1.100).

1.5.2 Construction of the Green's Tensor in Orthogonal Curvilinear Coordinate Systems

The constructing the tensor Green's function of the Helmholtz vector equation, developed in the monograph [89], is based on the possibility of representing its components as series expansions in the system of one longitudinal and two transverse Hansen vector functions, which are expressed through scalar eigenfunctions. This approach imposes several requirements to the coordinate systems that limit the number of systems in which the possibility of constructing Green's tensors exists. These requirements will be briefly analyzed bellow.

First of all, the coordinate system should ensure that the variable separations in the three-dimensional Helmholtz equation which allows determining the scalar eigenfunctions $\varphi(\vec{q})$, $\psi(\vec{q})$ and $\chi(\vec{q})$. There exist eleven types of coordinate systems which according to Robertson condition allow the variable separations. The scalar functions in each coordinate system are represented as products of functional multipliers, each of which depends only on one variable, for example, $\varphi(\vec{q}) = X_1(q_1)X_2(q_2)X_3(q_3)$.

However, in the Helmholtz vector equations, the variables cannot always be separated but even if they are separated, the components of the vector field may be admixed so that they are included in all three equations. To avoid this admixing, the method of vector field separation into one longitudinal component presented as a grad operator acting on the scalar potential and two transverse or solenoidal components given as rot operator acting on the vector potential [89]. Then, if the Lame coefficient $h_1 = 1$, the complete system of Hansen eigenvector functions can be obtained based on solutions of three scalar Helmholtz equation for the functions $\varphi(\vec{q})$, $\psi(\vec{q})$ and $\chi(\vec{q})$:

$$\vec{L} = grad\varphi(\vec{q}),$$
$$\vec{M} = \text{rot}(\vec{n}_1 w \psi(\vec{q})),$$
$$\vec{N} = \frac{1}{k}\text{rotrot}\,(\vec{n}_1 w \chi(\vec{q})), \tag{1.103}$$

where $\vec{n}_1$ is the unit vector of the coordinate q_1, $1/k$ is the coefficient introduced to ensure that the dimension of the components $\vec{L}$, $\vec{M}$, $\vec{N}$ is the same.

The requirements that $h_1 = 1$ and the ratio of the other two Lame coefficients do not depend upon the coefficient corresponding to this coordinate allows only six coordinate systems: rectangular coordinates in which x, y or z can be selected as the coordinate q_1, three cylindrical coordinate systems in which the coordinate z corresponds to q_1, spherical and conical coordinate systems in which the radius r is q_1. In the expressions (1.103), $w = 1$ in the first four cases and $w = r$ in the last two cases.

The system of the eigenvector vector functions (1.103) constructed in this way can ensure the fulfillment of boundary conditions for vector fields only if the boundary surfaces coincide with coordinate surfaces of the coordinate system. That is, the Green's functions can be constructed only for spatial domains with so-called coordinate boundaries. Thus, this requirement determines the choice of the coordinate system for the solutions of boundary-value problems of various geometries. The eigenfunctions of the vector Helmholtz equation can be defined based on (1.103) as

$$\vec{F}_n(\vec{q}) = \vec{L} + \vec{M} + \vec{N}, \tag{1.104}$$

where a four-digit subscript n is assign to each eigenvector $\vec{F}_n(\vec{q})$, so that the first digit of this index shows which of the three systems $\vec{L}$, $\vec{M}$ or $\vec{N}$ the vector belongs to, and the other three are numbers of the generating scalar eigenfunctions, $\varphi(\vec{q})$, $\psi(\vec{q})$ and $\chi(\vec{q})$. It can be shown that the eigenvector $\vec{F}_n(\vec{q})$ for all possible n form a complete orthogonal system of functions, that is

$$\iiint \vec{F}_n^*(\vec{q}) \vec{F}_m(\vec{q})\mathrm{d}v = \begin{cases} 0, m \neq n, \\ \lambda_n, m = n, \end{cases} \tag{1.105}$$

where integration is performed over the entire spatial domain, $\vec{F}_n^*(\vec{q})$ is a vector function, complex-conjugate to the function $\vec{F}_n(\vec{q})$, λ_n is the norm of the corresponding eigenfunctions.

The Green's tensor can be represented by using the operation of tensor multiplication of two vectors as an expansion in eigenvectors $\vec{F}_n(\vec{q})$

$$\widehat{G}(\vec{q}, \vec{q}') = 4\pi \sum_n \frac{\vec{F}_n^*(\vec{q}) \otimes \vec{F}_n(\vec{q}')}{\lambda_n(k_n^2 - k^2)}, \tag{1.106}$$

where the eigenvector $\vec{F}_n(\vec{q})$ satisfy the same boundary conditions as the components $\widehat{G}(\vec{q}, \vec{q}')$, k_n^2 are the corresponding eigenvalues, and $\otimes$ is the symbol of tensor multiplication. If $\vec{F}_n(\vec{q})$ is a complex vector, then the conjugate vector in the decomposition (1.106), therefore the tensor $\widehat{G}(\vec{q}, \vec{q}')$ is the Hermitian tensor. It can be also seen that the tensor $\widehat{G}(\vec{q}, \vec{q}')$ is symmetric relative to $\vec{q}$ and $\vec{q}'$.

The tensor $\widehat{G}(\vec{q}, \vec{q}')$ as a function of k has poles for all eigenvalues k_n, where the residue at the nth pole is equal to $2\pi \vec{F}_n^*(\vec{q}) \vec{F}_n(\vec{q}') / (\lambda_n k_n)$. This corresponds to an infinite amplitude of oscillations at the resonant frequency of the driving force in absence of attenuation.

Let us now construct using the described method, the Green's tensor for the most interesting, from a practical point of view, the generalized cylindrical coordinate systems.

1.5.3 Green's Tensor for Areas with Cylindrical Boundaries

Let us construct the tensor Green's function for the vector Helmholtz equation in the generalized cylindrical coordinate systems including rectangular coordinate systems by the proposed method which is also used in [90]. Let (q_1, q_2, z) and (q_1', q_2', z') be the coordinates of the observation and source points.

The tensor functions based on the formula (1.106) can be represented as series over the full system of vector functions (1.103). For example, the tensor $\widehat{G}^E(\vec{q}, \vec{q}')$ can be decompose as

$$\widehat{G}^E(\vec{q}, \vec{q}') = \sum_{n=0}^{\infty} \sum_{m=0}^{\infty} \left\{ \frac{1}{(k_{nm}^E)^2 \lambda_{nm}^E} \left[\vec{z}^\circ, \nabla\psi_{nm}(q_1, q_2) \right] \otimes \left[\vec{z}^\circ, \nabla\psi_{nm}^*(q_1', q_2') \right] h_{nm}^E \right.$$
$$+ \frac{1}{(k_{nm}^M)^2 \lambda_{nm}^M} \nabla\chi_{nm}(q_1, q_2) \otimes \chi_{nm}^*(q_1', q_2') h_{nm}^M$$
$$\left. + \frac{1}{\lambda_{nm}^M} \vec{z}^\circ \chi_{nm}(q_1, q_2) \times \vec{z}^\circ \chi_{nm}^*(q_1', q_2') h_{nm}^M \right\}, \tag{1.107}$$

where $\nabla u = \vec{q}_1^{\circ} \frac{1}{h_1} \frac{\partial u}{\partial q_1} + \vec{q}_2^{\circ} \frac{1}{h_2} \frac{\partial u}{\partial q_2}$, q_1, q_2 are curvilinear coordinates in the cross section of the considered area with Lame coefficients h_1 and h_2, $\vec{q}_1^{\circ}, \vec{q}_2^{\circ}$ are the unit vectors, and $\lambda_{nm}^{E(M)}$ are norms of the eigenfunctions.

In expression (1.107), $\psi_{nm}(q_1, q_2)$ and $\chi_{nm}(q_1, q_2)$ are mutually orthogonal eigenfunctions of the following two-dimensional Helmholtz equations

$$\Delta \chi_{nm}(q_1, q_2) + \left(k_{nm}^M\right)^2 \chi_{nm}(q_1, q_2) = 0,$$
$$\Delta \psi_{nm}(q_1, q_2) + \left(k_{nm}^E\right)^2 \psi_{nm}(q_1, q_2) = 0 \qquad (1.108)$$

with boundary conditions $\chi_{nm}(q_1, q_2) = \frac{\partial}{\partial n} \psi_{nm}(q_1, q_2) = 0$.

The functions h_{nm}^E and h_{nm}^M in (1.107) satisfy non-homogeneous ordinary differential equations

$$\frac{\partial^2}{\partial z^2} h_{nm}^{E(M)}(z, z') + \left(\gamma_{nm}^{E(M)}\right)^2 h_{nm}^{E(M)}(z, z') = -4\pi\delta(z - z'), \qquad (1.109)$$

with predefined boundary conditions at the ends of the interval z. In the Eq. (1.109) $\gamma_{nm}^{E(M)} = \sqrt{k^2 - \left(k_{nm}^{E(M)}\right)^2}$.

In the cylindrical coordinate systems, the form of Eq. (1.109) for $h_{nm}^{E(M)}(z, z')$ and their solutions do not depend on the field distribution in the region cross section. The solutions should be found in accordance with the geometry of the boundary value problem specified in the longitudinal direction.

The tensor Green's function $\widehat{G}^E(\vec{q}, \vec{q}')$ in the considered coordinate systems has the following form

$$\widehat{G}^E(\vec{q}, \vec{q}') = \begin{pmatrix} G_{11}^E(\vec{q}, \vec{q}') & G_{12}^E(\vec{q}, \vec{q}') & 0 \\ G_{21}^E(\vec{q}, \vec{q}') & G_{22}^E(\vec{q}, \vec{q}') & 0 \\ 0 & 0 & G_{33}^E(\vec{q}, \vec{q}') \end{pmatrix}, \qquad (1.110)$$

since only five nonzero components can be obtained from the tensor decomposition (1.107). For example, the explicit expression for the component of the tensor Green's function $G_{33}^E(\vec{q}, \vec{q}')$ can be written as

$$G_{33}^E(\vec{q}, \vec{q}') = G_{zz}^E(\vec{q}, \vec{q}') = \sum_{n=0}^{\infty} \sum_{m=0}^{\infty} \frac{1}{\lambda_{nm}^E} \chi_{nm}(q_1, q_2) \chi_{nm}^*(q_1, q_2) h_{nm}^E(z, z').$$

$$(1.111)$$

The tensor Green's functions of the magnetic type are also determined by the expression (1.107) with the following mutual substitutions: $\chi_{nm}(q_1, q_2) \Leftrightarrow \psi_{nm}(q_1, q_2)$, $h_{nm}^E(z, z') \Leftrightarrow h_{nm}^M(z, z')$, $k_{nm}^E \Leftrightarrow k_{nm}^M$, $\lambda_{nm}^E \Leftrightarrow \lambda_{nm}^M$.

The tensor Green's functions of electric and magnetic types for the spatial domains considered in the monograph, are presented in Appendix A.

References

1. Leontovich MA (1985) Theoretical physics. Selected works. Nauka, Moskow (in Russian)
2. Leontovich MA (1948) On approximate boundary conditions for an electromagnetic field on the surface of well-conducting bodies. Radio wave propagation studies, pp 5–12 (in Russian)
3. Miller MA, Talanov VI (1961) Using the concept of surface impedance in the theory of surface electromagnetic fields (review). Izv Vuz Radiophys 4(5):795–830 (in Russian)
4. Rytov SM (1940) Calculation of the skin-effect by the perturbation method. J Exp Theor Phys 10(2):180–190 (in Russian)
5. Ilyinsky AS, Kravtsov VV, Sveshnikov AG (1991) Mathematical models of electrodynamics. Vysshaya shkola, Moskow (in Russian)
6. Physical encyclopedia (1983) In: Prokhorov AM (ed) Soviet encyclopedia. Moskow (in Russian)
7. Feinberg EL (1999) Propagation of radio waves along the Earth's surface. Fizmatlit, Moskow (in Russian)
8. Ilyinsky AS, Slepyan GYa (1990) Impedance boundary conditions and their application for calculating the absorption of electromagnetic waves in conductive media. Radio Eng Electron 35(6):1121–1135 (in Russian)
9. Senior TBA, Volakis JL (1995) Approximate boundary conditions in electro-magnetics. IEE, London
10. Hoppe DJ, Rahmat-Samii Y (1995) Impedance boundary conditions in electro-magnetics. Taylor and Francis, Boca Raton
11. Tretyakov S (2003) Analytical modeling in applied electromagnetics. Artech House, Norwood, MA
12. Yuferev SV, Ida N (2009) surface impedance boundary conditions. A comprehensive approach. CRC Press, Boca Raton
13. Belkovich IV, Zheksenov MA, Kozlov DA (2011) Comparison of options for impedance boundary conditions when a plane electromagnetic wave is incident on a dielectric half-space. J Radio Electron (7) (in Russian)
14. Brekhovsky LM (1973) Waves in layered media. Nauka, Moskow (in Russian)
15. Alshits VI, Lyubimov VN (2009) Generalization of the Leontovich approximation for electromagnetic fields at the boundary of a dielectric-metal. Adv Phys Sci 179(8):865–871 (in Russian)
16. Nesterenko MV, Katrich VA, Penkin YuM, Berdnik SL (2008) Analytical and hybrid methods in theory of slot-hole coupling of electrodynamic volumes. Springer, New York
17. Nesterenko MV, Katrich VA, Penkin YuM, Dakhov VM, Berdnik SL (2011) Thin impedance vibrators. Theory and applications. Springer, New York
18. Weinstein LA (1988) Electromagnetic waves. Radio i svyaz', Moskow (in Russian)
19. Mende FF, Spitsyn AI (1985) Surface impedance of superconductors. Naukova dumka, Kiev (in Russian)
20. Rameau S, Winnery G (1948) Fields and waves in modern radio engineering. OGIZ, Moskow (in Russian)
21. Alexandrov BYa, Rybalchenko LF, Dukin VV (1970) Zone cleaning of copper. Izvest AN SSSR Met. Ser (4):68–75 (in Russian)
22. Reuter GEH, Sondheimer EH (1948) The theory of anomalous skin effect in metals. Proc Roy Soc A 195(1042):336–364
23. Chambers RS (1952) The anomalous skin effect. Proc Roy Soc A 215(1123):281–292
24. Hartmann LE, Luttsnger JM (1966) Exact solution of the integral equations for the anomalous skin effect and cyclotron resonance on metal. Phys Rev 151(2):430–433
25. Meissner W, Ochsenfeld R (1933) Ein neuer effekt bei eintritt der supraleitfaig-keit. Naturwissenchaften B 33(44):787–788 (in German)
26. Bardeen J, Cooper LN, Schieffer JR (1957) Theory of superconductivity. Phys Rev 108(5):1175–1204

27. Cooper LN (1956) Bound electron pairs in a degenerate Fermi gas. Phys Rev 108(4):1189–1190
28. Halbritter J (1969) Zur oberflachenimpdanz von supraleitern. Diss., Karlsruhe (in German)
29. Allison J, Benson FA (1955) Surface roughness and attenuation of precision drawn, chemically polished, electro polished, electroplated and electroformed waveguides. Proc IEE 102(1):251–259
30. Bass FG, Fuks IM (1972) Wave scattering on a statistically uneven surface. Nauka, Moskow (in Russian)
31. Mende FF, Spitsyn AI, Tereshchenko NA et al (1977) Measurement of the absolute value of reactive surface resistance and the depth of field penetration into superconducting lead. J Tech Phys 9(47):1916–1923 (in Russian)
32. Neganov VA, Nefedov EI, Yarovoy GP (1998) Modern design methods for transmission lines and resonators on ultra-high frequencies. Pedagogy-Press, Moskow (in Russian)
33. Efremov VN (2009) Mapping and monitoring of highly icy soils by radio-impedance sounding. Nauka i obrazovaniye 4(56):81–85 (in Russian)
34. Heymsfield SB, Lohman TG, Wang Z, Going SB (eds) (2005) Human body composition. Human Kinetics, Champaign, IL
35. Tichener JB, Willis SR (1991) The reflection electromagnetic waves form stratified anisotropic media. IEEE Trans Antennas Propag 39:35–40
36. Bagatskaya OV, Shulga SN (2001) Impedance approach to solving the problem of reflection of a plane electromagnetic wave from a multilayer plate of a uniaxial magnetodielectric. Vestnik Kharkov Univ 513:25–31 (in Russian)
37. Penkin D, Yarovoy A, Janssen G (2012) Surface impedance model for nano-scale device communications over an interface. In: Proceedings of the IEEE 19th symposium on communications and vehicular technology in the Benelux. Eindhoven, Netherlands, pp 1–5
38. King RWP, Smith GS (1981) Antennas in matter. MIT Press, Cambridge, MA
39. Pyatakov AP, Zvezdin AK (2012) Magnetoelectric materials and multiferroics. Adv Phys Sci 182(6):593–620 (in Russian)
40. Petit R (ed) (1980) Electromagnetic theory of grating. Springer, Berlin
41. Wu TK (ed) (2000) Frequency selective surfaces. Wiley, New York
42. Sidorchuk NV, Yachin VV (2005) Scattering of electromagnetic waves by a batch periodic magnetodielectric layer. Radiophys Radioastron 10(1):50–61 (in Russian)
43. Sidorchuk NV, Yachin VV, Prosvirnin SL (2002) The long-wave approximation in the problem of the propagation of electromagnetic waves in a two-period magnetodielectric layer. Radiophys Electron 7:208–212 (in Russian)
44. Yachin VV, Zinenko TL, Kiselev VK (2009) Quasistatic approximation for scattering of a plane wave by a two-period gyrotropic layer. Radiophys Radioastron 14(2):174–182 (in Russian)
45. Shevchenko VV (1998) Chiral electromagnetic objects and environments. Soros Educ J 2:109–114 (in Russian)
46. Lindell IV, Sihvola AH, Tretyakov SA, Viitanen AJ (1994) Electromagnetic waves in chiral and bi-isotropic media. Artech House, London
47. Katsenelenbaum BZ, Korshunova EN, Sivov AN, Shatrov AD (1997) Chiral electrodynamic objects. Adv Phys Sci 167(11):1201–1212 (in Russian)
48. Lakhtakia A, Varadan VK, Varadan VV (1989) Time-harmonic electromagnetic fields in chiral media. Lecture notes in physics. Springer, Berlin
49. Tretyakov SA (1994) Electrodynamics of complex media: chiral, biisotropic and some bianisotropic materials. Radioeng Electron 39(10):1457–1470 (in Russian)
50. Neganov VA, Osipov OV (2006) Reflective, waveguide and radiating structures with chiral elements. Radio i svyaz', Moskow (in Russian)
51. Lakhtakia A, Varadan VV, Varadan VK (1985) Scattering and absorption characteristics of lossy dielectric, chiral, nonspherical objects. Appl Opt 24(23):4146–4154
52. Lakhtakia A, Varadan VV, Varadan VK (1988) Field equations, Huygens's principle, integral equations, and theorems for radiation and scattering of electromagnetic waves in isotropic chiral media. J Opt Soc Am 5(2):175–184

53. Shevchenko VV (1995) Small chiral particle diffraction. Radioeng Electron 40(12):1777–1788 (in Russian)
54. Tretyakov SA, Mariotte F (1995) Maxwell Garnett modeling of uniaxial chiral composites with bianisotropic inclusions. J Electromagn Waves Appl 9(7/8):1011–1025
55. Prosvirnin SL (1999) Polarization transformation upon reflection of waves by a microstrip array of complex shape elements. Radioeng Electron 44(6):681–686 (in Russian)
56. Prosvirnin SL (1998) Analysis of electromagnetic wave scattering by plane periodical array of chiral strip elements. In: Proceedings of the 7th international conference on complex media, pp 185–188
57. Arnaut LR (1997) Mutual coupling in arrays of planar chiral structures. In: Advances in complex electromagnetic materials, vol 28. Kluwer Academic Publishers, Dordrecht; Boston; London, pp 293–309
58. Jaggard D, Engheta N, Kowarz MW, Pelet P, Liu JC, Kim Y (1989) Periodic chiral structures. IEEE Trans Antennas Propag 37:1447–1452
59. Vasilieva TD, Prosvirnin SL (1998) Diffraction of electromagnetic waves on a flat array of chiral strip elements of complex shape. Phys Wave Process Radio Eng Syst 1(4):5–9 (in Russian)
60. Prosvirnin SL, Zheludev NI (2005) Polarization effects in the diffraction of light by a planar chiral structure. Phys Rev E E71(3):3
61. Osipov OV, Panferova TA (2010) Approximate boundary conditions for thin chiral layers with a curved surface shape. Radioeng Electron 53(5):568–570 (in Russian)
62. Volobuev AN, Osipov OV, Panferova TA (2010) The method for determining the chirality parameter of artificial chiral media. Patent No. 2418292, dated March 22, 2010, IPC G01N23/02 (in Russian)
63. Amitay N, Galindo V, Wu CP (1972) Theory and analysis of phased array antennas. Wiley, New York
64. Gostyukhin VL, Trusov VN, Gostyukhin AV (2011) Active phased array antennas. Radiotekhnika, Moskow (in Russian)
65. Pringle LN, Harm PH, Blalock SP et al (2004) A reconfigurable aperture antenna based on switched links between electrically small metallic patches. IEEE Trans Antennas Propag 52:1434–1445
66. Prilutsky AA (2011) Reconfigurable antenna systems. Physical fundamentals of instrumentation 11:49–64 (in Russian)
67. Prilutsky AA (2005) 3rd generation smart antenna. Antennas 10:52–54 (in Russian)
68. Prilutsky AA (2011) Scanning reflective antenna with an impedance cylindrical reflector in the form of a layered semiconductor-dielectric-metal structure with optocoupler control. Adv Mod Electron 4:53–59 (in Russian)
69. Vardaxoglou JC (1997) Frequency selective surfaces, analysis and design. Research Studies Press, England
70. Munk BA (2000) Frequency selective surfaces: theory and design. Wiley, New York
71. King RWP, Wu T (1966) The imperfectly conducting cylindrical transmitting antenna. IEEE Trans Antennas Propag 14:524–534
72. Hanson GW (2008) Radiation efficiency of nano-radius dipole antennas in the microwave and far-infrared regimes. IEEE Antennas Propag Mag 50:66–77
73. Hanson GW (2005) Fundamental transmitting properties of carbon nanotube antennas. IEEE Trans Antennas Propag 53:3426–3435
74. Lerer AM, Makhno VV, Makhno PV (2010) Calculation of properties of carbon nanotube antennas. Int J Microwave Wirel Technol 2(5):457–462
75. Lagarkov AN, Kisel VN, Semenenko VN (2008) Wide-angle absorption by the use of a metamaterial plate. Prog Electromagn Res Lett 1:35–44
76. Lagarkov AN, Semenenko VN, Basharin AA, Balabukha NP (2008) Abnormal radiation pattern of metamaterial waveguide. PIERS Online 4(6):641–644
77. Vendik IB, Vendik OG (2013) Metamaterials and their application in microwave technology (overview). J Tech Phys 83(1):3–28 (in Russian)

78. Weinstein LA (1966) Diffraction theory and factorization method. Sov. Radio, Moskow (in Russian)
79. Nefedov EI (1979) Diffraction of electromagnetic waves by dielectric structures. Nauka, Moskow (in Russian)
80. Takizawa K, Hashimoto O (1999) Transparent wave absorber using resistive film at V-band frequency. IEEE Trans Microwave Theory Tech 47:1137–1141
81. Hillion P (2004) Impedance boundary conditions on a chiral film. Prog Electromagn Res 44:267–286
82. Potapov AA (2005) Fractals in radiophysics and radar. Universitetskaya kniga, Moskow (in Russian)
83. Felsen LB, Marcuvitz N (1973) Radiation and scattering of waves. Prentice-Hall Inc., Englewood Cliffs, NJ
84. Yoshitomi K (2001) Radiation from a slot in an impedance surface. IEEE Trans Antennas Propag 49:1370–1376
85. Katsenelenbaum BZ (1966) High frequency electrodynamics. Nauka, Moskow (in Russian)
86. Penkin YuM, Katrich VA (2003) Excitation of electromagnetic waves in the volumes with coordinate boundaries. Fakt, Kharkov (in Russian)
87. Feld YaN (1968) Basic equations, uniqueness theorem and boundary problems of electrodynamics. In: Ith all-union workshop on diffraction and wave propagation. VIRTA, Moskow, Kharkov, pp 93–109 (in Russian)
88. Penkin YuM (2001) Features of the uniqueness theorem and the duality principle for regions with impedance surfaces. Radio Eng 117:96–99 (in Russian)
89. Morse PM, Feshbach H (1953) Methods of theoretical physics. McGraw-Hill, New York
90. Panchenko BA (1970) Tensor Green's functions of Maxwell equations for cylindrical regions. Radio Eng 15:82–91 (in Russian)

Chapter 2
General Issues of the Theory of Thin Impedance Vibrators and Narrow Slots in a Spatial-Frequency Representation

2.1 Problem Formulation and Initial Integral Equations

Let us formulate in the most general form the problem of excitation, scattering and radiation of electromagnetic fields by material bodies of finite size in the presence of the coupling holes between the two electrodynamic volumes. Let there be an arbitrary volume V_1 bounded by a perfectly conducting, impedance, or partially impedance surface S_1, some parts of which can be removed to infinity. The volume V_1 is coupling with another arbitrary volume V_2 through holes $\Sigma_n (n = 1, 2 \ldots N)$ cut in the common surface between the two volumes. The boundary wall between the coupling volumes V_1 and V_2 has an infinitely small thickness in the region of the coupling holes. The permittivity and permeabilities of the media filling the volumes V_1 and V_2, are equal to ε_1, μ_1 and ε_2, μ_2. Let the volume V_1 contains material bodies enclosed in local volumes $V_{m_1} (m_1 = 1, 2 \ldots M_1)$, which are bounded by smooth closed surfaces S_{m_1}. The bodies are characterized by homogeneous material parameters: permittivity ε_{m_1}, permeability μ_{m_1}, and conductivity σ_{m_1}. Analogously, the volume V_2 contains in local volumes $V_{m_2} (m_2 = 1, 2 \ldots M_2)$, material bodies are limited by smooth closed surfaces S_{m_2}, which characterized by homogeneous material parameters: permittivity ε_{m_2}, permeability μ_{m_2} and conductivity σ_{m_2}(Fig. o).

Fields of extraneous sources can be defined in several different ways: (1) as fields of electromagnetic waves incident on bodies and holes in the scattering problem, (2) as fields of electromotive forces applied to the bodies which are different from zero only in certain areas of the volumes V_{m_1} and V_{m_2} in the radiation problem, and (3) as combination of these two fields. Without loss of generality, we will assume that the electromagnetic field $\{\vec{E}_0(\vec{r}), \vec{H}_0(\vec{r})\}$ of specified extraneous sources exist only inside the volume V_1 ($\vec{r}$ is the radius-vector of the observation point). The fields depend on time t as $e^{i\omega t}$ ($\omega = 2\pi f$ is the circular frequency, and f is the frequency in Hz). The problem is formulated as follows: to determine the total electromagnetic

© The Author(s), under exclusive license to Springer Nature Switzerland AG 2020
M. V. Nesterenko et al., *Combined Vibrator-Slot Structures: Theory and Applications*, Lecture Notes in Electrical Engineering 689,
https://doi.org/10.1007/978-3-030-60177-5_2

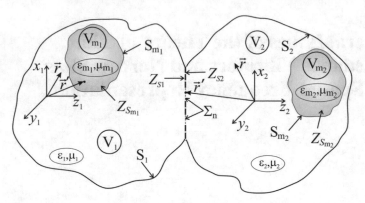

Fig. 2.1 The problem geometry and accepted notation

fields $\{\vec{E}_{V_1}(\vec{r}), \ \vec{H}_{V_1}(\vec{r})\}$ and $\{\vec{E}_{V_2}(\vec{r}), \ \vec{H}_{V_2}(\vec{r})\}$ in the volumes V_1 and V_2, satisfying the Maxwell equations and the boundary conditions on the surfaces $S_{m_1}, S_{m_2}, \Sigma_n, S_1$ and S_2.

The total electromagnetic fields in the volumes V_1 and V_2 can be expressed through the tangential field components on the surfaces S_{m_1}, S_{m_2} and Σ_n. In the Gaussian CGS system of units, the fields can be represented as Kirchhoff-Kotler integral equations [1, 2]

$$\vec{E}_{V_1}(\vec{r}) = \vec{E}_0(\vec{r}) + \frac{1}{4\pi i k \varepsilon_1}(\text{graddiv} + k_1^2)\sum_{m_1=1}^{M_1}\int\limits_{S_{m_1}} \hat{G}_{V_1}^e(\vec{r},\vec{r}'_{m_1})[\vec{n}_{m_1},\vec{H}_{V_1}(\vec{r}'_{m_1})]\mathrm{d}\vec{r}'_{m_1}$$

$$-\frac{1}{4\pi}\text{rot}\left\{ \begin{array}{l} \displaystyle\sum_{m_1=1}^{M_1}\int\limits_{S_{m_1}} \hat{G}_{V_1}^m(\vec{r},\vec{r}'_{m_1})[\vec{n}_{m_1},\vec{E}_{V_1}(\vec{r}'_{m_1})]\mathrm{d}\vec{r}'_{m_1} \\[4mm] \displaystyle+\sum_{n=1}^{N}\int\limits_{\Sigma_n} \hat{G}_{V_1}^m(\vec{r},\vec{r}'_{n})[\vec{n}_{n},\vec{E}_{V_1}(\vec{r}'_{n})]\mathrm{d}\vec{r}'_{n} \end{array}\right\},$$

$$\vec{H}_{V_1}(\vec{r}) = \vec{H}_0(\vec{r}) + \frac{1}{4\pi i k \mu_1}(\text{graddiv}$$

$$+ k_1^2)\left\{ \begin{array}{l} \displaystyle\sum_{m_1=1}^{M_1}\int\limits_{S_{m_1}} \hat{G}_{V_1}^m(\vec{r},\vec{r}'_{m_1})[\vec{n}_{m_1},\vec{E}_{V_1}(\vec{r}'_{m_1})]\mathrm{d}\vec{r}'_{m_1} \\[4mm] \displaystyle+\sum_{n=1}^{N}\int\limits_{\Sigma_n} \hat{G}_{V_1}^m(\vec{r},\vec{r}'_{n}[\vec{n}_{n},\vec{E}_{V_1}(\vec{r}'_{n})]\mathrm{d}\vec{r}'_{n} \end{array}\right\}$$

$$+\frac{1}{4\pi}\text{rot}\sum_{m_1=1}^{M_1}\int\limits_{S_{m_1}} \hat{G}_{V_1}^e(\vec{r},\vec{r}'_{m_1})[\vec{n}_{m_1},\vec{H}_{V_1}(\vec{r}'_{m_1})]\mathrm{d}\vec{r}'_{m_1}, \tag{2.1}$$

$$\vec{E}_{V_2}(\vec{r}) = \frac{1}{4\pi i k \epsilon_2}(\text{graddiv} + k_2^2) \sum_{m_2=1}^{M_2} \int_{S_{m_2}} \hat{G}^e_{V_2}(\vec{r}, \vec{r}'_{m_2})[\vec{n}_{m_2}, \vec{H}_{V_2}(\vec{r}'_{m_2})]d\vec{r}'_{m_2}$$

$$-\frac{1}{4\pi}\text{rot}\left\{\begin{array}{l} \sum_{m_2=1}^{M_2}\int_{S_{m_2}} \hat{G}^m_{V_2}(\vec{r},\vec{r}'_{m_2})[\vec{n}_{m_2}, \vec{E}_{V_2}(\vec{r}'_{m_2})]d\vec{r}'_{m_2} \\[1em] +\sum_{n=1}^{N}\int_{\Sigma_n} \hat{G}^m_{V_2}(\vec{r},\vec{r}'_n)[\vec{n}_n, \vec{E}_{V_2}(\vec{r}'_n)]d\vec{r}'_n \end{array}\right\},$$

$$\vec{H}_{V_2}(\vec{r}) = \frac{1}{4 up\pi i k \mu_2}(\text{graddiv} + k_2^2)\left\{\begin{array}{l} \sum_{m_2=1}^{M_2}\int_{S_{m_2}} \hat{G}^m_{V_2}(\vec{r},\vec{r}'_{m_2})[\vec{n}_{m_2}, \vec{E}_{V_2}(\vec{r}'_{m_2})]d\vec{r}'_{m_2} \\[1em] +\sum_{n=1}^{N}\int_{\Sigma_n} \hat{G}^m_{V_2}(\vec{r},\vec{r}'_n)[\vec{n}_n, \vec{E}_{V_2}(\vec{r}'_n)]d\vec{r}'_n \end{array}\right\}$$

$$+\frac{1}{4\pi}\text{rot}\sum_{m_2=1}^{M_2}\int_{S_{m_2}} \hat{G}^e_{V_2}(\vec{r},\vec{r}'_{m_2})[\vec{n}_{m_2}, \vec{H}_{V_2}(\vec{r}'_{m_2})]d\vec{r}'_{m_2},$$

where $k = 2\pi/\lambda$ is the wave number, λ is the free space wavelength, $k_1 = k\sqrt{\epsilon_1\mu_1}$, $k_2 = k\sqrt{\epsilon_2\mu_2}$ are wave numbers in the media filling the volumes, $\vec{r}'_{m_1,m_2,n}$ are radius-vectors of source points located on surfaces S_{m_1}, S_{m_2} and Σ_n, $\vec{n}_{m1,m2,n}$ are the unit vectors of external normals to the surfaces, $\hat{G}^e_{V_1,V_2}(\vec{r},\vec{r}')$ and $\hat{G}^m_{V_1,V_2}(\vec{r},\vec{r}')$ are the electric and magnetic tensor Green's functions for the vector Hertz potentials of the coupling volumes, satisfying the Helmholtz vector equation and corresponding boundary conditions on the surfaces S_1 and $S_)$. The boundary conditions for $\hat{G}^{e,m}_{V_1,V_2}(\vec{r},\vec{r}')$ at parts of the surfaces S_1 or S_2 displaced to infinity are transformed into the Sommerfeld radiation conditions.

The fields on the left-hand side of the Eq. (2.1) are interpreted depending on the position of the observation point $\vec{r}$. If the observation point is located on the surfaces S_{m_1}, S_{m_2} of the bodies V_{m_1} and V_{m_2} or on the hole apertures Σ_n, the fields $\vec{E}(\vec{r})$ and $\vec{H}(\vec{r})$ coincide with fields in the integrands of the right-hand side of the Eq. (2.1) which is the inhomogeneous linear integral Fredholm equation of the second kind that has a unique solution. If the observation point lies outside the volumes V_{m_1}, V_{m_2} and surfaces Σ_n, the Eq. (2.1) become equalities determining the total electromagnetic field by the field of the extraneous sources. These equalities allow us to solve, in the most general form, the excitation problem of electromagnetic fields by obstacles of finite size if the fields on the obstacle surfaces are known. Of course, these fields can be found if integral equations are solved.

The representation (2.1) can be also used to solve electrodynamic problems when some additional physical considerations for determining fields on the body surfaces are involved. For example, the current induced on well-conducting bodies ($\sigma \to \infty$), is concentrated near the body surface. Then, if the skin layer depth is small, the Shchukin-Leontovich approximate impedance boundary condition (1.17) can be used. If the impedance varies at the material body surface, the boundary condition can be presented as

$$[\vec{n}, \vec{E}(\vec{r})] = \bar{Z}_S(\vec{r})[\vec{n}, [\vec{n}, \vec{H}(\vec{r})]], \tag{2.2}$$

where $\bar{Z}_S(\vec{r}) = \bar{R}_S(\vec{r}) + i\bar{X}_S(\vec{r}) = Z_S(\vec{r})/Z_0$ is distributed variable surface impedance.

Based on the impedance boundary condition (2.2), the integral Eq. (2.1) can be converted to new variables, the density of the surface currents. Without loss of generality, let us convert the system (2.1) to the equation system relative to the surface currents for the configuration consisting of two material bodies in the volume V_1, coupling with the volume V_2 through a hole. If the observation point is placed on the impedance surfaces S_{1_1} and S_{2_1} (indices 1, 2) and the continuity condition for the tangential components of the magnetic fields on the hole Σ (index 3) are applied, the following system of integral equations can be obtained

$$Z_{S1}(\vec{r}_1)\vec{J}_1^e(\vec{r}_1) + \frac{k}{\omega}\mathrm{rot}\int_{\Sigma} \hat{G}_{V_1}^m(\vec{r}, \vec{r}_3')\vec{J}_3^m(\vec{r}_3')\mathrm{d}\vec{r}_3' = \vec{E}_0(\vec{r})$$

$$+ \frac{1}{i\omega\varepsilon_1}(\mathrm{graddiv} + k_1^2)\left\{\int_{S_{1_1}} \hat{G}_{V_1}^e(\vec{r}, \vec{r}_1')\vec{J}_1^e(\vec{r}_1')\mathrm{d}\vec{r}_1' + \int_{S_{2_1}} \hat{G}_{V_1}^e(\vec{r}, \vec{r}_2')\vec{J}_2^e(\vec{r}_2')\mathrm{d}\vec{r}_2'\right\}$$

$$+ \frac{1}{4\pi}\mathrm{rot}\left\{\int_{S_{1_1}} \hat{G}_{V_1}^m(\vec{r}, \vec{r}_1')Z_{S1}(\vec{r}_1')[\vec{n}_1, \vec{J}_1^e(\vec{r}_1')]\mathrm{d}\vec{r}_1'\right.$$

$$\left. + \int_{S_{2_1}} \hat{G}_{V_1}^m(\vec{r}, \vec{r}_2')Z_{S2}(\vec{r}_2')[\vec{n}_2, \vec{J}_2^e(\vec{r}_2')]\mathrm{d}\vec{r}_2'\right\}, \tag{2.3a}$$

$$Z_{S2}(\vec{r}_2)\vec{J}_2^e(\vec{r}_2) + \frac{k}{\omega}\mathrm{rot}\int_{\Sigma} \hat{G}_{V_1}^m(\vec{r}, \vec{r}_3')\vec{J}_3^m(\vec{r}_3')\mathrm{d}\vec{r}_3' = \vec{E}_0(\vec{r})$$

$$+ \frac{1}{i\omega\varepsilon_1}(\mathrm{graddiv} + k_1^2)\left\{\int_{S_{2_1}} \hat{G}_{V_1}^e(\vec{r}, \vec{r}_2')\vec{J}_2^e(\vec{r}_2')\mathrm{d}\vec{r}_2' + \int_{S_{1_1}} \hat{G}_{V_1}^e(\vec{r}, \vec{r}_1')\vec{J}_1^e(\vec{r}_1')\mathrm{d}\vec{r}_1'\right\}$$

$$+ \frac{1}{4\pi}\mathrm{rot}\left\{\int_{S_{2_1}} \hat{G}_{V_1}^m(\vec{r}, \vec{r}_2')Z_{S2}(\vec{r}_2')[\vec{n}_2, \vec{J}_2^e(\vec{r}_2')]\mathrm{d}\vec{r}_2' + \int_{S_{1_1}} \hat{G}_{V_1}^m(\vec{r}, \vec{r}_1')Z_{S1}(\vec{r}_1')[\vec{n}_1, \vec{J}_1^e(\vec{r}_1')]\mathrm{d}\vec{r}_1'\right\}, \tag{2.3b}$$

$$\vec{H}_0(\vec{r}) + \frac{1}{i\omega\mu_1}(\text{graddiv} + k_1^2)\int_\Sigma \hat{G}_{V_1}^m(\vec{r}, \vec{r}_3')\vec{J}_3^m(\vec{r}_3')\mathrm{d}\vec{r}_3'$$

$$+ \frac{1}{i\omega\mu_2}(\text{graddiv} + k_2^2)\int_\Sigma \hat{G}_{V_2}^m(\vec{r}, \vec{r}_3')\vec{J}_3^m(\vec{r}_3')\mathrm{d}\vec{r}_3'$$

$$= \frac{1}{i\omega\varepsilon_1}(\text{graddiv} + k_1^2)\left\{\begin{array}{l}\displaystyle\int_{S_{1_1}} \hat{G}_{V_1}^m(\vec{r}, \vec{r}_1')Z_{S1}(\vec{r}_1')[\vec{n}_1, \vec{J}_1^e(\vec{r}_1')]\mathrm{d}\vec{r}_1' \\[3mm] \displaystyle+ \int_{S_{2_1}} \hat{G}_{V_1}^m(\vec{r}, \vec{r}_2')Z_{S2}(\vec{r}_2')[\vec{n}_2, \vec{J}_2^e(\vec{r}_2')]\mathrm{d}\vec{r}_2'\end{array}\right\}$$

$$- \frac{k}{\omega}\text{rot}\left\{\int_{S_{1_1}} \hat{G}_{V_1}^e(\vec{r}, \vec{r}_1')\vec{J}_1^e(\vec{r}_1')\mathrm{d}\vec{r}_1' + \int_{S_{2_1}} \hat{G}_{V_1}^e(\vec{r}, \vec{r}_2')\vec{J}_2^e(\vec{r}_2')\mathrm{d}\vec{r}_2'\right\}. \qquad (2.3c)$$

with respect to electric surface current densities: $\vec{J}_{1,2}^e(\vec{r}_{1,2})$ on $S_{1_1,2_1}$ and equivalent magnetic $\vec{J}_3^m(\vec{r}_3)$ on Σ. These currents are equal to

$$\vec{J}_{1,2}^e(\vec{r}_{1,2}) = \frac{c}{4\pi}[\vec{n}_{1,2}, \vec{H}(\vec{r}_{1,2})], \quad \vec{J}_3^m(\vec{r}_3) = \frac{c}{4\pi}[\vec{n}_3, \vec{E}(\vec{r}_3)]. \qquad (2.4)$$

Thus, the problem of electromagnetic wave excitation by impedance bodies of finite size and coupling holes cut in the wall separating the electrodynamic volumes can be formulated as a boundary value problem of macroscopic electrodynamics which is reduced to integral equations for currents. The solution of this problem is an independent problem, often associated with significant mathematical difficulties. If the characteristic dimensions of the obstacles are much greater than the wavelength, that is, in the high-frequency approximation, the solution of the problem is usually sought in an inverse power of the wave number. In the low-frequency region, when the obstacles dimensions are much smaller than the wavelength, the representation of unknown functions in the form of power series expansion of the wavenumber reduces to solving a sequence of electrostatic problems. In the resonant region, where at least one of the dimensions of the obstacles is commensurable with the wavelength, the problem is the most difficult to analyze, since it requires a rigorous solution of the field equations. From a practical point of view, such obstacles as thin impedance vibrators and narrow slots in the resonant frequency domain are of particular interest.

2.1.1 Green's Function as Integral Equation Kernel

As noted above, the key problem of electromagnetic wave excitation by finite-size obstacles is to determine the fields induced by point sources. One of the approaches for solving this problem is the Green's function method. Analytical solutions of boundary-value problems of macroscopic electrodynamics can be found if the Green's functions for specific spatial domains are known [3–13]. As is known, the Green's function of the vector fields is a symmetric tensor of second rank (affinor), which is a function of the relative position of observation and source points defined by the radius-vectors $\vec{r}$ and $\vec{r}'$.

The Green's functions of inhomogeneous vector Eqs. (1.5) or (1.10), can be determined by solving one of the following tensor equations

$$\mathrm{rotrot}\hat{G}_{E(H)}(\vec{r}, \vec{r}') - k_1^2 \hat{G}_{E(H)}(\vec{r}, \vec{r}') = 4\pi\hat{I}\delta(|\vec{r} - \vec{r}'|), \tag{2.5}$$

$$\Delta\hat{G}_A^{e(m)}(\vec{r}, \vec{r}') + k_1^2 \hat{G}_A^{e(m)}(\vec{r}, \vec{r}') = -4\pi\hat{I}\delta(|\vec{r} - \vec{r}'|), \tag{2.6}$$

which satisfy the boundary conditions for the corresponding boundary value problem. In the Eqs. (2.5) and (2.6), $\hat{G}_{E(H)}(\vec{r}, \vec{r}')$ are the Green's functions for the electric and magnetic fields, and $\hat{G}_A^{e(m)}(\vec{r}, \vec{r}')$ are the Green's functions for the electric or magnetic vector potentials. The Eq. (2.6) are correct for coordinate systems where arbitrary electromagnetic fields can be represented as a sum of electric and magnetic fields. In the Cartesian coordinate system, $\delta(|\vec{r} - \vec{r}'|) = \delta(x - x')\delta(y - y')\delta(z - z')$ is the 3D Dirac delta-function, $\hat{I} = (\vec{e}_x \otimes \vec{e}_{x'}) + (\vec{e}_y \otimes \vec{e}_{y'}) + (\vec{e}_z \otimes \vec{e}_{z'})$ is the unit affinor, $\vec{e}_x, \vec{e}_y, \vec{e}_z$ are the unit vectors, Δ is the Laplace operator, and $\otimes$ is the sign of tensor multiplication.

The Green's functions allow us to obtain, in the closed form, the expression (2.1) defining the electromagnetic fields for arbitrary vector sources at any point in space. For example, if the electric current density $\vec{J}^e(\vec{r})$ is specified as the source, the electric field can be defined by one of the following expressions

$$\vec{E}(\vec{r}) = \frac{k_1^2}{i\omega} \int_V \hat{G}_E(\vec{r}, \vec{r}')\vec{J}^e(\vec{r}')d\vec{r}', \tag{2.7}$$

$$\vec{E}(\vec{r}) = \frac{1}{i\omega}(\mathrm{graddiv} + k_1^2) \int_V \hat{G}_A^e(\vec{r}, \vec{r}')\vec{J}^e(\vec{r}')d\vec{r}', \tag{2.8}$$

where the excitation field $\vec{E}_0(\vec{r}) = 0$ was used for simplicity. These two representations can formally be associated with a different potential calibration of electromagnetic fields. The Green's functions $\hat{G}_E(\vec{r}, \vec{r}')$ and $\hat{G}_A^e(\vec{r}, \vec{r}')$ are related as

$$\hat{G}_E(\vec{r}, \vec{r}') = \left(\hat{I} + \frac{1}{k_1^2}\mathrm{grad} \otimes \mathrm{grad}\right)\hat{G}_A^e(\vec{r}, \vec{r}'), \tag{2.9}$$

where operation {grad $\otimes$ grad} is the tensor product of the two symbolic vectors. The formula (2.9) is valid only in rectangular and cylindrical coordinate systems, where the expressions (1.6) are applicable, since they follow from the relation $\Delta \vec{F} = \text{graddiv} \vec{F} - \text{rotrot} \vec{F}$ in which the vector function $\vec{F}$ must satisfy the complete wave equation.

For infinite space, the only boundary condition imposed on the Green's function is the Sommerfeld radiation condition. In this case, the Green's function can be represented as

$$\hat{G}^{e(m)}(\vec{r}, \vec{r}') = \hat{I}\frac{e^{-ik_1|\vec{r}-\vec{r}'|}}{|\vec{r} - \vec{r}'|} = \hat{I}G_0^{e(m)}(\vec{r}, \vec{r}'). \qquad (2.10)$$

If $\vec{r} = \vec{r}'$, the functions (2.10) and (2.9) are singular and the integrals in (2.7), (2.8) cannot be defined as the limit of the integral sum, since it does not exist. Therefore, the formulas (2.7), (2.8) are valid for the points where the sources are absent. If the observation point coincides with one of the source points, the volume integrals in (2.7) and (2.8) are improper, that is, $\int_V = \lim_{\rho \to 0} \int_{V-v}$ where v is the excluded volume enclosed in a sphere of infinitely small radius ρ with center at the point $\vec{r}'$. The main difference between the two types of integrals is that the improper integral in (2.8) can be modified to absolutely convergent integral [9], whereas the integral in (2.7) is conditionally convergent [9], therefore, its value depends on the shape of the excluded region containing a singular point. Physically correct definition of the field in the source region require that principal integral value in (2.7) defining the main contribution to the integration result should be used. However, the singular part of the Green's functions can be isolated and the integrals may be interpreted in the usual sense. The behavior of the tensor Green's function (2.9) in the vicinity of the field source points is rather difficult to investigate. Rigorous treatment of these issues requires the application of generalized functions [11]. In this regard, it should be noted the advantage of using the expression for the field (2.8), based on the Green's function for the vector potential, compared to (2.7) and (2.9), namely: the field (2.8) has an feature and the calculation of the principal value the integral can be avoided.

The expression for the field (2.8) based on the Green's function for the vector potential has some advantages as compared with (2.7) and (2.9), since the Green's function is integrable.

This consideration fully applies to bounded regions, since the Green's function for a bounded region has the same singularity as in the free space [4]. We emphasize once again that the Green's function of a bounded region is the solution of Eqs. (2.5) or (2.6) defining the field (2.7) or (2.8), which must satisfy the corresponding boundary conditions on S_1 (S_2). In each of the volumes $V_{1(2)}$, the general solution of Eq. (2.6) can be represented as [1, 14, 15]

$$\hat{G}_{V_{1(2)}}^{e(m)}(\vec{r}, \vec{r}') = \hat{I}G_0^{e(m)}(\vec{r}, \vec{r}') + \hat{g}_{V_{1(2)}}^{e(m)}(\vec{r}, \vec{r}'), \qquad (2.11)$$

where $G_0^{e(m)}(\vec{r}, \vec{r}') = \frac{e^{-ik_{1(2)}|\vec{r}-\vec{r}'|}}{|\vec{r}-\vec{r}'|}$, and $\hat{g}_{V_{1(2)}}^{e(m)}(\vec{r}, \vec{r}')$ is a regular everywhere function satisfying a homogeneous equation

$$\Delta \hat{g}_{V_{1(2)}}^{e(m)}\left(\vec{r}, \vec{r}'\right) + k_{1(2)}^2 \hat{g}_{V_{1(2)}}^{e(m)}\left(\vec{r}, \vec{r}'\right) = 0 \tag{2.12}$$

which together with $\hat{I}G_0^{e(m)}\left(\vec{r}, \vec{r}'\right)$ provide fulfillment of boundary conditions on the surface $S_1(S_2)$ of the volume $V_1(V_2)$ for a point source field of electrical or magnetic type, located at the point $\vec{r}'$.

Representation (2.11) is fundamentally important for the solution of the equation system (2.1), which is quite natural if the Green's functions for specific volumes are constructed using the mirror image method. For example, the Green's functions for a free half space over a perfectly conducting plane $X0Y$ in a rectangular coordinate system (x, y, z) can be define as follows (Appendix A)

$$\hat{G}^{e(m)}\left(\vec{r}, \vec{r}'\right) = \hat{I}G_0^{e(m)}\left(\vec{r}, \vec{r}'\right) + \hat{g}^{e(m)}\left(\vec{r}, \vec{r}'\right) = \hat{I}G_0^{e(m)}\left(\vec{r}, \vec{r}'\right) - (+) \begin{vmatrix} g_{xx}(R_1) & 0 & 0 \\ 0 & g_{yy}(R_1) & 0 \\ 0 & 0 & -g_{zz}(R_1) \end{vmatrix},$$
$$\tag{2.13}$$

where $G_0^{e(m)}\left(\vec{r}, \vec{r}'\right) = \frac{e^{-ikR}}{R}$, $R = \sqrt{(x - x')^2 + (y - y')^2 + (z - z')^2}$, $g_{xx}(R_1) = g_{yy}(R_1) = g_{zz}(R_1) = \frac{e^{-ikR_1}}{R_1}$, $R_1 = \sqrt{(x - x')^2 + (y + y')^2 + (z - z')^2}$.

The Green's function components for the wedge region have a similar structure (Appendix A). That is, in any case, the Green's function representation of the will contain explicitly the term $\hat{I}G_0^{e(m)}\left(\vec{r}, \vec{r}'\right)$. Since this approach is not always possible to apply for more complex form of the volume $V_1(V_2)$, the Green's functions can be built by using other methods based on a basis of eigenfunctions for corresponding spatial domains.

Among these methods, constructing the electrical Green's function in the source-like representation (2.11) for circular waveguideы and resonator should be distinguished [16]. This method is based on solving the diffraction problem of tensor divergent spherical and quasispherical waves at the walls of electrodynamic volumes. However, the analytical expressions thus obtained turn out to be rather complicated.

An alternative approach is to construct the full Green's function $\hat{G}^{e(m)}\left(\vec{r}, \vec{r}'\right)$ by any method and then to extract its source-like component. the most common approach for such extraction consists in "forced formatting" when the Green's function is presented as

$$\hat{G}^{e(m)}\left(\vec{r}, \vec{r}'\right) = \hat{I}\frac{e^{-ik|\vec{r}-\vec{r}'|}}{|\vec{r} - \vec{r}'|} + \hat{G}^{e(m)}\left(\vec{r}, \vec{r}'\right) - \hat{I}G_0^{e(m)}\left(\vec{r}, \vec{r}'\right), \tag{2.14}$$

where the first term is the Green's function for infinite space in the source-like form, and the third term represent this function in the basis of coordinates and eigenfunctions in which the full function was constructed. The expressions of the Green's function $G_0^{e(m)}\left(\vec{r}, \vec{r}'\right)$ in the four basic orthogonal coordinate systems derived in the monograph [17], are given in Appendix B.

Next we should redefine the coefficients of the eigenfunctions in the expressions for the components $\hat{G}^{e(m)}(\vec{r}, \vec{r}')$, so as to satisfy the boundary condition for $\hat{g}^{e(m)}(\vec{r}, \vec{r}')$ on the surface S in the form $\hat{g}^{e(m)}(\vec{r}, \vec{r}')\big|_S = -\hat{I}G_0^{e(m)}(\vec{r}, \vec{r}')\big|_S$. Such a requirement ensures the equality to zero the sum of two surface integrals in (1.99) of corresponding to the source and regular components of the Green's function, taking into account the relation $\left(\Delta + k_{1(2)}^2\right)\left(\hat{G}^{e(m)}(\vec{r}, \vec{r}') - \hat{I}G_0^{e(m)}(\vec{r}, \vec{r}')\right) = 0$ (2.12). This requirement ensures that the sum of two surface integrals in (1.99) corresponding to the source-like and regular components were equal to zero.

The terms $\hat{G}^{e(m)}(\vec{r}, \vec{r}') - \hat{I}G_0^{e(m)}(\vec{r}, \vec{r}')$ in the identities (2.14) for the small distances $|\vec{r} - \vec{r}'|$ cannot be calculated independently to each other before applying the subtraction operation. This will inevitably lead to computational errors, since each of the terms is a slowly convergent series (or integral) that cannot be calculated with the same accuracy. Since the expression infinity minus infinity is indeterminate, the correct interpretation of the regular function infinity is impossible. Therefore, before numerical calculations, the terms in the expression $\hat{G}^{e(m)}(\vec{r}, \vec{r}') - \hat{I}G_0^{e(m)}(\vec{r}, \vec{r}')$ should first be combined within the framework of common basic decomposition.

Of course, such approach to the Green's function representation is applicable only for the volumes V_1 and V_2 for which the Green's functions $\hat{G}^{e(m)}(\vec{r}, \vec{r}')$ can be constructed. This approach is also applicable to rectangular resonators, infinite and semi-infinite rectangular waveguides discussed later in the monograph. However, the source-like representation of the Green's function $G_0^{e(m)}(\vec{r}, \vec{r}') = \frac{e^{-ik_{1(2)}|\vec{r}-\vec{r}'|}}{|\vec{r}-\vec{r}'|}$ can be used to solve the excitation problem for the complex electrodynamic volumes. In this case, both the volume and surface integrals should be taken into account in the relations (1.99).

2.1.2 Integral Equations for Electric and Magnetic Currents in Thin Vibrators and Narrow Slots

Solution of the equation system (2.3a) for material bodies with complex surface forms S and coupling holes Σ with arbitrary geometry certain mathematical difficulties may encounter. However, the equation system solution for cylinders and holes whose transverse dimensions are small compared to free space wavelength, that is, for thin vibrators and narrow slots is considerably simplified [15, 18]. In addition, the boundary condition (2.2) can be extended to the vibrator surfaces with an arbitrary distribution of complex impedance on their surfaces, regardless of the exciting field structure and electrical characteristics of the vibrator material [19].

Let us transform the equation system (2.3a) so that it can be applicable for thin vibrators made of circular cylindrical wires and rectilinear narrow slots. If the wire radii $r_{1,2}$ and lengths $2L_{1,2}$ and slot width d and length $2L_3$ satisfy the inequalities

$$\frac{r_{1,2}}{2L_{1,2}} \ll 1, \quad \frac{r_{1,2}}{\lambda_{1,2}} \ll 1, \quad \frac{d}{2L_3} \ll 1, \quad \frac{d}{\lambda_{1,2}} \ll 1, \tag{2.15}$$

($\lambda_{1,2}$ are the wavelengths in corresponding media), the induced currents on the vibrators and the equivalent magnetic current in the slot can be represented as follows

$$\vec{J}_{1(2)}^e(\vec{r}_{1(2)}) = \vec{e}_{s_{1(2)}} J_{1(2)}(s_{1(2)})\psi_{1(2)}(\rho_{1(2)}, \varphi_{1(2)}), \quad \vec{J}_3^m(\vec{r}_3) = \vec{e}_{s_3} J_3(s_3)\chi(\xi), \tag{2.16}$$

where $\vec{e}_{s_{1(2)}}$ and $\vec{e}_{s_3}$ are the unit vectors directed along the vibrator and slot axes, $s_{1(2)}$ and s_3 are local coordinates associated with the vibrator and slot axes, $\psi_{1(2)}(\rho_{1(2)}, \varphi_{1(2)})$ are the functions of transverse polar coordinates $(\rho_{1(2)}, \varphi_{1(2)})$ for vibrators, $\chi(\xi)$ is a function of a transverse coordinate ξ for the slot. The functions $\psi_{1(2)}(\rho_{1(2)}, \varphi_{1(2)})$ and $\chi(\xi)$ satisfy the normalization conditions

$$\int_{\perp_{1(2)}} \psi_{1(2)}(\rho_{1(2)}, \varphi_{1(2)})\rho_{1(2)}d\rho_{1(2)}d\varphi_{1(2)} = 1, \int_{\xi} \chi(\xi)d\xi = 1. \tag{2.17}$$

The unknown currents $J_{1(2)}(s_{1(2)})$ and $J_3(s_3)$ must satisfy the boundary conditions

$$J_{1(2)}(\pm L_{1(2)}) = 0, J_3(\pm L_3) = 0, \tag{2.18}$$

where superscripts $e(m)$ are omitted. Let us project the Eqs. (2.3a), (2.3b) and (2.3c) on the axes of the vibrators and slot. Then, since $[\vec{n}_{1(2)}, \vec{J}_{1(2)}(\vec{r}_{1(2)})] \ll 1$ according to (2.15), the system of integral equations for currents on the vibrators and in the slot, accounting for mutual interactions between them can be presented as

$$\left(\frac{d^2}{ds_1^2} + k_1^2\right)\left\{\int_{-L_1}^{L_1} J_1(s_1')G_{s_1}^{V_1}(s_1, s_1')ds_1' + \int_{-L_2}^{L_2} J_2(s_2')G_{s_2}^{V_1}(s_1, s_2')ds_2'\right\}$$

$$- ik\vec{e}_{s_1}\mathrm{rot}\int_{-L_3}^{L_3} J_3(s_3')G_{s_3}^{V_1}(s_1, s_3')ds_3' = -i\omega\varepsilon_1[E_{0s_1}(s_1) - z_{i1}(s_1)J_1(s_1)],$$

$$\left(\frac{d^2}{ds_2^2} + k_1^2\right)\left\{\int_{-L_2}^{L_2} J_2(s_2')G_{s_2}^{V_1}(s_2, s_2')ds_2' + \int_{-L_1}^{L_1} J_1(s_1')G_{s_1}^{V_1}(s_2, s_1')ds_1'\right\}$$

$$- ik\vec{e}_{s_2}\mathrm{rot}\int_{-L_3}^{L_3} J_3(s_3')G_{s_3}^{V_1}(s_2, s_3')ds_3' = -i\omega\varepsilon_1[E_{0s_2}(s_2) - z_{i2}(s_2)J_2(s_2)],$$

$$\frac{1}{\mu_1}\left(\frac{d^2}{ds_3^2} + k_1^2\right)\int_{-L_3}^{L_3} J_3(s_3')G_{s_3}^{V_1}(s_3, s_3')ds_3' + \frac{1}{\mu_2}\left(\frac{d^2}{ds_3^2} + k_2^2\right)\int_{-L_3}^{L_3} J_3(s_3')G_{s_3}^{V_2}(s_3, s_3')ds_3'$$

$$+ ik\vec{e}_{s_3}\mathrm{rot}\left\{\int\limits_{-L_1}^{L_1} J_1(s_1')G_{s_1}^{V_1}(s_3,s_1')ds_1' + \int\limits_{-L_2}^{L_2} J_2(s_2')G_{s_2}^{V_1}(s_3,s_2')ds_2'\right\} = -i\omega\, H_{0s_3}(s_3),$$

$$(2.19)$$

where $z_{i1(2)}(s_{1(2)})$ ([Ohm/m]) are the internal linear vibrator impedances $Z_{S1(2)}(\vec{r}_{1(2)}) = 2\pi r_{1(2)}z_{i1(2)}(\vec{r}_{1(2)})$, $E_{0s_{1(2)}}(s_{1(2)})$, and $H_{0s_3}(s_3)$ are the projections of the extraneous source fields on the vibrator and slot axes, and $G_{s_{1,2}}^{V_1}(s_{1,2,3},s_{1(2)}'),G_{s_3}^{V_{1(2)}}(s_{1,2,3},s_3')$ are the corresponding components of the tensor Green's functions of the volumes V_1 and V_2.

If the interaction between the vibrators and the slot is absent, the system of Eq. (2.19) is greatly simplified and has the following form

$$\left(\frac{d^2}{ds_1^2}+k_1^2\right)\left\{\int\limits_{-L_1}^{L_1} J_1(s_1')G_{s_1}^{V_1}(s_1,s_1')ds_1' + \int\limits_{-L_2}^{L_2} J_2(s_2')G_{s_2}^{V_1}(s_1,s_2')ds_2'\right\}$$

$$= -i\omega\varepsilon_1\left[E_{0s_1}(s_1) - z_{i1}(s_1)J_1(s_1)\right],$$

$$\left(\frac{d^2}{ds_2^2}+k_1^2\right)\left\{\int\limits_{-L_2}^{L_2} J_2(s_2')G_{s_2}^{V_1}(s_2,s_2')ds_2' + \int\limits_{-L_1}^{L_1} J_1(s_1')G_{s_1}^{V_1}(s_2,s_1')ds_1'\right\}$$

$$= -i\omega\varepsilon_1\left[E_{0s_2}(s_2) - z_{i2}(s_2)J_2(s_2)\right],$$

$$\frac{1}{\mu_1}\left(\frac{d^2}{ds_3^2}+k_1^2\right)\int\limits_{-L_3}^{L_3} J_3(s_3')G_{s_3}^{V_1}(s_3,s_3')ds_3'$$

$$+ \frac{1}{\mu_2}\left(\frac{d^2}{ds_3^2}+k_2^2\right)\int\limits_{-L_3}^{L_3} J_3(s_3')G_{s_3}^{V_2}(s_3,s_3')ds_3' = -i\omega H_{0s_3}(s_3). \qquad (2.20)$$

If configuration consists only of one vibrator and slot, the of equation system (2.20) splits into two independent equations

$$\left(\frac{d^2}{ds_1^2}+k_1^2\right)\int\limits_{-L_v}^{L_v} J_v(s_1')G_{s_1}^{V_1}(s_1,s_1')ds_1' = -i\omega\varepsilon_1\left[E_{0s_1}(s_1) - z_i(s_1)J_v(s_1)\right], \quad (2.21)$$

$$\frac{1}{\mu_1}\left(\frac{d^2}{ds_2^2}+k_1^2\right)\int\limits_{-L_{sl}}^{L_{sl}} J_{sl}(s_2')G_{s_2}^{V_1}(s_2,s_2')ds_2'$$

$$+ \frac{1}{\mu_2}\left(\frac{d^2}{ds_2^2}+k_2^2\right)\int\limits_{-L_{sl}}^{L_{sl}} J_{sl}(s_2')G_{s_2}^{V_2}(s_2,s_2')ds_2' = -i\omega H_{0s_2}(s_2), \qquad (2.22)$$

where indices v, 1, and sl, 2 are related to the vibrator and slot, respectively. The vibrator or slot may have a curved axial configuration, then, if the curvature radius of the vibrator axis and slot midline are large compared with their transverse dimensions, the equations for the electric current in the vibrator and the magnetic current in the slot based on (2.11) can be written as

$$
\int\limits_{L_v} \left\{ \left[\frac{\partial}{\partial s_1} \frac{\partial J_v(s_1')}{\partial s_1'} + k_1^2(\vec{e}_{s_1}\vec{e}_{s_1'}) J_v(s_1') \right] G_v(s_1, s_1') \right.
$$

$$
\left. + J_v(s_1') \left[\frac{\partial^2}{\partial s_1^2} + k_1^2 \right] \vec{e}_{s_1} \left(\hat{g}_{s_1}^{eV_1}(s_1, s_1')\vec{e}_{s_1'} \right) \right\} ds_1' = -i\omega\varepsilon_1 \left[E_{0s_1}(s_1) - z_i(s_1) J_v(s_1) \right],
\tag{2.23}
$$

$$
\int\limits_{L_{sl}} \left\{ \frac{1}{\mu_1} \left[\frac{\partial}{\partial s_2} \frac{\partial J_{sl}(s_2')}{\partial s_2'} + k_1^2(\vec{e}_{s_2}\vec{e}_{s_2'}) J_{sl}(s_2') \right] G_{sl}^{V_1}(s_2, s_2') \right.
$$

$$
+ \frac{1}{\mu_2} \left[\frac{\partial}{\partial s_2} \frac{\partial J_{sl}(s_2')}{\partial s_2'} + k_2^2(\vec{e}_{s_2}\vec{e}_{s_2'}) J_{sl}(s_2') \right] G_{sl}^{V_2}(s_2, s_2'),
\tag{2.24}
$$

$$
+ \frac{1}{\mu_1} J_{sl}(s_2') \left[\frac{\partial^2}{\partial s_2^2} + k_1^2 \right] \vec{e}_{s_2} \left(\hat{g}_{s_2}^{mV_1}(s_2, s_2')\vec{e}_{s_2'} \right)
$$

$$
\left. + \frac{1}{\mu_2} J_{sl}(s_2') \left[\frac{\partial^2}{\partial s_2^2} + k_2^2 \right] \vec{e}_{s_2} \left(\hat{g}_{s_2}^{mV_2}(s_2, s_2')\vec{e}_{s_2'} \right) \right\} ds_2' = -i\omega H_{0s_2}(s_2),
$$

where $\vec{e}_{s_1'}$ and $\vec{e}_{s_2'}$ are the unit vectors associated with the vibrator and slot axes at location of the sources,

$$
G_v(s_1, s_1') = \int\limits_{-\pi}^{\pi} \frac{e^{-ik_1\sqrt{(s_1-s_1')^2+[2r\sin(\varphi/2)]^2}}}{\sqrt{(s_1 - s_1')^2 + [2r\sin(\varphi/2)]^2}} \psi(r, \varphi) r \, d\varphi,
\tag{2.25a}
$$

$$
G_{sl}^{V_{1,2}}(s_2, s_2') = \int\limits_{-d/2}^{d/2} \frac{e^{-ik_{1,2}\sqrt{(s_2-s_2')^2+(\xi)^2}}}{\sqrt{(s_2 - s_2')^2 + (\xi)^2}} \chi(\xi) \, d\xi.
\tag{2.25b}
$$

The problem solution based on integral equation kernels (2.25a) may encounter serious difficulties, therefore we will use approximate expressions, known as "quasi-one-dimensional" kernels [20, 21]

$$
G_v(s_1, s_1') = \frac{e^{-ik_1\sqrt{(s_1-s_1')^2+r^2}}}{\sqrt{(s_1 - s_1')^2 + r^2}},
\tag{2.26a}
$$

$$
G_{sl}^{V_{1,2}}(s_2, s_2') = \frac{e^{-ik_{1,2}\sqrt{(s_2-s_2')^2+(d/4)^2}}}{\sqrt{(s_2 - s_2')^2 + (d/4)^2}},
\tag{2.26b}
$$

where the source points are assumed to be located on the geometric axes of the vibrator and slot. The observation points are placed on the physical vibrator surface and on the slot axis with the coordinates $\{s_2, \xi/2\}$. In this case, the functions $G_v(s_1, s'_1)$ and $G_{sl}^{V_{1,2}}(s_2, s'_2)$ are continuous everywhere, and the equations for the currents are significantly simplified without noticeable accuracy deterioration [22, 23].

Since the form of the Green's functions in the above equations was not specified, the Eqs. (2.19)–(2.24) are valid for any electrodynamic volumes if the corresponding Green's functions are known or can be constructed for these volumes.

Thus, the problem of the electromagnetic wave excitation by thin impedance vibrators and narrow slots connecting the two electrodynamic volumes has been reduced to the integral equations with respect to electric currents in the vibrators and equivalent magnetic currents in the slots. The solution of these equations is the final stage of the problem solution, since the total electromagnetic fields in the volumes can be easily found by using the expressions (2.1) and (2.4).

2.2 Methods for Solving Integral Equations for Currents

Starting with the fundamental publications of Poklington [24], Hallen [25], Brillouin [26], Leontovich and Levin [27], King [20], Weinstein [28, 29], Maya [30] up to the present time, a large number of publications is devoted to the theory of thin vibrators located in various electrodynamic volumes and material media. There also exist some publications related to studies of impedance vibrators [31–60] devoted to vibrator made of metal with finite conductivity. The authors of this monograph have also studied the vibrators with complex surface impedance [2, 15]. The publications [2, 6, 14, 21, 61–67] are fundamental in the theory of slot radiators and coupling holes in electrodynamic volumes. Finally, among the publications devoted to the study of combined vibrator-slot structures, the works [2, 23, 68–70] should be single out. An extended list of the authors' publications on the study of the vibrator and slot structures considered in this chapter is presented in Bibliography. As stated in the listed above and other known works, direct numerical, numerical-analytical, and analytical methods are used to solve the boundary-value problems.

Despite the fact that analytical methods are more vivid and effective as compared with numerical methods, various modifications the method of moment are used in the overwhelming majority of works. In this regard, the method of moments, as the most universal in the theory of thin vibrators, slots, and systems based on these elements will be detailly consider below.

2.2.1 Basics of the Method of Moments

The essence of the method of moments (MoM) consists in reducing the integral equation to a system of linear algebraic equations (SLAE) by substituting the unknown functions as a linear combination of predefined functions with an unknown complex amplitudes. Thus, the solution for the electric or magnetic currents can be sought in the form:

$$\vec{J}(s) = \sum_{m=-M}^{M} J_m \vec{\phi}_m(s), \qquad (2.27)$$

where s is a coordinate along the vibrator or slot axis, J_m is unknown complex amplitudes, $\vec{\phi}_m(s)$ are the vector basis functions. If the system of functions $\{\vec{\phi}_m(s)\}$ is complete in the corresponding functional space, then an exact solution of the problem can be obtained under condition $|M| \rightarrow \infty$. To find an approximate solution, M is chosen finite based on a preliminary justification.

The right-hand side of the original equation is also represented as a sum $\sum_{\mu=-L}^{L} E_\mu \vec{\psi}_\mu(s)$, where vector of the linearly independent functions $\vec{\psi}_\mu(s)$ are called by weighted or trial functions. The problem solution can be subdivided into several stages: (1) both sides of the equation are scalar multiplied by weight functions $\vec{\psi}_\mu(s)$ and integrated over the surface of the vibrator or slot; (2) the inner products, known as moments, are calculated, and the equations are reduced to the matrix form of the SLAE; (3) the solution of the SLAE for unknown complex amplitudes J_m included in the representation (2.27) are found. At the first stage of the solution, the SLAE consisting of (2L + 1) equations with (2M + 1) unknown amplitudes is obtained. When M = L, the SLAE matrix turns out to be quadratic and the amplitudes J_m can be easily found. If the weight functions coincide with the basis functions $\left(\vec{\psi}_\mu = \vec{\phi}_m\right)$, the method of moments is called the Galerkin method. The solution thus obtained has stationary properties, and the approach itself is equivalent to the variational Rayleigh-Ritz method.

In the theory of vibrator radiators, the Galerkin method is known as the generalized method of induced electromotive forces (EMF). This term is associated with the radio engineering, where the unknown function in the original integral equation corresponds to the electric current, and the equation kernel has the meaning of resistance. Therefore, the inner products, that is, the SLAE matrix coefficients, represent the EMF induced at some point on the antenna by the currents of other antenna sections or by other radiators. The right-hand side of the original equation usually corresponds to the extrinsic EMF. The term generalized method of the induced EMF reveals the physical meaning of the formulas used, therefore, the Galerkin method in the theory of slot radiators and coupling holes, is known as the generalized method of induced magnetomotive forces.

The SLAE can be solved numerically by using the Cramer's rule or matrix inversion, which is widely used for diffraction problem solutions. However, the matrix inversion may be inefficient and lengthy, especially when very large matrices are inverted or when a significant number of simulation options are required, for example, when studying the solution dependence upon many parameters. The inversion of ill-conditioned matrices, that is, matrices whose determinant is close to zero, can be difficult and can easily lead to large errors. Such situations can arise in problems related to perfectly conducting vibrators and slots in walls of electrodynamic volumes. These problems are based on Fredholm integral equations of the first kind. These problems are referred to as ill-posed electrodynamics problems whose numerical solution require additional procedures for its regularization.

In the method of moments, the choice of basic functions is of paramount importance. There exist an infinitely large number of basic and weighted function sets, however, for each specific electrodynamic problem, it is advisable to choose these functions so that they correspond to the physical entity and lead to efficient computational algorithms. The basic and weight functions, which for antenna radiators are called by spatial current harmonics, can be classified into two classes: (1) harmonics defined on the entire radiator, known as the basic functions of total domain; (2) harmonics that are non-zero only within some region of the radiator, known as the basic functions of subdomain or segment basic functions.

Among the most commonly used harmonics of the first group are:

- *harmonics of the Fourier series* $J(s) = \sum_m J_m e^{2\pi i m s/T}$. The Fourier series are often used for the approximation of functions, which properties are well studied. The formulas for SLAE coefficients are relatively simple, there are accuracy estimates based on a uniform and root-mean-square approximation. The harmonics of the Fourier series are often used to represent currents in rectilinear radiators, when the current on the radiator vanishes at its edges. In this case, the decomposition into sine or cosine functions can be quite natural if the coordinate origin coincides with the radiator center;
- *power polynomials* $J(s) = \sum_m J_m s^m$. This is also a class of approximating functions with known interpolation formulas and accuracy estimates. The power polynomials are often used with the Dirac delta functions as weighting functions, allowing to perform a pointwise function matching. This allows us to eliminate one of the two integrals in formulas for the mutual resistance. Polynomials $\left(1 - \frac{|s|}{L}\right)^m$ where L is the half length of the vibrator or slot can also be used;
- *Chebyshev polynomials* $J(s) = \sum_m J_m T_m(s)$. These polynomials are used as basis functions for representing currents in radiators with strong edge effect, for example, in strip vibrators or slots in infinitely thin walls of electrodynamic volumes. Since Chebyshev polynomials are characterized by the sharp increase at the edges of the interval, they allow us correctly describe the edge effect, that is, the diffraction at the radiator edge. This can accelerate convergence to the exact solution;
- *Bessel functions* $J(s) = \sum_m J_m I_m(s)$. They are usually used as basis functions for the radiators with axial symmetry, such as dielectric rods, spirals, etc.;

Consider some harmonics of the second group, which can be represented as
$\varphi_m(s) = \begin{cases} f_m(s), & s \in \Delta s_m; \\ 0, & s \notin \Delta s_m. \end{cases}$. The most commonly used types of such harmonics
are:

- *Dirac delta functions* $J(s) = \sum_m J_m \delta(s - s_m)$. These basis function can be used to satisfy the boundary conditions at separate points of the radiator. This procedure is also called pointwise stitching or the collocation method. Due to the properties of the delta function, calculation of the mutual resistances and the SLAE right-hand parts can be much-simplified. There are many examples of the successful application of this basis, however, it is often necessary to take into account quite a lot of stitching points and to solve a high order SLAE. A compromise between simplicity of the computational algorithms of the system matrix and the complexity of solving a SLAE with a large number of equations must be evaluated for specific problem solutions;
- *piecewise constant functions* $f_m(s) = J_m$. The properties of this basis functions is similar to that of the Dirac delta-functions. The integration universality and simplicity is compensated by slow convergence to the exact solution and the difficulty of solving high-order SLAE;
- *piecewise linear functions* $f_m(s)\Delta s_m = J_m(s_{m+1} - s) + J_{m+1}(s - s_m)$. This basis functions are characterized by comparative simplicity of integral calculating. It also accelerate convergence to the exact solution as compared with that for the piecewise constant functions;
- *piecewise sinusoidal functions* $f_m(s) \sin k\Delta s_m = J_m \sin k(s_{m+1} - s) + J_{m+1} \sin k(s - s_m)$. The advantage of this basis functions lies in the relatively small amount of computations, since the corresponding integrals can be presented in close forms allowing very fast convergence to the exact solution;
- *parabolic functions* $f_m(s) = A_m + B_m(s - s_m) + C_m(s - s_m)^2$. In this case, the subdomains can intersect. The properties of the approximation is close to the approximation by splines.

Recently, wavelets as basis and weight functions for numerical solution of electrodynamics problems, in particular, those concerning the theory of thin vibrators and narrow slots by the method of moments are increasingly used. In a number of cases, this approach can significantly improve the efficiency of numerical simulation, and, in addition, it optimizes the transition between the spatial-frequency and temporal representations of the solution. Wavelets are a generalized name for functions representing wave packets, localized along the axis of the independent variable, for example along the s coordinate in our case. The wave packets can be translated along the axis and scaled, that is, they can be compressed or stretched. These functions can be used in the method of moments both as basis functions of total domains and as segment basis functions. The Daubech orthogonal and Malvar wavelets are successfully used in the first and second cases.

Since in the monograph the approximate analytical methods will be used for solving the integral equations, we will detailly consider this method.

2.2.2 *Approximate Analytical Methods for Solving Current Integral Equations*

The rigorous solution of the integral equations for the electric current on the impedance vibrator and the magnetic current in the slot cannot be obtained in a closed form. Nevertheless, the current equation can be accurately solved by using approximate approaches. This can be done by applying the well-known methods developed earlier for perfectly conducting vibrators and slots cut in the infinite screen. The problems for the rectilinear vibrators in free space can be solved by the following approaches: a successive iteration method (further the iteration method) [20, 25], methods of a small-parameter powers series expansion of unknown function (further the expansion method) [27], variational method [20], the method of key equation searching [28, 29]. Similarly, the small parameter method [64, 65, 67], the variational method [66] or the iteration method [63] can be used for the slot in a thin infinite screen. To determine advantages and disadvantages of approximate analytical methods, let us find the solutions of the Eq. (2.19) by the expansion and iterations methods for the case of electromagnetic wave incidence on the perfectly conducting ($z_i = 0$) vibrator placed in a volume V_1 filled by a medium with material parameters $\varepsilon_1 = \mu_1 = 1$.

2.2.2.1 Series Expansion Method

The initial equation for the current on the perfectly conducting vibrator has the form

$$
\left(\frac{d^2}{ds^2} + k^2\right) \int_{-L}^{L} J(s') \frac{e^{-ikR(s,s')}}{R(s, s')} ds' = -i\omega E_{0s}(s) - f_0[s, J(s)], \tag{2.28}
$$

where $R(s, s') = \sqrt{(s - s')^2 + r^2}$ and

$$
f_0[s, J(s)] = \left(\frac{d^2}{ds^2} + k^2\right) \int_{-L}^{L} J(s') g_s^{V_1}(s, s') ds' \tag{2.29}
$$

is the projection of the regular part of the vibrator radiation field determined by the volume V_1 geometry on the vibrator axis when the vibrator is positioned along an arbitrarily axis of the Cartesian coordinate system.

If, the differential ds' is replaced by dR and the relations

$$
\left.\begin{array}{l}
s' = s - \sqrt{R^2 - r^2}, \quad \text{if} \quad s' \le s \\
s' = s + \sqrt{R^2 - r^2}, \quad \text{if} \quad s' \ge s
\end{array}\right\},
$$

are taken into account the Eq. (2.28) can be transform to

$$
\left(\frac{d^2}{ds^2} + k^2\right)\left\{-\int_{-L}^{s} J(s')e^{-ikR}d\ln[C(R + \sqrt{R^2 - r^2})]\right.
$$

$$
\left. + \int_{s}^{L} J(s')e^{-ikR}d\ln[C(R + \sqrt{R^2 - r^2})]\right\} = -i\omega E_{0s}(s) - f_0[s, J(s)], \quad (2.30)
$$

where C is an arbitrary constant. Integrating by parts in (2.30) and using the boundary conditions for the current (2.18), we obtained

$$
\left(\frac{d^2}{ds^2} + k^2\right)\left\{J(s)e^{-ikr}\ln Cr + \int_{-L}^{s}\ln[C(R + \sqrt{R^2 - r^2})]d[J(s')e^{-ikR}]\right.
$$

$$
\left. - \int_{s}^{L}\ln[C(R + \sqrt{R^2 - r^2})]d[J(s')e^{-ikR}]\right\} = i\omega E_{0s}(s) + f_0[s, J(s)]. \quad (2.31)
$$

Since the term $e^{-ikr} \simeq 1$ due to (2.15), the Eq. (2.28) can be converted to the following integral equation with the small parameter

$$
\frac{d^2 J(s)}{ds^2} + k^2 J(s) = \alpha\{i\omega E_{0s}(s) + f[s, J(s)] + f_0[s, J(s)]\}. \quad (2.32)
$$

where $\alpha = \frac{1}{2\ln[r/(2L)]}$ is the small parameter, and

$$
f[s, J(s)] = -\left(\frac{d^2}{ds^2} + k^2\right)\int_{-L}^{L}\text{sign}(s - s')\ln\frac{R + (s - s')}{2L}\frac{d}{ds'}[J(s')e^{-ikR}]ds'
$$

$$
(2.33)
$$

is self-field of the vibrator in free space. In the Eq. (2.31) the constant $C = 1/2L$ in contrast to [27], where the constant $C = k$.

Let us represent the current $J(s)$ as a power series in a small parameter $|\alpha| \ll 1$

$$
J(s) = J_0(s) + \alpha J_1(s) + \alpha^2 J_2(s) + \ldots \quad (2.34)
$$

After substituting (2.34) into (2.29) and (2.33), the function $f_\Sigma[s, J(s)]$ can be also expanded into a series

$$f_\Sigma[s, J(s)] = f_\Sigma[s, J_0(s)] + \alpha f_\Sigma[s, J_1(s)] + \alpha^2 f_\Sigma[s, J_2(s)] + \ldots, \quad (2.35)$$

where $f_\Sigma[s, J(s)] = f[s, J(s)] + f_0[s, J(s)]$ is the total vibrator field. Substituting (2.34) and (2.35) into the Eq. (2.32) and equating the terms of the same degrees in α we obtain the following system of differential equations:

$$\frac{d^2 J_0(s)}{ds^2} + k^2 J_0(s) = 0,$$

$$\frac{d^2 J_1(s)}{ds^2} + k^2 J_1(s) = i\omega E_{0s}(s) + f_\Sigma[s, J_0(s)],$$

$$\frac{d^2 J_2(s)}{ds^2} + k^2 J_2(s) = f_\Sigma[s, J_1(s)],$$

$$\ldots\ldots\ldots\ldots\ldots\ldots\ldots\ldots\ldots\ldots\ldots\ldots\ldots\ldots\ldots$$

$$\frac{d^2 J_n(s)}{ds^2} + k^2 J_n(s) = f_\Sigma[s, J_{n-1}(s)], \quad (2.36)$$

which can be solved by the method of successive approximations. The solution of each equation should satisfy the boundary conditions (2.18), namely: $J_0(\pm L) = 0$, $J_1(\pm L) = 0$, $J_2(\pm L) = 0, \ldots, J_n(\pm L) = 0$.

The solution of the first equation in (2.36) does not depend on the exciting field

$$J_0(s) = C_1 \cos ks + C_2 \sin ks, \quad (2.37)$$

which satisfies the boundary conditions only when

$$C_1 = 0 \text{ if } 2L = m\lambda; \quad C_2 = 0 \text{ if } 2L = (2n + 1)\frac{\lambda}{2}, \quad (2.38)$$

where m and n are integers. When the vibrator length $2L$ does not satisfy the conditions (2.38), $J_0 \equiv 0$, $f_\Sigma[s, J_0(s)] \equiv 0$ and, the current in the first approximation is equal to

$$J(s) = \alpha J_1(s) = -\alpha \frac{i\omega/k}{\sin 2kL}\left\{ \sin k(L - s) \int_{-L}^{s} E_{0s}(s') \sin k(L + s')ds' \right.$$

$$\left. + \sin k(L + s) \int_{s}^{L} E_{0s}(s') \sin k(L - s')ds' \right\}. \quad (2.39)$$

As can be seen, the expression for the current does not include the functions of the own vibrator fields $f[s, J(s)]$ and $f_0[s, J(s)]$, which determine resonant properties and energy characteristics of the vibrator. Obviously, the function $f_\Sigma[s, J(s)]$ can be taken into account only in the higher approximation, but its usage encounters significant mathematical difficulties. At present only the second approximation of the current $J_2(0)$ for the vibrator excited in the center by a voltage generator in free space is known [27].

As an example, let us consider a problem of H_{10}-wave scattering by a vibrator located in a cross-section plane of a standard rectangular waveguide parallel to its narrow wall. The extraneous field is

$$E_{0s}(s) = E_0 \sin(\pi x_0/a), \tag{2.40}$$

where E_0 is the amplitude of the incident wave H_{10}, a is the broad wall size, x_0 is the distance between the narrow wall and the vibrator axial line. Then the current induced on the vibrator according to (2.39) is equal to

$$J(s) = -\alpha E_0 \sin \frac{\pi x_0}{a} \frac{i\omega}{k^2} \frac{(\cos ks - \cos kL)}{\cos kL}. \tag{2.41}$$

And finally, the solution of the classical problem of the normal incidence of a plane electromagnetic wave $E_{0s}(s) = E_0$ on the vibrator in free space obtained by the small parameter method in the first approximation can be presented as

$$J(s) = -\alpha E_0 \frac{i\omega}{k^2} \frac{(\cos ks - \cos kL)}{\cos kL}. \tag{2.42}$$

It is quite evident, that the applicability condition of the formulas (2.41) and (2.42) is inequality [27]

$$|kL - n(\pi/2)| \gg |\alpha|, \tag{2.43}$$

which, together with relations (2.38), limits to a large extent the practical usage of the solution.

2.2.2.2 Successive Iterations Method

To eliminate the above disadvantages of the integral equation solution using the small parameter method, let us apply the method of successive iterations proposed by Hallen [25] and developed by King [20] to study vibrator characteristics in free space.

If the differential operator on the left-hand side of the Eq. (2.28) is inverted, the following integral equation can be obtained

$$\int_{-L}^{L} J(s')G_s^{\Sigma}(s, s')ds' = C_1 \cos ks + C_2 \sin ks - \frac{i\omega}{k} \int_{-L}^{s} E_{0s}(s') \sin k(s - s')ds',$$

$$\tag{2.44}$$

where the relation $G_s^\Sigma(s, s') = G_s(s, s') + g_s^{V_1}(s, s')$ is taken into account. To find one of the arbitrary constants C_1 and C_2, the symmetry conditions [20] uniquely related to the vibrator excitation mode should be used. In other words, to solve the original equation using the iteration method the field of extraneous sources $E_{0s}(s)$ should be specified at this stage of solution. In other words, already at this stage of solving the original equation by the iteration method, it is necessary to specify the fields of the extraneous sources. Let us assume that in accordance with (2.40) the identity $E_{0s}(s) = E_0$ is valid, that is, the vibrator in the rectangular waveguide is excited by the H_{10}-wave when the vibrator axis is located at $x_0 = a/2$. Then,

$$\int_{-L}^{L} J(s')G_s^\Sigma(s, s')ds' = C_1 \cos ks + \frac{i\omega}{k^2} E_0(\cos ks \ \cos kL - 1). \qquad (2.45)$$

If the kernel of the Eq. (2.45) is $G_s^\Sigma(s, s') = e^{-ikR}/R$, the equation is similar to the linearized Hallen integral equation [20, 25], which is the basis of many works devoted to the theory of thin dipole antennas.

The kernel of the integral Eq. (2.45) on the vibrator surface, has a quasi-stationary singularity. To isolate the singularity, we will use the inequality (2.15) and rewrite the left-hand side of the Eq. (2.45) in the following way:

$$\int_{-L}^{L} J(s')G_s^\Sigma(s, s')ds' = \int_{-L}^{L} J(s')\frac{e^{-ikR(s,s')}}{R(s, s')}ds' + \int_{-L}^{L} J(s')g_s^{V_1}(s, s')ds'. \qquad (2.46)$$

Then

$$\int_{-L}^{L} J(s')\frac{e^{-ikR(s,s')}}{R(s, s')}ds' = \Omega(s)J(s) + \int_{-L}^{L}\left[J(s')\frac{e^{-ikR(s,s')}}{R(s, s')} - \frac{J(s)}{R(s, s')}\right]ds', \qquad (2.47)$$

where

$$\Omega(s) = \int_{-L}^{L} \frac{ds'}{\sqrt{(s - s')^2 + r^2}}. \qquad (2.48)$$

The first term on the right-hand side of expression (2.47) is logarithmically large compared to the second regular term. The function $\Omega(s)$ differs from its mean value $\overline{\Omega}(s) = 2\ln(2L/r) - 0.614$ only at the vibrator ends where the current $J(\pm L) = 0$, therefore, the Eq. (2.45) can be converted to

$$J(s) = -\alpha\left[C_1 \cos ks + \frac{i\omega}{k^2} E_0(\cos ks \ \cos kL - 1)\right]$$

$$+ \alpha \int\limits_{-L}^{L} \left[J(s')G_s^{\Sigma}(s, s') - \frac{J(s)}{R(s, s')} \right] ds', \tag{2.49}$$

where $\alpha = \frac{1}{2\ln[r/(2L)]}$ is a small parameter coinciding with that obtained in Sect. 2.2.2.1 when the integration constant $C = 1/2L$.

Then, using the approach proposed in [20, 25], we substitute $s = L$ into (2.49) and subtract the result from the initial Eq. (2.49). In fact, the subtrahend is 0, since $J(L) \equiv 0$. In this case, Eq. (2.49) is transformed as follows

$$J(s) = -\alpha \left[C_1(\cos ks - \cos kL) + \frac{i\omega}{k^2} E_0 \cos kL(\cos ks - \cos kL) \right]$$

$$+ \alpha \left\{ \int\limits_{-L}^{L} \left[J(s')G_s^{\Sigma}(s, s') - \frac{J(s)}{R(s, s')} \right] ds' - \int\limits_{-L}^{L} J(s')G_s^{\Sigma}(L, s')ds' \right\}. \tag{2.50}$$

When the first term in the right-hand side of Eq. (2.50) is used as the zero approximation of the current $J_0(s)$, the following expression can be obtained

$$J_0(s) = -\alpha E_0 \frac{i\omega}{k^2} \frac{(\cos ks - \cos kL)}{\cos kL}, \tag{2.51}$$

where the condition (2.18) was used to define the constant C_1. As can be seen, the expression (2.51) is identical to (2.41) when $x_0 = a/2$. Substituting (2.51) into (2.50), we find the first approximation for the current with accuracy up to α^2

$$J_1(s) = -\alpha E_0 \frac{i\omega}{k^2} \frac{(\cos ks - \cos kL)}{\cos kL + \alpha F(kr, kL)}, \tag{2.52}$$

where

$$F(kr, kL) = \int\limits_{-L}^{L} \left[(\cos ks' - \cos kL)G_s^{\Sigma}(L, s') \right] ds' \tag{2.53}$$

is the function of the vibrator's self-field, allowing to analyze both tuned ($\cos kL = 0$) and untuned ($\cos kL \neq 0$) vibrators by using the formula of the first approximation. The vibrator is tuned or untuned if the frequency of the external field is close to the natural vibrator frequency or significantly differ from it. Certainly, the tuned and untuned vibrators cannot be analyzed by using the small parameter method.

Thus, solving the quasi-one-dimensional integral equation for electric currents on thin vibrators using the small parameter method allows us to defined current on the tuned and the untuned vibrators. The first approximation solution for the untuned vibrator can be obtained for the arbitrary excitation. Thus, the solution of the integral equation for the current by the iteration method can be obtained in the form of a single formula, suitable for the tuned and untuned vibrators, however, it can only be used if the extraneous source fields are predefined at the initial stage of the analysis. The solutions of the integral Eq. (2.22) for the magnetic current in the slotted coupling holes by using the small parameter and iteration methods give similar results [18, 64, 67, 70]. In the next Subsection, a general analytical expression for the vibrator current will be derived by an asymptotic averaging method. The formula is valid for both tuned and untuned vibrators, it does not require that the field of extraneous sources and the form of electrodynamic volumes in which they are located were defined explicitly.

2.2.2.3 Mathematical Aspects of the Asymptotic Averaging Method

A rigorous justification of the asymptotic averaging method is a purely mathematical problem, which has been studied in detail in monographs [71, 72], where the averaging theorems were proved. Let us briefly discuss the principles that will be needed below.

Let a system of ordinary differential equations be written in standard form when the first derivatives $\frac{dx}{ds}$ are proportional to the small parameter

$$\frac{dx}{ds} = \alpha X(s, x) \tag{2.54}$$

where x is the n-dimensional vector, $0 < \alpha \ll 1$ is the small parameter. There exist several approaches to reduce the initial integral equations to the form (2.54), but the method of variation of arbitrary constants is most commonly used [72]. If the reduction of the initial equation system to the standard form has already been fulfilled, change of variables can be made in (2.54)

$$x = \xi + \alpha \tilde{X}(s, \xi), \tag{2.55}$$

where ξ is the new unknowns. Let $\frac{\partial \tilde{X}}{\partial s} = X(s, \xi) - \overline{X}(\xi)$, where the bar denotes averaging over the explicit variable s, that is

$$\overline{X}(\xi) = \lim_{l \to \infty} \frac{1}{l} \int_0^l X(s, \xi) ds. \tag{2.56}$$

Then, after a series of transformations, we obtained [71]

$$\frac{d\xi}{ds} = \alpha \overline{X}(\xi) + \alpha^2 \dots . \tag{2.57}$$

That is, if the variable ξ satisfies Eq. (2.57), which the right-hand side differs from the right-hand side of the averaged equations

$$\frac{d\xi}{ds} = \alpha \overline{X}(\xi) \tag{2.58}$$

by the terms proportional to α^2, then expression (2.55) is an exact solution of the original Eq. (2.54). Therefore, the equality $x = \xi$ can be accepted as the first approximation, when ξ is the solution of Eq. (2.58), satisfying Eq. (2.54) up to the second order of smallness. Note, that the expression (2.55), satisfying the Eq. (2.57), is called the improved first approximation.

Thus, the first approximation Eq. (2.58) can be obtained from the exact Eq. (2.54) by averaging over the variable s, while ξ is constant. This formal process, consisting in the replacement of exact equations by averaged ones, is called the averaging principle. The essence of this approach was revealed by N. Bogolyubov and Y. Mitropolsky, who showed that the averaging method is associated with the existence of a certain change of variables, which allows to exclude the variable s from the right-hand sides of the equations with any degree of accuracy relative to the small parameter α[71]. Thus, one can construct not only the first approximation system (2.58), but also to find higher-order averaged systems whose solutions approximate solutions (2.54) with an arbitrarily specified accuracy, although only the first approximation can practically be effectively used due to the rapid complication of the higher order formulas. The proof that the error of the first approximation is small was also obtained in [71], where it was found that under very general conditions the difference $x(s) - \xi(s)$ can be made arbitrarily small for a sufficiently small α at arbitrarily large finite interval $0 < s < l$. Thus, the following averaging theorem holds [71, 72].

Theorem *Let the continuous function $X(s, x)$ be defined in the domain Q ($s \geq 0$, $x \in D$) where the following conditions hold:*

1. *the function $X(s, x) \in \mathrm{Lip}_x(\lambda, Q)$, i.e., it satisfies the Lipschitz condition over x with constants λ;*
2. *the inequality $||X(s, x)|| < M$ holds, i.e. $X(s, x)$ is bounded function;*
3. *there exists the limit (2.56) for every point $x \in D$;*
4. *the solution $\xi(s)$ of the averaged system (2.58) is defined for all $s \geq 0$; it lies in the domain D within a ρ neighborhood.*

Then for any arbitrarily small $\eta > 0$ and arbitrarily large $L > 0$, such α_0 can be specified that if $0 < \alpha < \alpha_0$, the inequality

$$||x(s) - \xi(s)|| < \eta,$$

holds when $0 \leq s \leq L\alpha^{-1}$. The functions $x(s)$ and $\xi(s)$ are the solutions of systems (2.54) and (2.58) coincide when $s = 0$.

This theorem can be generalized for systems of integral-differential equations in standard form

$$\frac{dx}{ds} = \alpha X\left(s, x, \int_0^s \phi(s, s', x(s'))ds'\right),\tag{2.59}$$

for which the following averaging scheme is possible [71]. First let us compute the integral

$$\psi(s, x) = \int_0^s \phi(s, s', x)ds'\tag{2.60}$$

over the explicit variable s', while s and x are considered to be parameters. Then, along with (2.59), let us consider the system of differential equations

$$\frac{dy}{ds} = \alpha X(s, y, \psi(s, y)),\tag{2.61}$$

which can be averaged if the limit

$$\lim_{l \to \infty} \frac{1}{l} \int_0^l X(s, y, \psi(s, y))ds = \overline{X}(y)\tag{2.62}$$

exists. Then, the following system of differential equations can be obtained

$$\frac{d\xi}{ds} = \alpha \overline{X}(\xi),\tag{2.63}$$

which is averaged system, corresponding to the system of integral-differential Eq. (2.59). Thus, the averaging in the systems (2.59) consists in approximating the solutions of this system by that of the differential equation system (2.63). Obviously, this system can be solved much easier than the original system of integral-differential equations. The conditions under which the solutions (2.59), (2.61) and (2.63) are close can be found in the monograph [72]. In contrast to the differential equation system, the systems of integral-differential equations allow various averaging options. In the general case, one system of integral-differential equations can be associated with several different systems of averaged equations. Some of these averaged systems can be the differential equation systems, while the others represent the systems of integral-differential equations. The possibility of choosing the most acceptable averaged system determines the high efficiency of the averaging method for solving applied problems.

Consider the following system of integral-differential equations

$$\frac{dx}{ds} = \alpha X\left(s, x, \frac{dx}{ds}, \int_0^s \phi\left(s, s', x(s'), \frac{dx(s')}{ds'}\right)ds'\right),$$ (2.64)

which is a standard system not resolved relative to the derivative. In practice, an attempt to resolve a system relative to the derivative $\frac{dx}{ds}$ often encounters a need to perform cumbersome and time-consuming calculations, therefore, appropriate averaging schemes have been developed to avoid this drawback [72]. For example, in the first approximation, instead of (2.64), we can consider a simplified system

$$\frac{dy}{ds} = \alpha X\left(s, y, 0, \int_0^s \phi(s, s', y(s'), 0)ds'\right),$$ (2.65)

since the derivatives on the right-hand side of (2.64) begin to manifest itself in the second and subsequent approximations, as well as in the improved first approximation. Under sufficiently general conditions and small α, solutions (2.64) and (2.65) are arbitrarily close on an interval of $L\alpha^{-1}$ order.

The systems of differential and integral-differential equations allow various variants of partial averaging. For example, only some terms or separate equations in the system can be averaged [72]. In particular, if the source system is representable as

$$\frac{dx}{ds} = \alpha X_1(s, x) + \alpha X_2(s, x)$$ (2.66)

and the limit

$$\lim_{l\to\infty} \frac{1}{l} \int_0^l X_1(s, x)ds = \overline{X}_1(x),$$ (2.67)

exists, then the partially averaged system can be put in correspondence to the system (2.67)

$$\frac{d\xi}{ds} = \alpha\overline{X}_1(\xi) + \alpha X_2(s, \xi).$$ (2.68)

The partial averaging schemes are very diverse; they can be also applied to the equation systems (2.59) and (2.64).

2.2.2.4   Application of the Generalized Method of Induced EMMF for Multi-Element Vibrator-Slot Structures

The integral equations for the multi-element vibrator-slot structures can be quite complicated, which significantly limits the application of various modifications of the numerical method of moments considered in Sect. 2.2.1. A promising method for analyzing vibrator-slot systems consists in the numerical-analytical implementation of the generalized method of induced electromagnetic motive forces (EMMF) based on the previously built asymptotic solutions for currents are used as basic functions for the system elements.

Initially, the method of induced electromotive forces (EMF), as a special case of the EMMF was proposed for analyzing a system of linear perfectly conducting vibrators. In this approximations, the current on each m-th vibrator of the system can be described by basic functions $\vec{\varphi}_m(s_m) = \vec{e}_{s_m} \sin k(L_m - |s_m|)$, while functions $\vec{\psi}_\mu(s) = \vec{\varphi}_m(s)$ are used as weight functions. The number of equations in the system is equal to the number of vibrators, the complex amplitudes in the antinodes of the basic functions are unknown, the right-hand sides of the equations are source voltages applied at the center of each vibrator which are re-calculated to the current in the antinodes. Such a choice of weight functions is associated with the following physical considerations. If the current is approximated by one continuous basis function, the correspondence between the current at an antinode determining the vibrator radiation, and the current at the input of the vibrator defining the power supplied by the power source can be inevitably disrupted. The complex amplitude of the current can be chosen so that either the radiation power or other power calculated from this value is equal to or at least close to its true value. However, in this case, the balance between the power radiated by the vibrator and the power coming from the source can be violated. The choice of the weight functions $\vec{\psi}_\mu(s) = \vec{\varphi}_m(s)$ corresponds to such a selection of complex amplitudes that the balance of these powers is achieved, although both powers are determined with some error. The weight functions thus build cannot be optimal in all cases, although it is always desirable to use functions $\vec{\varphi}_m(s)$ and $\vec{\psi}_\mu(s)$ that do not differ substantially from exact solutions, especially if the vibrators have distributed surface impedance.

The most vulnerable point of the approach is that in real vibrator systems, due to the interaction of vibrators through surrounding space or asymmetrical self-excitation, the current distribution in each of them can be asymmetric. Therefore, using only the symmetric basis and weight functions considered above becomes inappropriate. That is, functions asymmetric relative to the vibrator center should be included into the system of basic functions representing the current in each vibrator; this require further generalizing the electromotive forces (EMF) method to use complex functional current distributions as adequate basic and weight functions. Similar arguments can be made for the induced magnetomotive forces (MMF) method, and, in general, for the hybrid method of induced EMMF.

2.3   New Aspects in the Development of the Theory of Thin Impedance Vibrators

Currently, the theory of thin vibrators based on a huge number of published articles and monographs is considered to be classical, both for perfectly conducting and impedance vibrators. However, this problem still attracts the researches attention, since the vibrator structures are constantly applied to ensure the necessary excitation mode of waves in various devices and systems. Therefore, complex boundary value problems should be solved taking into account non-coordinate boundaries of spatial regions, the presence of scattering inhomogeneities, and the heterogeneity of filling medium, etc. An experimental optimization of new devices being developed is practically impossible due to a multiparameter nature of the problem. The main point in carrying out physically adequate mathematical modeling of such structures consists in searching the vibrator current distributions, which can be greatly simplified by correctly choosing basis current distributions. This choice should be made taking into account the complex electrodynamic environment of the vibrator radiator that could not always be done by analyzing publications available in the literature. Therefore, generalizing the results of thin impedance vibrator theory concerning the influence of the electro-dynamic environment on the current distribution in the radiating vibrator is very actual problem.

This Subsection presents generalizations of some results concerning the theory of thin impedance vibrators when the vibrator radiators with concentrated excitation are placed in free space or in electrodynamic volumes with coordinate boundaries. Such a generalization based on the methodological aspects of the problem can be conveniently made by formulating two theorems. On the one hand, this approach allows us to systematically evaluate already known results, and, on the other hand, to extend the knowledge gained to solving new boundary-value problems with complex boundaries. Let us list the requirements imposed on the boundary surface S: it must be a Lyapunov-type surface; it should not cross external current sources; it must be passive and do not allow mutual transformations of electric and magnetic fields.

2.3.1   Presentation of Two Fundamental Theorems of the Vibrator Theory

One of the approaches allowing such a generalization is based on using the natural small parameter (2.33) in the analytical solution of the integral equation for the vibrator current. For example, an influence functional of hollow rectangular waveguide walls included into an analytical solution for the current in a linear dissipative vibrator located inside the waveguide is proportional to a small natural parameter [1]. A similar situation is also observed when solving the problem related to radial monopole current located on an perfectly conducting sphere [13]. A distinctive feature of these solutions consists in fact that the total field could be represented as waves of the electric and magnetic type. The assessment of this factor influence on boundary value problems with arbitrary boundaries that do not possess the properties of mutual transformation of electric and magnetic field types is implemented in the formulation of the first theorem.

Theorem 1 *Let a thin radiating impedance vibrator, represented by a segment of a circular cylinder excited by a point source is placed in an electrodynamic volume V uniformly filled by a medium with material parameters (ε_1, μ_1). The vibrator radius r and length 2L are such that inequality $\frac{r}{2L} \ll 1$ and $\frac{r\sqrt{\varepsilon_1\mu_1}}{\lambda} \ll 1$ hold (λ is the wavelength in free space); the volume V is bounded by the Lyapunov surface S which do not crosses extraneous current sources. Then the influence of the volume boundaries on the vibrator current distribution is proportional to the natural small parameter $\alpha = \frac{1}{2Ln[r/(2L)]}$.*

The Theorem 1 was formulated for sufficiently general conditions; it allows to establish a series of physically interesting corollaries which determines its fundamental nature.

Corollary 1.1
Since no restrictions were imposed on magnitude of the vibrator constant impedance, the theorem is valid for a perfectly conducting vibrator, $z_i = 0$.

Corollary 1.2
Since no restrictions were imposed on the material parameter of the medium (ε_1, μ_1) the theorem is valid both for unbounded space and hollow closed volumes with $\varepsilon_1 = \mu_1 = 1$.

Corollary 1.3
Since the formulation and proof of the theorem does not require specifying the coordinates of a feed point s_0 of the voltage δ-generator $E_{0s}(s') = V_0\delta(s'-s_0)$, the theorem is valid for arbitrary choice of the feed point. Here V_0 is the voltage amplitude, $\delta(s' - s_0)$ is one-dimensional Dirac delta-function.

Commentary. Really, the expression for the first approximation of the current (11) was obtained for an arbitrary extraneous excitation field $E_{0s}(s')$. The term "arbitrary" applies only to the choice of the feed point coordinates and model describing the local excitation source. It is related to the fact that the vibrator excitation by the incident extraneous field, formed in the spatial domain with account of the borders, which influence through structure of the field $E_{0s}(s')$ will have an indirect reflection in the first approximation for the current. This certainly violates the logic of the theorem proof.

Corollary 1.4
One of the Theorem 1 conditions are the following requirements on the surface: (1) it must be a Lyapunov surface; (2) it should not cross current sources; (3) in a more general sense, it has to be passive, i.e., without generation of extraneous fields, and

(4) it does not possess the properties of mutual transformation of spatial harmonics of the electric and magnetic fields. Since the type of electrodynamic boundary surfaces is not defined and do not exclude the possibility of its composite presentation, the theorem is valid both for different types of surfaces such as perfectly conducting, impedance, partially impedance, etc. and for the different scattering inhomogeneities in the spatial domain, whose surfaces can be interpreted as parts of the total surface.

Commentary. Parts of the boundary surface can be presented by impedance surfaces of coupling holes between coupling volumes. Separate areas of piecewise inhomogeneous of the magneto-dielectric filling of the electrodynamic volume can be thought as scattering irregularities.

This approach allows us to formulate the Theorem 2. It concerns the fundamental possibility to compensate the influence of the spatial domain boundaries the current distribution on a perfectly conducting vibrator by "applying" impedance to the vibrator surface. In this case, we will use the generalized concept, when the impedance $z_i(\vec{r}) = z_i(s)$ can be presented as continues functions of the coordinate s or as lumped impedance inclusions in some points on the vibrator; the superposition of both options is also possible.

Theorem 2 *Let a thin perfectly conducting vibrator, represented by a segment of a circular cylinder excited by a point source is placed in an electrodynamic volume V uniformly filled by a medium with material parameters (ε_1, μ_1). The vibrator radius r and length 2L are such that inequality $\frac{r}{2L} \ll 1$ and $\frac{r\sqrt{\varepsilon_1\mu_1}}{\lambda} \ll 1$ hold (λ is the wavelength in free space); the electrodynamic volume V is bounded by the Lyapunov surface S which do not crosses extraneous current sources. Then the influence of the volume boundaries on the vibrator current distribution can be compensated by coating the vibrator surface with complex impedance $z_i(s) = \frac{F_S(s, J(s))}{J(s)}$, variable along the vibrator axis, where $J(s)$ is the current distribution in the perfectly conducting vibrator and $F_S(s, J(s))$ is the functional of the boundary reaction defined by the expression (D.17):*

$$F_S\left(\vec{r}, \vec{J}(\vec{r})\right) = (\mathrm{graddiv} + k_1^2) \oint_S \left\{ \mathrm{div}\,\vec{\Pi}^e(\vec{r}')\left(\hat{I}G_0^e(\vec{r}, \vec{r}')\right)\vec{n} - \mathrm{div}\left(\hat{I}G_0^e(\vec{r}, \vec{r}')\right)\vec{\Pi}^e(\vec{r}')\vec{n} \right.$$

$$\left. + \left[\vec{n}, \left(\hat{I}G_0^e(\vec{r}, \vec{r}')\right)\right]\mathrm{rot}\,\vec{\Pi}^e(\vec{r}') - \left[\vec{n}, \vec{\Pi}^e(\vec{r}')\right]\mathrm{rot}\left(\hat{I}G_0^e(\vec{r}, \vec{r}')\right) \right\} \mathrm{d}s'.$$

Corollary 2.1
Since the surfaces of scattering inhomogeneities located in the electrodynamic volume can be interpreted as components of the common surface S, the theorem is also valid for this case.

Corollary 2.2
The input resistance of the vibrator with an impedance distribution (D.21) $z_i(s) = \frac{F_S(s, J(s))}{J(s)}$, located in the electrodynamic volume V, is equivalent to the input resistance of the perfectly conducting vibrator of the same geometry located in free space.

The main difficulty in applying the Theorem 2 consists in finding the functional $F_S(s, J(s))$. However, if the Green's function for the electromagnetic volume is known, the problem is considerably simplified, since the fields on the boundary surface in expression (D.21) can be easily found. The proofs of the Theorem 1 and Theorem 2 are given in Appendix D.

2.3.2   Alternative Green's Function of the Electric Field on the Vibrator

As it follows from the Sect. 2.1.1, the field Green's functions allowing direct calculations of the electric field at the observation point can be conveniently used for numerical solving integral equations in boundary value problems. However, the field Green's functions are seldom used in commercial computer-aided design (CAD) systems. In modern CAD systems, various mathematical methods are implemented. Among them, direct numerical methods for solving boundary value problems can be distinguished: the finite element (FE) and the Finite Difference Time Domain (FDTD) methods, which are characterized by considerable versatility and capability to analyze structures of a rather large complexity. Modeling of microwave structures using the software packages may require significant expenditures of computer resources and computation time due to a high degree of discretization of the spatial domains in the FE and FDTD methods. The number of discrete determines the problem dimension. The FE and FDTD methods are characterized by the highest discretization degree among all currently known methods. The FE and FDTD methods are used in the HFSS and CST MWS software packages.

An alternative approach to solving computational electrodynamics problems are indirect methods, among which the moment method should be distinguished. The MoM differ from direct numerical approaches, since the field at the observation point is determined by an analytical solution of some key problem, namely, a problem of a structure excitation by an elementary current source. Such a solution is uniquely determined by the Green's function. The MoM is effective if the Green's function can be written analytically in a simple form. In this case, only the surface can be discretized, hence the problem dimension is greatly reduced. Unfortunately, the Green's function can be simply found only for a limited number of structures which include flat-layered structures and free space. The CAD systems based on MoM include following software packages: Microwave Office, ADS, and FEKO.

It can be stated that one of the factors that hinder the further expansion of the functionality of commercial software packages is the lack of a unified methodological approach to representation of Green's functions for structures and regions with complex boundaries. Therefore, the search and justification of such a methodical platform is an urgent and important problem for engineering practice. For vibrator problems when boundary reaction can be presented in an explicit form, such an approach can be implemented by using the source-like representation of Green's functions (2.11), which allows single-type formalization of the Green's functions in computational algorithms for domains of various geometry. This subsection is aimed at development of a general approach to representing the Green's function of the electric field on a thin impedance vibrator surface located in an arbitrary electrodynamic volume based on the use of the relation (2.11). In this case, the term of the Green's function responsible for the field response of the spatial domain boundaries can be analyzed in the impedance approach framework [15, 73].

Let us consider the current integral Eq. (2.21) for a single vibrator radiator, valid in an arbitrary electrodynamic volume filled with an isotropic homogeneous medium

$$
\left(\frac{d^2}{ds^2} + k_1^2\right) \int_{-L}^{L} J(s') \frac{e^{-ikR(s,s')}}{R(s,s')} ds' + f_{reg}[s, J(s)] - z_i(s)i\omega\varepsilon_1 J(s) = -i\omega\varepsilon_1 E_{0s}(s),
$$

(2.69)

where (s, s') are local coordinates along the vibrator axis, $R(s, s') = \sqrt{(s - s')^2 + r^2}$, and

$$
f_{reg}[s, J(s)] = \left([graddiv + k_1^2] \int_{-L}^{L} J(s')\vec{e}_s \cdot \hat{g}_V^e(\vec{r}, \vec{e}_s \cdot s')ds', \vec{e}_s\right)
$$

(2.70)

is the projection on the vibrator axis of the regular part of radiated electric field determined by the volume V geometry.

As can be seen from the Eq. (2.69), the first integral term is the left-hand side of the well-known Poklington equation [24] for the perfectly conducting vibrator in free space. This term can be interpreted as the vibrator own field, which can be rewritten by using quasi-one-dimensional kernel in the form suggested by Richmond [74]

$$
\int_{-L}^{L} J(s')\left[\frac{\partial^2}{\partial s^2} + k_1^2\right]\frac{e^{-ik_1 R}}{R} ds'
$$

$$
= \int_{-L}^{L} J(s')\frac{e^{-ik_1 R}}{R^5}\left[(1 + ik_1 R)(2R^2 - 3r^2) + k_1^2 r^2 R^2\right]\Bigg|_{R=R(s,s')} ds'.
$$

(2.71)

The second integral term (2.70) in (2.69) is a functional taking into account the reaction of electrodynamic volume boundaries on the vibrator current also needs a correct physical interpretation. The Theorem 2 formulated in Sect. 2.3.1 states that the influence of the spatial domain boundaries on the current distribution on the perfectly conducting radiator can be compensated by covering the vibrator surface by complex impedance $z_i(s)$. In the notation (2.69), this possibility can be realized when the identity

$$z_i(s) = \frac{-i f_{reg}[s, J(s)]}{\omega \varepsilon_1 J(s)} \tag{2.72}$$

holds.

Since in physical processes similar can be compensated only be similar, it can be argued that, according to (2.72), the physical analysis of the influence functional of boundaries should be carried out using the impedance concept. That is, the influence of boundaries on the internal characteristics of the vibrator, can be taken into account by introducing the concept of induced effective external impedance

$$z_e(s) = \frac{i}{\omega \varepsilon_1 J(s)} \left([\text{graddiv} + k_1^2] \int_{-L}^{L} J(s') \vec{e}_s \cdot \hat{g}_V^e(\vec{r}, \vec{e}_s \cdot s') ds', \vec{e}_s \right). \tag{2.73}$$

Therefore, the requirement of the Theorem 2 can be reduced to the requirement $z_i(s) = -z_e(s)$. In contrast to the internal impedance $z_i(s)$, the external induced impedance should be determined as effective, since fulfillment of the condition $\text{Re} z_e(s) \geq 0$ cannot be guaranteed. The condition $\text{Re} z_i(s) \geq 0$ imposed from energy considerations defines the internal vibrator impedance $z_i(s)$ as the real physical quantity. However, introduction of a functional quantity is methodologically expedient, since it allows, even at the level of the initial equation formulation to localize the factor influence of external scattering surfaces upon the vibrator radiator. Thus, Eq. (2.69) can be represented as follows

$$\frac{1}{i\omega \varepsilon_1} \int_{-L}^{L} J(s') \frac{e^{-ik_1 R}}{R^5} \left[(1 + ik_1 R)(2R^2 - 3r^2) + k_1^2 r^2 R^2 \right] \bigg|_{R=\sqrt{(s-s')^2 + r^2}} ds' \tag{2.74}$$

$$- [z_e(s) + z_i(s)] J(s) = -E_0(s).$$

Based on a comparison of the left-hand side of Eq. (2.74) and the right-hand side of relation (2.7), we can write the final representation of the Green's function for the electric field on the generatrix of a thin impedance vibrator

$$G_E(s, s') = \frac{e^{-ik_1 R}}{k_1^2 R^5} \left[(1 + ik_1 R)(2R^2 - 3r^2) + k_1^2 r^2 R^2 \right] \bigg|_{R=\sqrt{(s-s')^2 + r^2}} \tag{2.75}$$

$$- \frac{i\omega \varepsilon_1}{k_1^2} [z_e(s) + z_i(s)] \delta(s - s'),$$

where $\delta(s - s')$ is delta-function.

The expression (2.75) has a number of methodological features. First, it combines the own field of the vibrator current and the secondary fields defined by the external boundaries of the electrodynamic volume and the internal impedance of the vibrator as a simple sum. This makes it possible to realize an autonomous calculation of these terms using different analysis methods. Second, the representation of the field of boundary reaction in the form of the effective external induced impedance $z_e(s)$ makes it possible to perform the same type formalization of its definition for regions with boundaries of different geometries. Third, the current Eq. (2.69) can be easily solved based on Corollaries of the Theorem 1 from Sect. 2.3.1 by using the method of successive approximations, in which the secondary fields in (2.75) are defined by the vibrator current found in the previous approximation.

These features prove that the format for representing the field Green's function in the form (2.75) is optimal for use in commercial software packages in order to expand their functional capabilities. Let us consider the advantages of the proposed methodological approach by considering the problem of a horizontal vibrator located above an infinite perfectly conducting plane.

2.3.3 Radiation of a Vibrator Located Above a Plane

The boundary value problem for a thin symmetric horizontal vibrator radiating into a uniformly filled half-space over a perfectly conducting plane is classical in the electrodynamics theory. Over the past hundred years, the problem has been intensively studied by various authors for both perfectly conducting vibrators and vibrators whose surface is characterized by internal impedance (for example, [1, 3, 15, 20, 74–76]). In the thin-wire approximations, a quasi-one-dimensional kernel can be used

$$\int_S \hat{G}^e(\vec{r}, \vec{r}') \vec{J}(\vec{r}') d\vec{r}' \approx \int_{-L}^{L} J(s') \left[\frac{e^{-ik_1\sqrt{(s-s')^2+r^2}}}{\sqrt{(s-s')^2+r^2}} - \frac{e^{-ik_1\sqrt{(s-s')^2+(2h+r)^2}}}{\sqrt{(s-s')^2+(2h+r)^2}} \right] ds',$$

(2.76)

where h is the distance between the vibrator axis and the plane. The relation (2.76), which was obtained based on the Green's function for a half-space (Appendix A), makes it possible to identify the effect of the perfectly conducting plane on a vibrator radiator as influence of a vibrator mirror image. Historically, this approach was adopted in literature as the most common. Therefore, the plane influence on the vibrator radiation field was traditionally interpreted by using a concept of intrinsic and mutual resistance of the real and mirror image vibrators. In other words, functional values involved in the interpretation of the physical processes of the boundary-value problem, which appeared in the course of its solution by the chosen method, but did not figure in any way in the original problem. The mathematical correctness of the results thus obtained does not raise any objections, but from a physical point of view, this approach looks like a methodological case. The case is not only that the physical processes in the radiating system are interpreted by non-existent values, but also that such an interpretation cannot be generalized to other cases, for example, for an impedance plane or a boundary surface of a more complex geometry. The proposed impedance interpretation of extraneous scatterer influence on the radiating vibrator is devoid of these shortcomings. This Subsection is aimed at the numeric estimate

Fig. 2.2 The problem
geometry

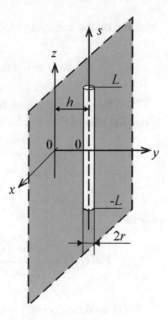

of the induced effective impedance $z_e(s)$ for various distances between the vibrator
and plane and also at analyzing the feasibility to implement the distributions internal
vibrator impedance, which can compensate the plane influence.

The structure and the adopted notation are presented in Fig. 2.2, where the Cartesian coordinate system $\{x, y, z\}$ which axis is directed along the vibrator axis. Let
us consider the integral equation for the current identical to the Eq. (2.69):

$$\left(\frac{d^2}{ds^2} + k_1^2\right) \int_{-L}^{L} J(s') \frac{e^{-ik_1 R(s,s')}}{R(s, s')} ds'$$

$$= -i\omega\varepsilon_1 E_{0S}(s) + \left(\frac{d^2}{ds^2} + k_1^2\right) \int_{-L}^{L} J(s') \frac{e^{-ik_1\sqrt{(s-s')^2+(2h+r)^2}}}{\sqrt{(s-s')^2 + (2h+r)^2}} ds' + i\omega\varepsilon_1 z_i(s) J(s),$$

$$(2.77)$$

which can be used to write the explicit expression for the induced external effective
impedance on the vibrator, located in a free half-space above the plane

$$z_e(s) = -\frac{i}{\omega J(s)} \left(\frac{d^2}{ds^2} + k^2\right) \int_{-L}^{L} J(s') \frac{e^{-ik\sqrt{(s-s')^2+(2h+r)^2}}}{\sqrt{(s-s')^2 + (2h+r)^2}} ds'. \quad (2.78)$$

If the vibrator in the infinite free space is excited in the center by a lumped voltage
generator with an amplitude V_0, the vibrator current can be obtained by solving the

problem by the averaging method [15]:

$$J(s) = -\alpha V_0 \left(\frac{ik}{60}\right) \frac{\sin k(L - |s|) + \alpha P_\delta^s(kr, ks)}{\cos kL + \alpha P_L^s(kr, kL)}. \tag{2.79}$$

Then the final expression for the distribution of the induced effective impedance along the axis of the vibrator can be written as follows

$$z_e(s) = -\frac{30i}{k Fi(s)} \int_{-L}^{L} Fi(s') Fe(s, s') ds', \tag{2.80}$$

where

$$Fe(s, s') = \frac{e^{-ikR_1(s,s')}}{(R_1(s, s'))^4} \left[\begin{array}{c} (s - s')^2 \left(3ik - \frac{k^2-3}{R_1(s,s')}\right) - R_1(s, s') \\ -ik(R_1(s, s'))^2 + k^2(R_1(s, s'))^3 \end{array} \right],$$

$Fi(s) = \sin k(L - |s|) + \alpha P_\delta^s(kr, ks)$, $R_1(s, s') = \sqrt{(s - s')^2 + (2h + r)^2}$, $\alpha = \frac{1}{2\ln[r/(2L)]}$, $P_\delta^s(kr, ks) = P^s[kr, k(L + s)] - (\sin ks + \sin k|s|) P_L^s(kr, kL)$, and the expressions $P^s[kr, k(L + s)]$ and $P_L^s(kr, kL)$ are defined by

$$P^s[kr, k(L + s)] = \int_{-L}^{s} \left[\frac{e^{-ikR(s',-L)}}{R(s', -L)} + \frac{e^{-ikR(s',L)}}{R(s', L)} \right] \sin k(s - s') ds' \Bigg|_{s=L}$$

$$= \frac{e^{ikL}}{2i} \int_{-L}^{L} \left[\frac{e^{-ikR(s',-L)}}{R(s', -L)} + \frac{e^{-ikR(s',L)}}{R(s', L)} \right] e^{-iks'} ds'$$

$$- \frac{e^{-ikL}}{2i} \int_{-L}^{L} \left[\frac{e^{-ikR(s',-L)}}{R(s', -L)} + \frac{e^{-ikR(s',L)}}{R(s', L)} \right] e^{iks'} ds',$$

$$P_L^s(kr, kL) = \left[2\ln 2 - \gamma(L) - \frac{\mathrm{Cin}(4kL) + i\,\mathrm{Si}(4kL)}{2} \right] \cos kL$$

$$+ \frac{\mathrm{Si}(4kL) - i\,\mathrm{Cin}(4kL)}{2} \sin kL,$$

where $\mathrm{Si}(x)$ and $\mathrm{Cin}(x)$ are the integral sine and cosine complex argument and

$$\gamma(s) = \ln \frac{[(L + s) + \sqrt{(L + s)^2 + r^2}][(L - s) + \sqrt{(L - s)^2 + r^2}]}{4L^2}.$$

The numerical simulations carried out for the half-wave vibrator have shown that the distribution of the induced effective impedance (2.80) along the vibrator has a

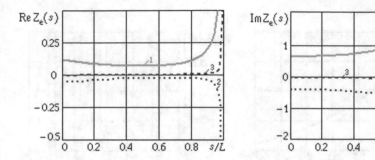

Fig. 2.3 The plots of the effective impedance as functions of the vibrator arm coordinate, $2L = 0.5\lambda$, $r = L/100$: 1—$h = 1.25\lambda$, 2—$h = 2\lambda$, 3—$h = 20\lambda$)

U-shaped character if $h \geq \lambda$. The plots of the effective impedance normalized to the free space resistance $Z_e(s) = z_e(s)/(120\pi)$ as function of the ratio s/L are shown in Fig. 2.3 for three fixed distances h. Due to the problem symmetry the curves are plotted only for one arm of the vibrator. Such shape of the curves is expected, since the denominator at the right-hand side of the expression (2.78) includes the current distribution on the vibrator, which obeys the boundary conditions $J(\pm L) = 0$. That is, the function $Z_e(s)$ has singular points at the edges of the vibrator, however, the Eq. (2.74) depends upon the product $z_e(s)J(s)$ that is zero at the edges of the vibrator. This eliminates the need to accurately determine the impedance $z_e(\pm L)$ at the end points by using approximate relation $z_e(\pm L) \approx z_e(\pm L \mp r/2)$ as proposed in the monograph [74].

As can be seen from Fig. 2.3, the flat curve pieces at the plots $\mathrm{Re}\,Z_e(s)$ and $\mathrm{Im}\,Z_e(s)$ are widening and become more uniform as the distance between the plane and vibrator increases. In addition, the curve level decreases and at distances $\infty > h \geq \lambda$, where the plane no longer has a significant effect on the vibrator and its influence becomes small. A more complete characterization of the plane influence upon the distance h can be obtained from Fig. 2.4, where the levels of flat curve pieces at the plots of impedance at the point $s = L/2$, the middle of the vibrator arm, are shown.

According to the physical assumptions, the effective impedance modulus $|Z_e(L/2)|$ is an exponentially decaying function of the parameter h/λ. The argument at the interval $[-\pi; \pi]$ is represented by a saw-tooth periodic function which period equal to $\lambda/2$. Therefore, the curves $\mathrm{Re}\,Z_e(s)$ and $\mathrm{Im}\,Z_e(s)$ are damping periodic functions (Fig. 2.4c and Fig. 2.4d). These curves are phase shifted, and the effective impedance in the vicinity of $h = n\lambda/2$ $(n = 1, 2, 3\ldots)$ is imaginary, since its real part is close to zero. Whereas the impedance in the vicinity $h = n\lambda/4$ $(n = 1, 3, 5\ldots)$ is of real character and its imaginary part is close to zero. In this case, the radiating vibrator, tuned to resonance, remains resonant, but its radiating capacity depends upon the real part of the effective impedance. Since the $\mathrm{Re}\,Z_e(s)$ can be positive or negative, the radiation resistance of the vibrator can decrease or increase. At distances $0 \leq h \leq 1.5\lambda$, the influence of the plane (Fig. 2.5) leads to

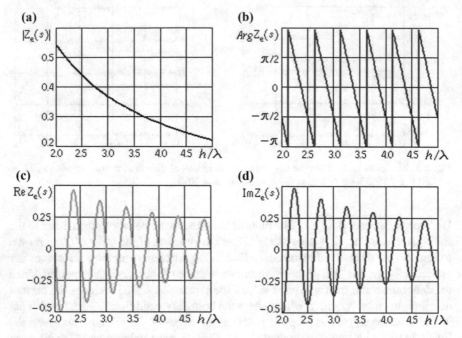

Fig. 2.4 The plots of effective impedance induced on the vibrator at the point $s = L/2$ as function of the distance h ($2L = 0.5\lambda$ and $r = L/100$)

abrupt changes of the values $\mathrm{Re}Z_e(s)$ and $\mathrm{Im}Z_e(s)$ along the vibrator can exceed the resistance of free space by tens and hundreds of times.

Under these circumstances, the value of $\mathrm{Re}Z_e(s)$ and $\mathrm{Im}Z_e(s)$ at the local point on the vibrator cannot be estimated. However, tendencies similar to those identified above can also be observable at small distances h. As an example, consider the points $h_1 \approx 0.356\lambda$ and $h_2 \approx 0.61\lambda$, in which vicinity the sign of imaginary part of the effective impedance $\mathrm{Im}Z_e(s)$ changes to the opposite. The plots of the effective impedance for $h_2 \approx 0.61\lambda$ is shown in Fig. 2.6. As shown in [15], the first two resonances are observed precisely at distances $h_1 \approx 0.356\lambda$ and $h_2 \approx 0.61\lambda$. The simulation results have also shown that the impedance is not very susceptible to the choice of the vibrator radius r and at the interval $L/50 \leq r \leq L/500$ is almost constant.

The effect of the perfectly conducting plane on the vibrator radiator allows us within the framework of the impedance concept based on the condition $z_i(s) = -z_e(s)$ to consider the problem of direct compensation of the plane's influence by introducing the internal impedance on the vibrator $z_i(s)$. The problem can be divided into two levels: physical and technological. At the first level, one should keep in mind that, in accordance with the energy laws, the real part of internal impedance $z_i(s)$ should always be positive, i.e., $\mathrm{Re}z_e(s) \geq 0$. Since any restrictions on the sign of the imaginary part of the impedance are not imposed, the full compensation of the induced effective impedance by the impedance $z_i(s)$ is possible only when

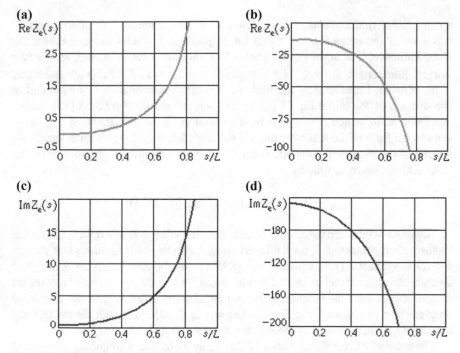

Fig. 2.5 The distribution of the effective impedance along the vibrator arm ($2L = 0.5\lambda$, $r = L/100$): **a** $h = 0.5\lambda$, **b** $h = 0.25\lambda$

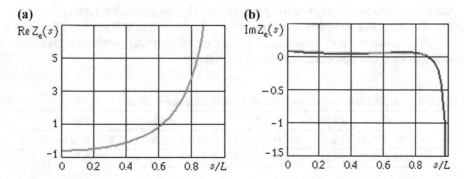

Fig. 2.6 The distribution of the effective impedance along the vibrator arm ($2L = 0.5\lambda$, $r = L/100$, $h = 0.61\lambda$)

$\mathrm{Re} z_e(s) \leq 0$. Otherwise, partial compensation is possible only for the imaginary part of the effective impedance.

The technological side of the issue is associated with the difficulties of practical vibrator manufacturing, which surfaces can be characterized by a certain profile of the internal impedance distribution [15]. From this point of view, the vibrators with a constant internal impedance are simplest to realize. Therefore, a natural question arises about the correctness of using the impedance vibrators to compensate the plane influence. One of the possible ways to studry this issue is to compare the two integral functionals: $J_1 = \int_{-L}^{L} Fi^2(s)z_e(s)ds$ and $J_2 = z_c \int_{-L}^{L} Fi^2(s)ds$, where z_c is the constant impedance of the vibrator. The first functional J_1 can be found in the solution of the initial Eq. (2.77) by the generalized EMF method [15], which describes the influence of the plane on the vibrator. The second functional J_2 is the asymptotic form of the first functional if the impedance z_c is correctly selected. As a result of the numerical simulations for a half-wave vibrator, we propose to fix the constant impedance as follows

$$z_c = \text{Re}(z_e(0.05\lambda)) + i\text{Im}(z_e(0.115\lambda)). \tag{2.81}$$

Although such a representation of effective impedance is not optimal from the mathematical point of view, but it allows to obtain an acceptable accuracy at distances $h \geq 2\lambda$ (see Table 2.1). Moreover, for large distances, the calculated values of J_1 and J_2 coincide. Thus, it can be argued that at distances $h \geq 2\lambda$, the effect of a plane on the perfectly conducting vibrator can be compensated by using an impedance vibrator characterized by constant impedance $z_i(s) = -z_c$ (2.81). At shorter distances, only direct compensation based on the relation $z_i(s) = -z_e(s)$ can be valid.

The induced effective impedance (2.80) is, by its nature, a frequency-dependent quantity. However, the estimates obtained above for the fixed wavelength can also be used to analyzing the vibrator radiation at the operating wavelengths. That is, the data obtained for the average wavelength can be generalized to the entire operating band of the vibrator. This possibility is confirmed by the simulation results shown in Fig. 2.7, according to which, changing of the vibrator length at the interval $0.23\lambda \leq L \leq 0.25\lambda$, causes small variations of the real part the effective vibrator impedance, while its imaginary part remains almost unchanged.

Table 2.1 Dependence of functionals J_1 and J_2 on the distance to plane

h	2.25λ	2.5λ	5λ	10.25λ	20λ
J_1	$-0.206 + i1.819$	$0.195 - i1.641$	$0.117 - i0.822$	$-0.062 + i0.401$	$0.033 - i0.205$
J_2	$-0.232 + i1.821$	$0.215 - i1.642$	$0.122 - i0.823$	$-0.063 + i0.401$	$0.033 - i0.205$

Fig. 2.7 The distribution of the effective impedance along the vibrator arm ($h = 2.1\lambda, r = L/100$): 1—$L = 0.25\lambda$, 2—$L = 0.24\lambda$, 3—$L = 0.23\lambda$)

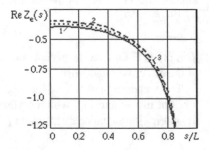

References

1. Khizhnyak NA (1986) Integral equations of macroscopical electrodynamics. Naukova dumka, Kiev (in Russian)
2. Penkin YM, Katrich VA, Nesterenko MV, Berdnik SL, Dakhov VM (2019) Electromagnetic fields excited in volumes with spherical boundaries. Springer Nature Swizerland AG, Cham, Swizerland
3. Felsen LB, Marcuvitz N (1973) Radiation and scattering of waves. Prentice-Hall Inc., Englewood Cliffs, NJ
4. Morse PM, Feshbach H (1953) Methods of theoretical physics. McGraw-Hill, New York
5. Panchenko BA (1970) Tensor Green's functions of Maxwell equations for cylindrical regions. Radio Engineering. 15:82–91 (in Russian)
6. Levin H, Schwinger J (1950) On the theory of electromagnetic wave diffraction by an aperture in an infinite plane conducting screen. Commun Pure Appl Math 355–391
7. Collin RE (1960) Field theory of guided waves. McGraw-Hill, New York
8. Tai CT (1971) Dyadic Green's function in electromagnetic theory. Intex Educ Publ, Scranton
9. Tikhonov AN, Samarsky AA (1977) Equations of mathematical physics. Nauka, Moscow (in Russian)
10. Tikhonov AN, Samarsky AA (1947) On the excitation of radio waveguides. J Tech Phys 47:1283–1296 (in Russian)
11. Van Bladel J (1961) Some remarks on Green's dyadic for infinite space. IRE Trans Antennas Propag AP-9:563–566
12. Markov GG, Panchenko BA (1964) Tensor Green's functions of rectangular waveguides and resonators. Izv Vuzov Radiotek 1:34–41 (in Russian)
13. Penkin YuM, Katrich VA (2003) Excitation of electromagnetic waves in the volumes with coordinate boundaries. Fakt, Kharkov (in Russian)
14. Repin VM (1971) A numerical method for solving the problem of electromagnetic coupling of volumes through holes. J Comput Math Math Phys 11(1):152–163 (in Russian)
15. Nesterenko MV, Katrich VA, Penkin YuM, Dakhov VM, Berdnik SL (2011) Thin impedance vibrators. Theory and Applications. Springer Science + Business Media, New York
16. Priymenko SD, Bondarenko LA (2007) The source-like Green's function of a circular resonator. Radiotekhnika 149:C, 22–26 (in Russian)
17. Markov GG, Chaplin AF (1983) Excitation of electromagnetic waves. Radio i svyaz' Moskow (in Russian)
18. Nesterenko MV, Katrich VA, Penkin YuM, Berdnik SL (2008) Analytical and hybrid methods in theory of slot-hole coupling of electrodynamic volumes. Springer Science + Business Media, New York
19. Miller MA, Talanov VI (1961) Using the concept of surface impedance in the theory of surface electromagnetic fields (review). Izv Vuz Radiophys 4(5):795–830 (in Russian)
20. King RWP (1956) The theory of linear antennas. Harvard University Press, Cambridge
21. Leontovich M (1946) On a theorem of the theory of diffraction and its application to diffraction by a narrow slot of arbitrary length. J Exp Theor Phys 16(6):474–479 (in Russian)
22. King RWP, Aronson EA, Harrison CW (1966) Determination of the admittance and effective length of cylindrical antennas. Radio Sci 1:835–850
23. Naiheng Y, Harrington R (1983) Electromagnetic coupling to an infinite wire through a slot in a conducting plane. IEEE Trans Antennas Propag 31:310–316
24. Pocklington HC (1897) Electrical oscillations in wires. Proc Cambridge Philos Soc 9(pt VII):324–332
25. Hallen E (1938) Theoretical investigations into the transmitting and receiving qualities of antenna. Nova Acta Reg Soc Sci Ups Ser IV 11:1–44
26. Brillouin L (1943) The antenna problem. Quar. Appl Math 201–214
27. Leontovich M, Levin M (1944) To the theory of oscillations excitation in antennas' vibrators. J Tech Phys 14(9):481–506 (in Russian)

28. Vineshtein LA (1959) The current waves in a thin cylindrical conductor. J Tech Phys 29(6):65–80 (in Russian)
29. Vineshtein LA (1959) Current waves in a thin cylindrical conductor. II. Passive vibrator current and transmit vibrator radiation. J Tech Phys 31(6):81–91 (in Russian)
30. Mei KK (1965) On the integral equation of thin wire antennas. IEEE Trans Antennas Propag 13:374–378
31. Hanson GW (2008) Radiation efficiency of nano-radius dipole antennas in the microwave and far-infrared regimes. IEEE Antennas Propag Mag 50:66–77
32. Hanson GW (2005) Fundamental transmitting properties of carbon nanotube antennas. IEEE Trans Antennas Propag 53:3426–3435
33. Lewin L (1951) Advanced theory of waveguides. Iliffe, London
34. Miller MA (1954) Application of homogeneous boundary conditions in the theory of thin antennas. J Tech Phys 24(8):1483–1495 (in Russian)
35. Govorun NN (1959) Integral equations for antenna—body of revolution with an impedance surface. Rep USSR Acad Sci 126(1):49–52 (in Russian)
36. Cassedy ES, Fainberg J (1960) Back scattering cross sections of cylindrical wires of finite conductivity. IEEE Trans Antennas Propag 8:1–7
37. Vineshtein LA (1961) Current waves in a thin cylindrical conductor. III. The variational method and its application to the theory of ideal and impedance wires. J Tech Phys 33(1):110–125 (in Russian)
38. Harrison CW, Heinz RO (1963) On the radar cross section of rods, tubes, and strips of finite conductivity. IEEE Trans Antennas Propag 11:459–468
39. Wu TT, King RWP (1965) The cylindrical antenna with nonreflecting resistive loading. IEEE Trans Antennas Propag 13:369–373
40. King RWP, Wu TT (1966) The imperfectly conducting cylindrical transmitting antenna. IEEE Trans Antennas Propag 14:524–534
41. King RWP, Harrison CW, Aronson EA (1966) The imperfectly conducting cylindrical transmitting antenna: numerical results. IEEE Trans Antennas Propag 14:535–542
42. Glushkovsky EA, Levin BM, Rabinovich EY (1967) Integral equation for current in a thin impedance vibrator. Radiotekhnika 22(12):18–23 (in Russian)
43. Glushkovsky EA, Isralit AB, Levin BM, Rabinovich EY (1967) Variable surface impedance linear antennas. Antennas 2:154–165 (in Russian)
44. Shen L-C (1967) An experimental study of the antenna with nonreflecting resistive loading. IEEE Trans Antennas Propag 15:606–611
45. Lamensdorf D (1967) An experimental investigation of dielectric-coated antennas. IEEE Trans Antennas Propag 15:767–771
46. Richmond J (1967) Scattering by imperfectly conducting wires. IEEE Trans Antennas Propag 15:802–806
47. Glushkovsky EA, Isralit AB, Levin BM, Rabinovich EY (1968) Linear impedance vibrator calculation methods. Radiotekhnika 23(1):40–46 (in Russian)
48. Taylor CD (1968) Cylindrical transmitting antenna: tapered resistivity and multiple impedance loadings. IEEE Trans Antennas Propag 16:176–179
49. Rao BLJ, Ferris JE, Zimmerman WE (1969) Broadband characteristics of cylindrical antennas with exponentially tapered capacitive loading. IEEE Trans Antennas Propag 17:145–151
50. Einarsson O, Plato T (1969) Electromagnetic scattering by a thin resistive wire. Electron Lett 5(25):637–638
51. Inagaki N, Kukino O, Sekiguchi T (1972) Integral equation analysis of cylindrical antennas characterized by arbitrary surface impedance. IEICE Trans Commun 55-B(6):683–690 (1972)
52. Senior T (1979) Backscattering from resistive strips. IEEE Trans Antennas Propag 27:808–813
53. Richmond J (1980) Green's function technique for near-zone scattering by cylindrical wires with finite conductivity. IEEE Trans Antennas Propag 28:114–117
54. Garb HL, Friedberg PS, Yakover IM (1982) Scattering of electromagnetic waves by a thin impedance film in a rectangular waveguide. Radioengineering Electron 27(4):690–695 (in Russian)

55. Garb HL, Friedberg PS, Yakover IM (1983) Distribution of surface electric current on an ideally conducting plate and impedance film in a rectangular waveguide. Lith Phys Collect 23(1):34–40 (in Russian)
56. Yakover IM (1984) Scattering of an H_{10}-wave by a thin resonant impedance rod in a rectangular waveguide. Lith Phys Collect 24:41–46 (in Russian)
57. Levin BM, Yakovlev AD (1985) Antenna with loads as an impedance vibrator with variable impedance. Radioengineering Electron 30(1):25–33 (in Russian)
58. Garb HL, Friedberg PS, Yakover IM (1985) Diffraction of an H_{10}-wave on a thin resistive film with a stepwise change of surface impedance in a rectangular waveguide. Radioengineering Electron 30(1):41–48 (in Russian)
59. Bretones AR, Martín RG, García IS (1995) Time-domain analysis of magnetic-coated wire antennas. IEEE Trans Antennas Propag 43:591–596
60. Mäkinen RM (2004) An efficient surface-impedance boundary condition for thin wires of finite conductivity. IEEE Trans Antennas Propag 52:3364–3372
61. Bethe HA (1944) Theory of diffraction by small holes. Phys Rev 66:163–182
62. Watson WH (1946) Resonant slots. Proc IEEE 93(4):747–777
63. Stevenson AF (1948) Theory of slots in rectangular wave-guides. J Appl Phys 19(1):24–38
64. Feld YN (1948) Fundamentals of the Theory of Slot Antennas. Sov Radio Moskow (in Russian)
65. Levin ML (1951) To the conclusion of the basic equation of the theory of slot antennas. J Tech Phys 21(7):787–794 (in Russian)
66. Suzuki M (1956) Diffraction of plane electromagnetic waves by a rectangular aperture. IRE Trans Antennas Propag. 4:149–156
67. Feld YN, Benenson LS (1959) Antenna-feeder devices. Publishing House VVIA them. In: Zhukovsky (ed) Moskow (in Russian)
68. Levin ML (1951) Slot antenna with guide device. J Tech Phys 21(7):795–801 (in Russian)
69. Yamaguchi S, Aramaki Y, Takahashi T, Otsuka M, Konishi Y (2012) A slotted waveguide array antenna covered by a dielectric slab with a post-wall cavity. In: Proceedings of IEEE APS URSI international symposium. Chicago, USA, pp 559–560
70. Nesterenko MV, Berdnik SL, Katrich VA, Penkin YM (2015) Electromagnetic waves excitation by thin impedance vibrators and narrow slots in electrodynamic volumes. In: Bashir SO (ed) Advanced electromagnetic waves, chapter 6. InTech, Rijeka, pp 147–175
71. Bogoliubov NN, Mitropolsky YA (1961) Asymptotic methods in the theory of nonlinear oscillations. Gordon Breach Science Publishers, New York
72. Philatov AN (1974) Asymptotic methods in the theory of differential and integral-differential equations. PHAN, Tashkent (in Russian)
73. Berdnik SL, Penkin DY, Katrich VA, Penkin YM, Nesterenko MV (2014) Using the concept of surface impedance in problems of electrodynamics (75 years later) (review). Radiophys Radioastronomy 19(1):57–80 (in Russian)
74. Mittra R (1973) Computer techniques for electromagnetics. Pergamon Press, Oxford
75. Feinberg EL (1999) Propagation of radio waves along the Earth's surface. Fizmatlit, Moskow (in Russian)
76. King RWP, Smith GS (1981) Antennas in matter. MIT Press, Cambridge

Chapter 3
Solution of Current Equations for Isolated Vibrator and Slot Scatterers

3.1 Vibrator with Variable Surface Impedance in Free Space

As is known, vibrators characterized by variable surface impedance can be used as an additional tool for controlling the antenna electrodynamic characteristics [1–9]. We will not analyze the methods outlined in the above references which were used for the problem solution, but we simply note that all of them are devoted to theoretical and experimental study of radiating vibrators excited in the center by lamped EMF. However, the induced current on a scattering vibrator excited by an incident electromagnetic wave should be known to carry out the analysis of receiving antennas. This problem has also independent applied worth to study scattering characteristics of material bodies with irregular shapes [10]. The results presented in this Subsection have independent importance, besides, they will also be used through the monograph.

3.1.1 Solving the Current Equation by the Averaging Method

Consider the integral Eq. (2.21) defining the electric current on the vibrator located in free space ($\varepsilon_1 = \mu_1 = 1$)

$$\left(\frac{d^2}{ds^2} + k^2\right) \int_{-L}^{L} J(s') \frac{e^{-ik\sqrt{(s-s')^2+r^2}}}{\sqrt{(s-s')^2+r^2}} ds' = -i\omega[E_{0s}(s) - z_i(s)J(s)], \qquad (3.1)$$

M. V. Nesterenko et al., *Combined Vibrator-Slot Structures: Theory and Applications*, Lecture Notes in Electrical Engineering 689, https://doi.org/10.1007/978-3-030-60177-5_3

where $R(s, s') = \sqrt{(s - s')^2 + r^2}$. First, we single out the logarithmic singularity of the above equation kernel as it was done for the Eq. (2.47). Then the vibrator current can be written as

$$\int_{-L}^{L} J(s') \frac{e^{-ikR(s,s')}}{R(s, s')} ds' = \Omega(s) J(s) + \int_{-L}^{L} \frac{J(s') e^{-ikR(s,s')} - J(s)}{R(s, s')} ds', \qquad (3.2)$$

where

$$\Omega(s) = \int_{-L}^{L} \frac{ds'}{\sqrt{(s - s')^2 + r^2}} = \Omega + \gamma(s), \qquad (3.3)$$

$\Omega = 2 \ln \frac{2L}{r}$ is the large parameter, $\gamma(s) = \ln \frac{[(L+s)+\sqrt{(L+s)^2+r^2}][(L-s)+\sqrt{(L-s)^2+r^2}]}{4L^2}$ is a function equal to zero in the vibrator center and reaches the greatest value at the vibrator ends, where the current is zero according to the boundary conditions (2.18). Then, based on (3.3), the Eq. (3.1) can be transformed to the integral-differential equation with a small parameter

$$\frac{d^2 J(s)}{ds^2} + k^2 J(s) = \alpha\{i\omega E_{0s}(s) + F[s, J(s)] - i\omega z_i(s) J(s)\} \qquad (3.4)$$

where $\alpha = \frac{1}{2 \ln[r/(2L)]}$ is the natural small parameter ($|\alpha| \ll 1$), and

$$F[s, J(s)] = -\frac{dJ(s')}{ds'} \frac{e^{-ikR(s,s')}}{R(s, s')} \Bigg|_{-L}^{L} + \left[\frac{d^2 J(s)}{ds^2} + k^2 J(s) \right] \gamma(s)$$

$$+ \int_{-L}^{L} \frac{\left[\frac{d^2 J(s')}{ds'^2} + k^2 J(s') \right] e^{-ikR(s,s')} - \left[\frac{d^2 J(s)}{ds^2} + k^2 J(s) \right]}{R(s, s')} ds' \qquad (3.5)$$

is the vibrator eigenfield in free space.

The approximate analytical solution of the Eq. (3.4) will be obtained by the asymptotic averaging method, described in the Sect. 2.2.2.3. To reduce (3.4) the equation system to the standard form (2.64) with a small parameter, we make the change of variables and obtain by the method of variation of arbitrary constants

$$J(s) = A(s) \cos ks + B(s) \sin ks,$$

$$\frac{dJ(s)}{ds} = -A(s)k \sin ks + B(s)k \cos ks, \quad \left(\frac{dA(s)}{ds} \cos ks + \frac{dB(s)}{ds} \sin ks = 0 \right),$$

$$\frac{d^2 J(s)}{ds^2} + k^2 J(s) = -\frac{dA(s)}{ds} \sin ks + \frac{dB(s)}{ds} \cos ks, \qquad (3.6)$$

where $A(s)$ and $B(s)$ are new unknown functions. Then, the Eq. (3.4) is transformed to the following system of integral-differential equations:

$$
\frac{dA(s)}{ds} = -\frac{\alpha}{k}\left\{
\begin{array}{c}
i\omega E_{0s}(s) + F\left[s, A(s), \dfrac{dA(s)}{ds}, B(s), \dfrac{dB(s)}{ds}\right] \\
-i\omega z_i(s)[A(s)\cos ks + B(s)\sin ks]
\end{array}
\right\}\sin ks,
$$

$$
\frac{dB(s)}{ds} = +\frac{\alpha}{k}\left\{
\begin{array}{c}
i\omega E_{0s}(s) + F\left[s, A(s), \dfrac{dA(s)}{ds}, B(s), \dfrac{dB(s)}{ds}\right] \\
-i\omega z_i(s)[A(s)\cos ks + B(s)\sin ks]
\end{array}
\right\}\cos ks. \quad (3.7)
$$

These equations are equivalent to the system of integral-differential Eqs. (3.4) not resolved with respect to the derivative which are presented in the standard form (2.64). Since the right-hand sides of these equations are proportional to the small parameter α, the functions $A(s)$ and $B(s)$, on the right-hand sides of Eqs. (3.7) are slowly varying functions, and the asymptotic method of averaging can be used to solve the equation system (3.7). The equation system (3.7) can be put into one-to-one correspondence with the simplified system (2.65), in which the derivatives $\frac{dA(s)}{ds}$ and $\frac{dB(s)}{ds}$ on right-hand sides of the equations are zeroes. After partial averaging the system (2.65) over the explicit variable s, we obtain the equations of the first approximation. The term partial means that the averaging operator (2.56) acts on all terms, except those that contain the incident field $E_{0s}(S)$. Such averaging is valid for equation systems in the form (3.7) [11]:

$$
\frac{d\bar{A}(s)}{ds} = -\alpha\left\{\frac{i\omega}{k}E_{0s}(s) + \bar{F}[s, \bar{A}(s), \bar{B}(s)]\right\}\sin ks + \chi_a\bar{A}(s) + \chi_s\bar{B}(s),
$$

$$
\frac{d\bar{B}(s)}{ds} = +\alpha\left\{\frac{i\omega}{k}E_{0s}(s) + \bar{F}[s, \bar{A}(s), \bar{B}(s)]\right\}\cos ks - \chi_s\bar{A}(s) - \chi_a\bar{B}(s), \quad (3.8)
$$

where

$$
\bar{F}[s, \bar{A}(s), \bar{B}(s)] = [\bar{A}(s')\sin ks' - \bar{B}(s')\cos ks']\left.\frac{e^{-ikR(s,s')}}{R(s, s')}\right|_{-L}^{L} \quad (3.9)
$$

is the vibrator eigenfield averaged over its length. In the Eq. (3.8)

$$
\chi_s = \lim_{2L\to\infty}\frac{1}{2L}\frac{\alpha i\omega}{k}\int_{-L}^{L}z_i^s(s)\sin^2 ks\{\cos^2 ks\}ds,
$$

$$
\chi_a = \lim_{2L\to\infty}\frac{1}{2L}\frac{\alpha i\omega}{k}\int_{-L}^{L}\frac{1}{2}z_i^a(s)\sin 2ks\,ds, \quad (3.10)
$$

where $z_i^s(s)$ and $z_i^a(s)$ are the symmetric and antisymmetric components of the impedance $z_i(s) = z_i^s(s) + z_i^a(s)$ distributed over the vibrator length. The solution of the equations system of (3.8) can be seek in the following form [12]:

$$\bar{A}(s) = \chi_s[C_1(s) \sin \chi s + C_2(s) \cos \chi s],$$
$$\bar{B}(s) = -\chi_a[C_1(s) \sin \chi s + C_2(s) \cos \chi s]$$
$$+ \chi[C_1(s) \cos \chi s - C_2(s) \sin \cos \chi s], \tag{3.11}$$

where $\chi^2 = \chi_s^2 - \chi_a^2$, $C_1(s)$ and $C_2(s)$ are new unknown functions. The functions $C_1(s)$, $C_2(s)$, and, hence, $\bar{A}(s)$, $\bar{B}(s)$ can be found by substituting the expressions (3.11) into the equation system (3.8). Thus, we obtain:

$$\bar{A}(s) = \frac{1}{\chi}\Big\{\bar{A}(-L)[\chi \cos \chi(L+s) + \chi_a \sin \chi(L+s)] + \bar{B}(-L)[\chi_s \sin \chi(L+s)]$$
$$+ \alpha \int_{-L}^{s} \left\{ \begin{array}{c} \frac{i\omega}{k} E_{0s}(s') + \bar{F}[s', \bar{A}, \bar{B}] \\ \times[\chi_s \cos ks' \sin \chi(s-s') - \chi \sin ks' \cos \chi(s-s') + \chi_a \sin ks' \sin \chi(s-s')]ds' \end{array} \right\} \Big\},$$

$$\bar{B}(s) = \frac{1}{\chi}\Big\{-\bar{A}(-L)[\chi_s \sin \chi(L+s)] + \bar{B}(-L)[\chi \cos \chi(L+s) - \chi_a \sin \chi(L+s)]$$
$$+ \alpha \int_{-L}^{s} \left\{ \begin{array}{c} \frac{i\omega}{k} E_{0s}(s') + \bar{F}[s', \bar{A}, \bar{B}] \\ \times[\chi_s \sin ks' \sin \chi(s-s') + \chi \cos ks' \cos \chi(s-s') - \chi_a \cos ks' \sin \chi(s-s')]ds' \end{array} \right\} \Big\}.$$
$$\tag{3.12}$$

Substituting the functions $\bar{A}(s)$ and $\bar{B}(s)$ as approximating functions for $A(s)$ and $B(s)$ in the formula $J(s) = A(s) \cos ks + B(s) \sin ks$, we obtain the most general asymptotic expression relative the parameter α for the current in the thin vibrator with impedance variable along its axis. The formula valid for arbitrary vibrator excitation can be written as

$$J(s) = \bar{A}(-L)[\cos ks \cos \chi(L+s) - (\bar{\chi}_s \sin ks - \bar{\chi}_a \cos ks) \sin \chi(L+s)]$$
$$+ \bar{B}(-L)[\sin ks \cos \chi(L+s) + (\bar{\chi}_s \cos ks - \bar{\chi}_a \sin ks) \sin \chi(L+s)]$$
$$+ \alpha \int_{-L}^{s} \left\{ \frac{i\omega}{k} E_{0s}^s(s') + \bar{F}[s', \bar{A}(\pm L), \bar{B}(\pm L)] \right\} \{\sin k(s-s') \cos \chi(s-s')$$
$$+ [\bar{\chi}_s \cos k(s-s') - \bar{\chi}_a \sin k(s+s')] \sin \chi(s-s')\}ds' \tag{3.13}$$

where $\bar{\chi}_s = \chi_s/\chi$, $\bar{\chi}_a = \chi_a/\chi$.

When the vibrator impedance is constant, $\chi = \alpha \frac{i\omega}{2k} z_i$, $\bar{\chi}_s = 1$, and $\bar{\chi}_a = 0$, the expression (3.13) is reduced to

$$J(s) = \overline{A}(-L) \cos(\tilde{k}s + \chi L) + \overline{B}(-L) \sin(\tilde{k}s + \chi L)$$

$$+ \alpha \int_{-L}^{s} \left\{ \frac{i\omega}{k} E_{0s}(s') + \overline{F}[s', \overline{A}, \overline{B}] \right\} \sin \tilde{k}(s - s') ds', \tag{3.14}$$

where $\tilde{k} = k + \chi = k + i(\alpha/r)\overline{Z}_S$.

The constants $\overline{A}(\pm L)$ and $\overline{B}(\pm L)$ can be find by using the boundary conditions (2.18) for the current and symmetry conditions [13], that are uniquely related to the vibrator excitation method. If $E_{0s}(S) = E_{0s}^s(S)$, then $J(s) = J(-s) = J^s(s)$ and $\overline{A}(-L) = \overline{A}(+L)$, $\overline{B}(-L) = -\overline{B}(+L)$; if $E_{0s}(s) = E_{0s}^a(s)$, then $J(s) = -J(-s) = J^a(s)$ and, $\overline{A}(-L) = -\overline{A}(+L)$, $\overline{B}(-L) = \overline{B}(+L)$. Then, for arbitrary vibrator excitation we obtain the following expression

$$J(s) = J^s(s) + J^a(s) = \alpha \frac{i\omega}{k} \left\{ \int_{-L}^{s} E_{0s}(s') \sin \tilde{k}(s - s') ds' \right.$$

$$- \frac{\sin \tilde{k}(L + s) + \alpha P^s[kr, \tilde{k}(L + s)]}{\sin 2\tilde{k}L + \alpha P^s(kr, 2\tilde{k}L)} \int_{-L}^{L} E_{0s}^s(s') \sin \tilde{k}(L - s') ds'$$

$$\left. - \frac{\sin \tilde{k}(L + s) + \alpha P^a[kr, \tilde{k}(L + s)]}{\sin 2\tilde{k}L + \alpha P^a(kr, 2\tilde{k}L)} \int_{-L}^{L} E_{0s}^a(s') \sin \tilde{k}(L - s') ds' \right\}, \tag{3.15}$$

where P^s and P^a are the vibrator eigenfield functions

$$P^s[kr, \tilde{k}(L + s)] = \int_{-L}^{s} \left[\frac{e^{-ikR(s', -L)}}{R(s', -L)} + \frac{e^{-ikR(s', L)}}{R(s', L)} \right] \sin \tilde{k}(s - s') ds' \bigg|_{s=L}$$

$$= P^s(kr, 2\tilde{k}L), \tag{3.16a}$$

$$P^a[kr, \tilde{k}(L + s)] = \int_{-L}^{s} \left[\frac{e^{-ikR(s', -L)}}{R(s', -L)} - \frac{e^{-ikR(s', L)}}{R(s', L)} \right] \sin \tilde{k}(s - s') ds' \bigg|_{s=L}$$

$$= P^a(kr, 2\tilde{k}L). \tag{3.16b}$$

Let the vibrator be excited by a normally incident plane electromagnetic wave with amplitude E_0 so that $E_{0s}^s(s) = E_0$, $E_{0s}^a(s) = 0$. Then, we obtain based on (3.15), (3.16a) the first approximation of the averaging method

$$J(s) = -\alpha \frac{i\omega}{k\tilde{k}} E_0 \frac{(\cos \tilde{k}s - \cos \tilde{k}L) + \alpha \left\{ \begin{array}{l} \sin \tilde{k}L P^s[kr, \tilde{k}(L + s)] \\ -[1 - \cos \tilde{k}(L + s)]P^s[kr, \tilde{k}L] \end{array} \right\}}{\cos \tilde{k}L + \alpha P^s[kr, \tilde{k}L]},$$

$$\tag{3.17}$$

where $P^s[kr, \tilde{k}L] = \int_{-L}^{L} \frac{e^{-ik\sqrt{(L-s)^2+r^2}}}{\sqrt{(L-s)^2+r^2}} \cos \tilde{k}s \, ds$.

Since the final expression for the current obtained from (3.13) is rather cumbersome and unsuitable for practical use, we will construct the solution of Eq. (3.1) by the generalized method of EMF by using formula (3.17).

3.1.2 Solving the Current Equation by the Generalized Method of Induced EMF

Let us apply the generalized method of induced EMF for the approximate analytical solution of Eq. (3.1) under symmetric vibrator excitation, $E_{0s}(s) = E_{0s}(-s)$, provided that $z_i(s) = z_i(-s)$. First, we approximate the distribution of electric current on the vibrator by

$$J(s) = J_0 f(s), \quad f(\pm L) = 0, \tag{3.18}$$

where J_0 is unknown current amplitude, $f(s)$ is the predefined function. Then, we multiply the left- and right-hand sides of Eq. (3.1) by $f(s)$ and integrate the resulting expressions over the vibrator length, and obtain

$$J_0 = \frac{-\frac{i\omega}{2k} \int_{-L}^{L} f(s) E_{0s}(s) ds}{\frac{1}{2k} \int_{-L}^{L} f(s) \left[\left(\frac{d^2}{ds^2} + k^2 \right) \int_{-L}^{L} f(s') \frac{e^{-ikR(s,s')}}{R(s,s')} ds' \right] ds - \frac{i\omega}{2k} \int_{-L}^{L} f^2(s) z_i(s) ds}. \tag{3.19}$$

As is known, the accuracy of the integral equation solution for the vibrator current by the induced EMF method substantially depends on the choice of approximating functions for the current [14, 15]. If the impedance depends on the vibrator length, the most adequate to the real physical process, from our point of view, is a function containing information about the average value of the impedance along the vibrator. Such approach is quite obvious if the small fluctuations of the impedance around the average value are observed. The type of the functions defining impedance variation over the vibrator length considerable influences the current amplitude and, hence, the electrodynamic characteristics of the vibrator.

Consider normal incidence of the plane electromagnetic wave $E_{0s}(s) = E_0$ on the vibrator. Then, the first term in the numerator of (3.17) can be used as the function $f(s)$, i.e.,

$$f(s) = \cos \tilde{k}s - \cos \tilde{k}L, \tag{3.20}$$

where $\tilde{k} = k - \frac{i2\pi z_i^{av}}{Z_0\Omega}$, $z_i^{av} = \frac{1}{2L} \int_{-L}^{L} z_i(s) ds$ is the averaged internal impedance along the vibrator length, and $\Omega = 2\ln(2L/r)$. The expression (3.20) contains information concerning the function $f(s)$ that is significantly different from the functions used in the literature [1, 2]. Argumentation for such function choice is presented in Sect 3.1.3, were the function $f(s)$ is also proposed for the case $E_{0s}(s) = E_{0s}(-s)$, $z_i(s) \neq z_i(-s)$. After substituting (3.20) into (3.19) and calculating J_0, we obtain, based on (3.18), the expression

$$J(s) = -\frac{i\omega}{k\tilde{k}} E_0 \frac{(\sin \tilde{k}L - \tilde{k}L \cos \tilde{k}L)(\cos \tilde{k}s - \cos \tilde{k}L)}{Z(kr, \tilde{k}L) + F_z(\tilde{k}r, \tilde{k}L)}, \qquad (3.21)$$

where

$$Z(kr, \tilde{k}L) = \frac{1}{2k} \int_{-L}^{L} (\cos \tilde{k}s - \cos \tilde{k}L)\left(\frac{d^2}{ds^2} + k^2\right) F_f(s) ds, \qquad (3.22)$$

$$F_f(s) = \int_{-L}^{L} (\cos \tilde{k}s' - \cos \tilde{k}L) \frac{e^{-ikR(s,s')}}{R(s, s')} ds', \qquad (3.23)$$

$$F_z(\tilde{k}r, \tilde{k}L) = -\frac{i}{r} \int_{-L}^{L} (\cos \tilde{k}s - \cos \tilde{k}L)^2 \bar{Z}_S(s) ds, \qquad (3.24)$$

$\bar{Z}_S(s) = \frac{2\pi r z_i(s)}{Z_0}$ is the normalized surface complex impedance of the vibrator $\bar{Z}_S(s) = \bar{R}_S + i\bar{X}_S\phi(s)$, and $\phi(s)$ is the predefined function.

3.1.3 Justification of the Choice of Approximating Functions for Current

Let the vibrator with the distributed surface symmetric impedance, $z_i(s) = z_i(-s) = z_i^s(s)$, be excited by the symmetric field $E_{0s}(s) = E_{0s}(-s) = E_{0s}^s(s)$ created by external sources. Then we substitute the equalities $\chi_a = 0$, $\chi = \chi_s$, $J(s) = J(-s)$, $\bar{A}(-L) = \bar{A}(L) = C_1$, $\bar{B}(-L) = -\bar{B}(L) = C_2$ in the formula (3.13) and obtain that the current is

$$J(s) = C_1 \cos[(k + \chi_s)s + \chi_s L] + C_2 \sin[(k + \chi_s)s + \chi_s L]$$
$$+ \alpha \int_{-L}^{s} \left\{\frac{i\omega}{k} E_{0s}^s(s') + \bar{F}[s', C_1, C_2]\right\} \sin[(k + \chi_s)(s - s')] ds'. \qquad (3.25)$$

The constants C_1 and C_2 are determined from the boundary conditions $J(\pm L) = 0$. Assuming $E_{0s}^s(s) = E_0$ in (3.25), we obtain

$$J(s) = J_c\{\alpha[\cos(k + \chi_s)s - \cos(k + \chi_s)L] + \alpha^2 P^s[kr, (k + \chi_s)s]\}, \quad (3.26)$$

where J_c is the current amplitude, $P^s[kr, (k + \chi_s)s]$ is the function defined in (3.16a). Assume that $z_i(s) = z_i\phi(s)$ where $\phi(s) = e^{-\beta\frac{|s|}{L}}$ Since the impedance $\bar{Z}_S = 2\pi r z_i/Z_0$, we obtain

$$k + \chi_s = k + i\frac{\alpha}{r}\bar{Z}_S\frac{1 - e^{-\beta}}{\beta}. \quad (3.27)$$

Using the expression for $\tilde{k}$ defined in (3.20), we have

$$\tilde{k} = k - \frac{i2\pi z_i}{Z_0\Omega}\frac{1}{2L}\int_{-L}^{L}\phi(s)\mathrm{d}s = k + i\frac{\alpha}{r}\bar{Z}_S\frac{1 - e^{-\beta}}{\beta}. \quad (3.28)$$

As can be seen, expressions (3.27) and (3.28) are identically equal. When terms proportional to α^2 are discarded, the functions $f(s)$ in formulas (3.20) and (3.26) are also identical. The above conclusions confirm the validity of approximation (3.20) if the vibrator is symmetrically and the function $f(s)$ is also symmetric with respect to the vibrator feeding point.

The generalized method of induced EMF with adequate approximating functions can also be used to solve a more specific problem as compared to that defined in (3.17), namely: $E_{0s}(s) = E_{0s}^s(s)$, $z_i(s) = z_i^s(s) + z_i^a(s)$. In this case, the vibrator current will consist of symmetric and asymmetric parts

$$J(s) = J^s(s) + J^a(s) = J_0^s f^s(s) + J_0^a f^a(s). \quad (3.29)$$

The expression for the function $f^a(s)$ can be obtained from (3.13), assuming that $z_i(s) = -z_i(-s) = z_i^a(s)$. Finally, it can be written as

$$\begin{cases} f^a(s) = \sin ks \mathrm{sh}\chi_a L - \sin kL \mathrm{sh}\chi_a s, & \chi_a \neq 0, \\ f^a(s) = \sin 2ks - 2\sin ks \cos kL, & \chi_a = 0. \end{cases} \quad (3.30)$$

3.2 Vibrator with Variable Surface Impedance in a Rectangular Waveguide

3.2.1 Solving the Current Equation by the Averaging Method

Let a thin rectilinear impedance vibrator be placed in a cross-sectional plane of infinite or semi-infinite hollow ($\varepsilon_1 = \mu_1 = 0$) rectangular waveguide. We will

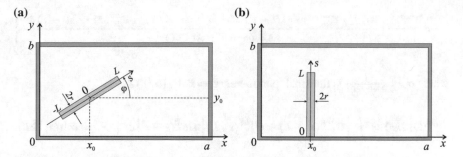

Fig. 3.1 The structure configuration with: **a** the symmetric vibrator oriented at the angle φ between the vibrator axis and the vibrator broad wall; **b** monopole on the waveguide broad wall

consider symmetric (Fig. 3.1a) and asymmetric known as the monopole (Fig. 3.1b): the first does not contact with the waveguide walls, while the second touches the waveguide broad wall by one of its ends. Internal dimensions of the waveguide cross-section is $\{a \times b\}$. The vibrator radius r and length $2L$ satisfy the thin wire approximation, $r/(2L) \ll 1$, $r/(L) \ll 1$, where λ is wavelength in free space. Let us introduce the Cartesian coordinate system $\{x, y, z\}$ related to the waveguide, and local coordinate systems $\{0s'\}$ and $\{0s\}$ connected with the vibrator axis and surface.

When impedance distribution of the rectilinear vibrator is defined by a constant function, $z_i(s) = const$, the structure analysis can be carried out by using the integral Eq. (2.23) written as

$$\left(\frac{d^2}{ds^2} + k^2\right) \int_{-L}^{L} J(s') \frac{e^{-ikR(s,s')}}{R(s,s')} ds' = -i\omega[E_{0s}(s) - z_i J(s)]$$

$$- \left(\frac{d^2}{ds^2} + k^2\right) \int_{-L}^{L} J(s') g_s^{Wg}(s, s') ds', \tag{3.31}$$

where $R(s, s') = \sqrt{(s - s')^2 + r^2}$, $E_{0s}(s)$ is projection of the electric field induced by extrinsic sources at the vibrator axis, and $g_s^{Wg}(s, s')$ is the component of the regular part of the waveguide electric Green's function defined in (2.11).

The solution of the Eq. (3.31) will be obtained by the averaging method described in Sect. 3.1. The first approximation expression for the current in the structure with the symmetric impedance vibrator under arbitrary excitation, $E_{0s}(s) = E_{0s}^s(s) + E_{0s}^a(s)$, can be written as

$$J^{s,a}(s) = -\alpha \frac{i\omega}{k}$$

$$\times \left\{ \frac{\sin \tilde{k}(L - s) \int_{-L}^{s} E_{0s}^{s,a}(s') \sin \tilde{k}(L + s') ds' + \sin \tilde{k}(L + s) \int_{s}^{-L} E_{0s}^{s,a}(s') \sin \tilde{k}(L - s') ds'}{\sin 2\tilde{k}L + \alpha W^{s,a}(kr, 2\tilde{k}L)} \right.$$

$$+\alpha \frac{W^{s,a}(kr, 2\tilde{k}s) \int_{-L}^{L} E_{0s}^{s,a}(s') \sin \tilde{k}(L-s') \, ds' - W^{s,a}(kr, 2\tilde{k}L) \int_{-L}^{s} E_{0s}^{s,a}(s') \sin \tilde{k}(s-s') \, ds'}{\sin 2\tilde{k}L + \alpha W^{s,a}(kr, 2\tilde{k}L)}\Bigg\},$$

$$(3.32)$$

where $\alpha = \frac{1}{2\ln(r/2L)}$ is the small parameter, $\tilde{k} = k + i(\alpha/r)\bar{Z}_S$,

$$W^s(kr, 2\tilde{k}s) = \int_{-L}^{s} \left[G_s^{Wg}(s', -L) + G_s^{Wg}(s', L) \right] \sin \tilde{k}(s-s') \, ds'\Bigg|_{s=L} = W^s(kr, 2\tilde{k}L),$$

$$(3.33a)$$

$$W^a(kr, 2\tilde{k}s) = \int_{-L}^{s} \left[G_s^{Wg}(s', -L) - G_s^{Wg}(s', L) \right] \sin \tilde{k}(s-s') \, ds'\Bigg|_{s=L} = W^a(kr, 2\tilde{k}L)$$

$$(3.33b)$$

are the functions of the vibrator eigenfield in the waveguide, $G_s^{Wg}(s, s') = \frac{e^{-ikR(s,s')}}{R(s,s')} + g_s^{Wg}(s, s')$ is s-component of the total electrical Green's function of the rectangular waveguide (Appendix A).

If the $H_{10}-$ wave propagates in the infinite rectangular waveguide ($a < \lambda < 2a$, $a > 2b$) from the region $z = -\infty$, we can write for the structure shown in Fig. 3.1a

$$E_{0s}(s) = E_0 \sin \varphi \left(\sin \frac{\pi x_0}{a} \cos k_\varphi s + \cos \frac{\pi x_0}{a} \sin k_\varphi s \right), \qquad (3.34)$$

where E_0 is the H_{10}-wave amplitude, $(k_\varphi = (\pi/a) \cos \varphi$. The expression for the current (3.32) with accuracy up to α^2 terms, i.e., defined by the first term in curly brackets of formula (3.32)) are

$$J(s) = J_0[f^s(s) + f^a(s)] = -\alpha E_0 \frac{2(i\omega/k\tilde{k})\sin \varphi}{[1 - (k_\varphi/\tilde{k})^2][\sin 2\tilde{k}L + \alpha W_\varphi^{sa}(kr, 2\tilde{k}L)]}$$

$$\times \left[\begin{array}{l} \sin \dfrac{\pi x_0}{a} \sin \tilde{k}L (\cos \tilde{k}s \cos k_\varphi L - \cos \tilde{k}L \cos k_\varphi s) \\[2mm] + \cos \dfrac{\pi x_0}{a} \cos \tilde{k}L (\sin \tilde{k}s \sin k_\varphi L - \sin \tilde{k}L \sin k_\varphi s) \end{array} \right], \qquad (3.35)$$

where J_0 is the current amplitude, $f^s(s)$ and $f^a(s)$ are symmetric and antisymmetric terms of the current distribution functions, $W_\varphi^{sa}(kr, 2\tilde{k}L)$ are the functions defining the vibrator eigenfield which takes into account in (3.33a) both the symmetric and antisymmetric components; these functions are determined by the corresponding components of the electric Green function of the infinite rectangular waveguide, which take into account all types of oscillations in the local region near vibrator.

If the angle $\varphi = 90^O$, we obtain from (3.34) that $E_{0s}(s) = E_{0s}^s(s) = E_0 \sin \frac{\pi x_0}{a}$. Hence, the current on the monopole (Fig. 3.1b) with accuracy up to the terms α^2 is equal to

$$J(s) = J_0[f(s) + \alpha F(s)]$$

$$= -\alpha E_0 \left(\frac{i\omega}{k\tilde{k}} \right) \sin \frac{\pi x_0}{a} \frac{(\cos \tilde{k}s - \cos \tilde{k}L) + \alpha[W(kr, \tilde{k}s) - W(kr, \tilde{k}L)]}{\cos \tilde{k}L + \alpha W(kr, \tilde{k}L)},$$

$$(3.36)$$

where

$$W(kr, \tilde{k}s) = \frac{4\pi}{ab} \sum_{m=1}^{\infty} \sum_{n=0}^{\infty} \frac{\varepsilon_n e^{-k_z r}}{k_z (\tilde{k}^2 - k_y^2)} \sin^2 k_x x_0 \cos k_y L$$

$$\times (\tilde{k} \sin \tilde{k}L \cos k_y s - k_y \sin k_y L \cos \tilde{k}s), \qquad (3.37a)$$

$$W(kr, \tilde{k}L) = W(kr, \tilde{k}s) \Big|_{s=L}, \qquad (3.38b)$$

$$\varepsilon_n = \begin{cases} 1, & n = 0 \\ 2, & n \neq 0 \end{cases}, k_x = \frac{m\pi}{a}, k_y = \frac{n\pi}{b}, k_z = \sqrt{k_x^2 + k_y^2 - k^2}, m \text{ and } n \text{ are integers,}$$

and x_0 is the distance from the narrow waveguide wall to the monopole axis.

3.2.2 Solving the Current Equation by the Generalized Method of Induced EMF

Consider the structure with symmetric vibrator shown in Fig. 3.1a. Since the impedance boundary condition $E_s(s) = z_i(s) J(s)$ should be fulfilled on the vibrator surface, the following integral equation for the current $J(s)$ can be written

$$\left(\frac{d^2}{ds^2} + k^2 \right) \int_{-L}^{L} J(s') G_s(s, s') ds' = -i\omega E_{0s}(s) + i\omega z_i(s) J(s), \qquad (3.39)$$

where $G_s(s, s')$ is $s-$ component of the electrical Green's function of the rectangular waveguide.

Consider the case of symmetric vibrator excitation, $E_{0s}(s) = E_{0s}(-s)$. The approximate analytical solution of Eq. (3.39) under conditions $\varphi = 90°$ and $z_i(s) = z_i(-s)$ can be obtained by the generalized method of induced EMF, as it was done in Sect. 3.1.2. Let us select the basis approximating function $f(s) = \cos \tilde{k}s - \cos \tilde{k}L$ defined (3.36) and take into account that the field of extrinsic sources created by incident $H_{10}-$ wave is $E_{0s}(s) = E_0 \sin \frac{\pi x_0}{a}$. Then the expression for the current according to (3.19) can be written as

$$J(s) = -\frac{i\omega}{k\tilde{k}} E_0 \sin\frac{\pi x_0}{a} \frac{(\sin\tilde{k}L - \tilde{k}L\cos\tilde{k}L)(\cos\tilde{k}s - \cos\tilde{k}L)}{Z^W(kr, \tilde{k}L) + F_z^W(\tilde{k}r, \tilde{k}L)}, \qquad (3.40)$$

where $\tilde{k} = k - \frac{i2\pi z_i^{av}}{Z_0\Omega}$, $z_i^{av} = \frac{1}{2L}\int_{-L}^{L} z_i(s)ds$ is the internal impedance averaged over the vibrator length,

$$Z^W(kr, \tilde{k}L) = \frac{1}{2k}\int_{-L}^{L} f(s)\left(\frac{d^2}{ds^2} + k^2\right)F_f^W(s)ds, \qquad (3.41)$$

$$F_f^W(s) = \int_{-L}^{L} f(s')G_s(s, s')ds', \qquad (3.42)$$

$$F_z^W(\tilde{k}r, \tilde{k}L) = -\frac{i}{r}\int_{-L}^{L} f^2(s)\bar{Z}_S(s)ds. \qquad (3.43)$$

Since the Green's function according to the formula (A.4) can be presented as

$$G_s(s, s') = \frac{4\pi}{ab}\sum_{m=1}^{\infty}\sum_{n=0}^{\infty}\frac{\varepsilon_n}{k_z}e^{-k_z r}\sin^2 k_x x_0 \cos k_y(y_0 + s)\cos k_y(y_0 + s'), \quad (3.44)$$

the impedance $Z^W(kr, \tilde{k}L)$ is equal

$$Z^W(kr, \tilde{k}L) = \frac{8\pi}{ab}\sum_{m=1}^{\infty}\sum_{n=0}^{\infty}\frac{\varepsilon_n(k^2 - k_y^2)\tilde{k}^2}{kk_z(\tilde{k}^2 - k_y^2)^2}e^{-k_z r}\sin^2 k_x x_0 \cos^2 k_y y_0$$
$$\times [\sin\tilde{k}L\cos k_y L - (\tilde{k}/k_y)\cos\tilde{k}L\sin k_y L]^2. \qquad (3.45)$$

The expression for the current on the monopole shown in Fig. 3.1b can be obtain by using the formula (3.40) after substitutions $y_0 = 0$, $b \to 2b$ in the formula (3.45). Then we get

$$Z^W(kr, \tilde{k}L) = \frac{4\pi}{ab}\sum_{m=1}^{\infty}\sum_{n=0}^{\infty}\frac{\varepsilon_n(k^2 - k_y^2)\tilde{k}^2}{kk_z(\tilde{k}^2 - k_y^2)^2}e^{-k_z r}\sin^2 k_x x_0$$
$$\times [\sin\tilde{k}L\cos k_y L - (\tilde{k}/k_y)\cos\tilde{k}L\sin k_y L]^2. \qquad (3.46)$$

Now, we will study the energy characteristics of the monopole. The field reflection and transmission coefficients, S_{11} and S_{12}, in the waveguide excited by the $H_{10}-$ wave are:

$$S_{11} = -\frac{4\pi i}{abkk_g}\left(\frac{k}{\tilde{k}}\sin\frac{\pi x_0}{a}\right)^2 \frac{(\sin \tilde{k}L - \tilde{k}L\cos\tilde{k}L)^2\, e^{2ik_g z}}{Z^W(kr,\tilde{k}L) + F_z^W(\tilde{k}r,\tilde{k}L)}, \tag{3.47}$$

$$S_{12} = 1 + \frac{4\pi i}{abkk_g}\left(\frac{k}{\tilde{k}}\sin\frac{\pi x_0}{a}\right)^2 \frac{(\sin \tilde{k}L - \tilde{k}L\cos\tilde{k}L)^2}{Z^W(kr,\tilde{k}L) + F_z^W(\tilde{k}r,\tilde{k}L)}, \tag{3.48}$$

where $k_g = \sqrt{k^2 - k_c^2}$ is the propagation constant of the $H_{10}-$ wave, $k_c = \pi/a$.

Consider three distribution functions of the imaginary part of vibrator surface impedance: (1) constant, $\phi_0(s) = 1$; (2) $\phi_1(s) = 2[1 - (s/L)]$, linearly decreasing towards the vibrator end; (3) $\phi_1(s) = 2[(s/L)]$, linearly increasing towards the vibrator end. All functions have the equal averaged values over the vibrator length. The analytical expressions for $F_{z0,1,2}^W$ obtained by using the formula (3.43) have the form

$$F_{z0}^W = -\frac{2i(\bar{R}_S + i\bar{X}_S)}{\tilde{k}^2 Lr}\left[\left(\frac{\tilde{k}L}{2}\right)^2\left(2 + \cos 2\tilde{k}L\right) - \frac{3}{8}\tilde{k}L\sin 2\tilde{k}L\right]$$

$$= \tilde{F}_z^W(\bar{R}_S + i\bar{X}_S)\Phi \tag{3.49}$$

for the function (1),

$$F_{z1}^W = \tilde{F}_z^W\left\{\bar{R}_S\Phi + i\bar{X}_S\left[\left(\frac{\tilde{k}L}{2}\right)^2\left(2 + \cos 2\tilde{k}L\right) - \frac{7}{4}\sin^2\tilde{k}L - 2\left(\cos\tilde{k}L - 1\right)\right]\right\} \tag{3.50}$$

for the function (2), and

$$F_{z2}^W = \tilde{F}_z^W\left\{\bar{R}_S\Phi + i\bar{X}_S\left[\left(\frac{\tilde{k}L}{2}\right)^2\left(2 + \cos 2\tilde{k}L\right) + \frac{7}{4}\sin^2\tilde{k}L\right.\right.$$

$$\left.\left. - \frac{3}{4}\tilde{k}L\sin 2\tilde{k}L + 2(\cos\tilde{k}L - 1)\right]\right\} \tag{3.51}$$

for the function (3).

As can be seen, the formulas for $F_{z0,1,2}^W$ differ from each other, despite the equal averaged values of the functions $\phi_{0,1,2}(s)$. Therefore, the current amplitudes and, hence, the energy characteristics of the vibrator differ significantly from each other, although the functions $f(s) = \cos\tilde{k}s - \cos\tilde{k}L$ coincide for all three impedance distribution functions

Consider a corrugated metal conductor as an example of constant surface impedance implementation shown in Fig. 3.2b. As evident, the size of a unit cell along the axis $\{0s\}$ should be much smaller than the operating wavelength. If conductivity of the vibrator metal is not taken into account, its surface impedance determined by

Fig. 3.2 The possible implementations of impedance vibrators

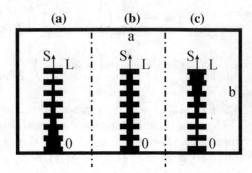

the formula (1.64) with $\mu = 1$ is purely inductive

$$\bar{Z}_S = i\bar{X}_S = ikr\ln(r/r_i),\tag{3.52}$$

where r and r_i are the external and internal corrugation radii. Then, if the surface impedance varies over the vibrator length, we obtain that $\bar{Z}_S^{av} = ikr\ln(r/r_i)$ and $\bar{Z}_S(s) = ikr\ln(r/r_i)\,\phi_n(s)$, $(n = 1, 2)$, where r_i corresponds to the case $\phi(s) = 1$. In this case, the variable impedance can be implement at by varying the conductor internal radius as $r_i(s) = re^{-\ln(r/r_i)\phi_n(s)}$, as shown in Fig. 3.2a, b.

3.2.3 *Numerical and Experimental Results*

Consider the influence of the functions $\phi_1(s)$ (Fig. 3.2c) and $\phi_2(s)$ (Fig. 3.2a) upon the vibrator resonant wavelength λ_{res} determined by the condition arg $S_{11} = 0$, i.e., by the energy of reactive electric and magnetic fields for all types of waves in the vibrator vicinity averaged over the wave period [7, 16]. This effect, as in the case of vibrator located in free space, can be explained by the fact that the maximal amplitude of the current distribution (3.40) is equal to zero at the vibrator end. As expected, the impedance for all three distribution for the perfectly conducting vibrator ($\phi(s) = 0$, $r = 2.0$ mm).

The vibrator with fixed parameters a, b, c and x_0 can be resonantly tuned at any wavelength in the waveguide operating band by varying the vibrator length, magnitude and type of its surface impedance, and function $\phi(s)$, as shown in Fig. 3.3. As can be seen from the plots in Fig. 3.4, the shape of the resonance curve $|S_{11}| = f(\lambda)$ depends mainly on the vibrator length and only weakly on the parameters of surface impedance. If the vibrator length is increased, the width of the resonance curve at the level $0.5|S_{11}|$ is also increased.

The proposed mathematical model was verified by comparing the simulation and experimental results shown in Fig. 3.5. The data were obtained for structures shown in Figs. 3.2b, c. The corrugated brass vibrator with cell width equal to 1.0 mm +

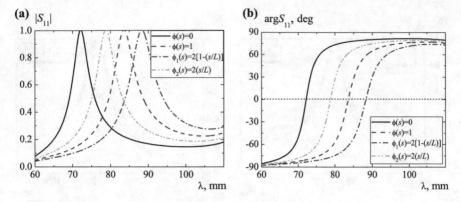

Fig. 3.3 The reflection coefficient as a function of the wavelength for corrugated metal monopole with parameters $L = 15.0$ mm, $r = 2.0$ mm, $r_i = 0.5$ mm, $x_0 = a/8$: **a**—$|S_{11}|$; **b**—arg S_{11}

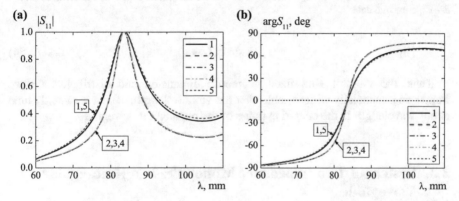

Fig 3.4 The reflection coefficient as a function of the wavelength for impedance vibrator with parameters $r = 2.0$ mm, $x_0 = a/8$: **a** $-|S_{11}|$; **b** $-$ arg S_{11}; **1** $-L = 18.5$ mm, $\phi(s) = 0$; **2** $-L = 15.0$ mm $\tilde{Z}_S = ikr \ln(4.0)$, $\phi(s) = 1$; **3** $-L = 15.0$ mm, $\tilde{Z}_S = ikr \ln(2.65)$, $\phi_1(s) = 2[1-(s/L)]$; **4** $-L = 15.0$ mm, $\tilde{Z}_S = ikr \ln(11.5)$, $\phi_2(s) = 2(s/L)$; **5** $-L = 20.0$ mm, $\tilde{Z}_S = -i0.014/kr$, $\phi(s) = 1$

1.0 mm $= 2.0$ mm, the external radius $r = 2$ mm, and the internal radii defined by the functions $\phi(s) = 1$ and $\phi(s) = 2[1 - (s/L)]$.

In conclusion, we note that if symmetric vibrator located at the angle φ to the waveguide broad wall, the approximating function can be presented as

$$f^s(s) + f^a(s) = \left[\begin{array}{c} \sin \frac{\pi x_0}{a} \sin \tilde{k}L(\cos \tilde{k}s \cos k_\varphi L - \cos \tilde{k}L \cos k_\varphi s) \\ + \cos \frac{\pi x_0}{a} \cos \tilde{k}L(\sin \tilde{k}s \sin k_\varphi L - \sin \tilde{k}L \sin k_\varphi s) \end{array} \right], \quad (3.53)$$

according to (3.35). The formula for the monopole, excited at the base by the voltage $\delta-$ generator can be obtained by using (3.32) as

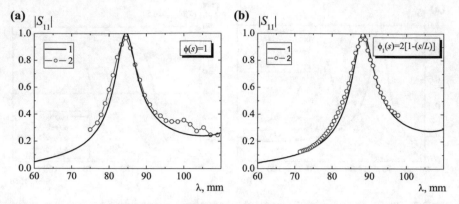

Fig. 3.5 The reflection coefficient as a function of the wavelength for the monopole with various distribution functions of the surface impedance $L = 15.0\,\text{mm}$, $r = 2.0\,\text{mm}$, $r_i = 0.5\,\text{mm}$, $x_0 = a/8$: **a** $\phi(s) = 1$, **b** $\phi(s) = 2[1 - (s/L)]$, **1**—numerical results based on the formula (3.47), **2**—experimental data

$$f(s) = \sin \tilde{k}(L - s). \tag{3.54}$$

Thus, the vibrator with fixed geometric dimension and distributed surface impedance, especially with variable over the vibrator length, can be resonantly tune at any wavelength in the operating range of the rectangular waveguide.

3.3   System of Two Impedance Monopoles in a Rectangular Waveguide

3.3.1   Problem Formulation and Solution the System of Equations for Currents

Let a system of two monopoles with variable surface impedances be located in a hollow infinite rectangular waveguide with perfectly conducting walls. The $H_{10}-$ wave propagates in the waveguide from the region $z = -\infty$. The waveguide cross-section is $\{a \times b\}$, the monopole radii and lengths are $r_{1,2}$ and $L_{1,2}$. The monopole axes are placed parallel to the narrow waveguide walls at the distances x_{01} and x_{02} from the left wall and are displaced along the waveguide axis at z_0 as shown in Fig. 3.6.

In this case, analysis of the system can be carried out by using the system of integral equations for the monopole currents, $J_1(s_1)$ and $J_2(s_2)$, which can be written according to (2.20) as:

Fig. 3.6 The system of two monopoles in the rectangular waveguide

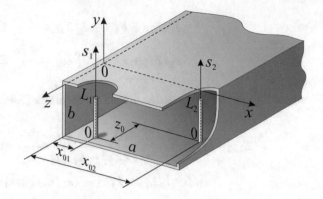

$$\left(\frac{d^2}{ds_1^2} + k^2\right)\left\{\int_{-L_1}^{L_1} J_1(s_1')G_{s_1}^{Wg}(s_1, s_1')ds_1' + \int_{-L_2}^{L_2} J_2(s_2')G_{s_2}^{Wg}(s_1, s_2')ds_2'\right\}$$
$$= -i\omega\left[E_{0s_1}(s_1) - z_{i1}(s_1)J_1(s_1)\right], \tag{3.55a}$$

$$\left(\frac{d^2}{ds_2^2} + k^2\right)\left\{\int_{-L_2}^{L_2} J_2(s_2')G_{s_2}^{Wg}(s_2, s_2')ds_2' + \int_{-L_1}^{L_1} J_1(s_1')G_{s_1}^{Wg}(s_2, s_1')ds_1'\right\}$$
$$= -i\omega\left[F_{0s_2}(s_2) - z_{i2}(s_2)J_2(s_2)\right]. \tag{3.55b}$$

Solution of the equations system (3.55) will be obtained by the generalized method of induced EMF, using the approximating expressions for the currents $J_{1(2)}(s_{1(2)}) = J_{1(2)}^0 f_{1(2)}(s_{1(2)})$ in which $J_{1(2)}^0$ are unknown current amplitudes and $f_{1(2)}(s_{1(2)})$ are predefined distribution functions. If the system is excited by the $H_{10}-$ wave the functions $f_{1(2)}(s_{1(2)})$ can be presented in accordance with (3.20) as

$$f_1(s_1) = \cos\tilde{k}_1 s_1 - \cos\tilde{k}_1 L_1, \tag{3.56a}$$

$$f_2(s_2) = \cos\tilde{k}_2 s_2 - \cos\tilde{k}_2 L_2, \tag{3.56b}$$

where $\tilde{k}_{1(2)} = k - \frac{i2\pi z_{i1(2)}^{av}}{Z_0\Omega_{1(2)}}$, $z_{i1(2)}^{av} = \frac{1}{2L_{1(2)}}\int_{-L_{1(2)}}^{L_{1(2)}} z_{i1(2)}(s_{1(2)})ds_{1(2)}$ are the values of intrinsic impedance averaged over the monopole lengths, and $\Omega_{1(2)} = 2\ln(2L_{1(2)}/r_{1(2)})$.

To solve the problem, we first multiply Eqs. (3.55a) and (3.55b) by the functions $f_1(s_1)$ and $f_2(s_2)$ and then integrate Eqs. (3.55a) and (3.55b) over the monopole lengths. As a result, we obtain the SLAE, which solution allows us to find the current amplitudes

$$J_1^0(Z_{11} + F_1^{\bar{Z}}) + J_2^0 Z_{12} = -\frac{i\omega}{2k}E_1,$$

$$J_2^0(Z_{22} + F_2^{\bar{Z}}) + J_1^0 Z_{21} = -\frac{i\omega}{2k}E_2, \qquad (3.57)$$

where $Z_{11(22)}$ and $F_{1(2)}^{\bar{Z}}$ are determined by formulas (3.46) and (3.43),

$$
\begin{aligned}
Z_{12(21)} = \frac{4\pi}{ab} \sum_{m=1}^{\infty} \sum_{n=0}^{\infty} & \frac{\varepsilon_n(k^2 - k_y^2)\tilde{k}_1 \tilde{k}_2 e^{-k_z z_0}}{kk_z(\tilde{k}_1^2 - k_y^2)(\tilde{k}_2^2 - k_y^2)} \sin k_x x_{01} \sin k_x x_{02} \\
& \times [\sin \tilde{k}_1 L_1 \cos k_y L_1 - (\tilde{k}_1/k_y)\cos \tilde{k}_1 L_1 \sin k_y L_1] \\
& \times [\sin \tilde{k}_2 L_2 \cos k_y L_2 - (\tilde{k}_2/k_y)\cos \tilde{k}_2 L_2 \sin k_y L_2], \qquad (3.58)
\end{aligned}
$$

$$E_{1(2)} = 2H_0 \frac{k}{k_g \tilde{k}_{1(2)}} \sin \frac{\pi}{a} x_{01(02)} f(\tilde{k}_{1(2)} L_{1(2)}),$$

$$f(\tilde{k}_{1(2)} L_{1(2)}) = \sin \tilde{k}_{1(2)} L_{1(2)} - \tilde{k}_{1(2)} L_{1(2)} \cos \tilde{k}_{1(2)} L_{1(2)},$$

The analytical solution of the equation system (3.57) has the following form

$$J_1^0 = -\frac{i\omega}{2k} \frac{E_1(Z_{22} + F_2^z) - E_2 Z_{12}}{(Z_{11} + F_1^z)(Z_{22} + F_2^z) - Z_{21} Z_{12}} = -\frac{i\omega}{2k} \tilde{J}_1^0,$$

$$J_2^0 = -\frac{i\omega}{2k} \frac{E_2(Z_{11} + F_1^z) - E_1 Z_{21}}{(Z_{11} + F_1^z)(Z_{22} + F_2^z) - Z_{21} Z_{12}} = -\frac{i\omega}{2k} \tilde{J}_2^0. \qquad (3.60)$$

The final expressions for the currents on the monopoles based on the formulas (3.56) and (3.60) can be presented as

$$J_{1(2)}(s_{1(2)}) = -\frac{i\omega}{2k} \tilde{J}_{1(2)}^0 (\cos \tilde{k}_{1(2)} s_{1(2)} - \cos \tilde{k}_{1(2)} L_{1(2)}). \qquad (3.61)$$

The energy characteristics of the structure, the field reflection and transmission coefficients, S_{11} and S_{12}, if $H_0 = 1$ will be defined as:

$$
\begin{aligned}
S_{11} = -\frac{4\pi i}{abkk_g} \Bigg\{ & \frac{k^2}{\tilde{k}_1} \tilde{J}_1^0 \sin\left(\frac{\pi x_{01}}{a}\right) f(\tilde{k}_1 L_1) e^{-ik_g z_0} \\
& + \frac{k^2}{\tilde{k}_2} \tilde{J}_2^0 \sin\left(\frac{\pi x_{02}}{a}\right) f(\tilde{k}_2 L_2) \Bigg\} e^{2ik_g z}, \qquad (3.62)
\end{aligned}
$$

$$
\begin{aligned}
S_{12} = 1 + \frac{4\pi i}{abkk_g} \Bigg\{ & \frac{k^2}{\tilde{k}_1} \tilde{J}_1^0 \sin\left(\frac{\pi x_{01}}{a}\right) f(\tilde{k}_1 L_1) e^{ik_g z_0} \\
& + \frac{k^2}{\tilde{k}_2} \tilde{J}_2^0 \sin\left(\frac{\pi x_{02}}{a}\right) f(\tilde{k}_2 L_2) \Bigg\}, \qquad (3.63)
\end{aligned}
$$

and the voltage standing wave ratio, $\text{VSWR} = (1 + |S_{11}|)/(1 - |S_{11}|)$.

3.3.2 Numerical and Experimental Results

Let us considered as in the Sect. 3.2.2, the following impedance distribution functions: $\phi_0(s_{1(2)}) = 1$, $\phi_1(s_{1(2)}) = 2[1 - (s_{1(2)}/L_{1(2)})]$ and $\phi_2(s_{1(2)}) = 2(s_{1(2)}/L_{1(2)})$. The formulas defining $F_{1(2)}^{\bar{Z}}$ are given in (3.49)–(3.51).

The plots of the reflection coefficient modulus as function of the wavelength for the single monopole in a rectangular waveguide, $|S_{11}| = f(\lambda)$, are presented in Fig. 3.7. The monopole parameters are given in Table 3.1. The other common parameter of the structure for the plots in Fig. 3.7 and other plots in this Subsection are as follows: $a = 58$mm, $b = 25$mm, $r_{1,2} = 2.0$mm, $L_{1,2} = 15$mm, $r_{1,2}/2L_{1,2} = 0.07$, and the ratio $r_{1,2}/\lambda$ is in range from 0.02 to 0.03. The vibrator lengths were selected so as to decrease the probability of electrical breakdown when operating at high powers.

There exist several possibilities to realize the surface impedance which can be used to resonant vibrator tuning. The active impedance $\overline{R}_s > 0$ can be obtained by using a dielectric cylinder with a metal coating which thickness is less than the skin layer depth, formula (1.59), the inductive impedance $\overline{X}_s > 0$ can be realized by a metal cylinder coated by a magnetodielectric or as a corrugated metal cylinder, formulas (1.64), (1.65), The capacitive impedance $\overline{X}_s < 0$ can be obtained by layered metal-dielectric cylinder, formula (1.62). As can be seen from the plots in Fig. 3.7, the distributed surface impedance of the above listed types allows resonant vibrator tuning ($|S_{11}| = f(\lambda)$, $\arg S_{11} = 0$) to any wavelength from the band of single mode waveguide operation. The vibrator tuning also depends on the type of the impedance distribution function. As expected, the capacitive impedance decreases and inductive impedance increases the resonant wavelength λ_{res}, while the active impedance $\overline{R}_S$ decreases the $|S_{11}|_{max}$ value, but practically does not affects λ_{res}

Fig. 3.7 The reflection coefficient $|S_{11}|$ for the system with the single monopole as the function of wavelength:
1—$\bar{X}_S = -0.03/(kr)$,
2—$\bar{Z}_S = 0$, 3—$\bar{Z}_S = 0.01$,
4—$\bar{Z}_S = 0.03$,
5—$\bar{X}_S = kr\ln(4.0)\phi_2(s)$,
6—$\bar{X}_S = kr\ln(4.0)$,
7—$\bar{X}_S = kr\ln(2.7)\phi_1(s)$,
8—$\bar{X}_S = kr\ln(4.0)\phi_1(s)$,
9—$\bar{X}_S = kr\ln(8.0)\phi_1(s)$

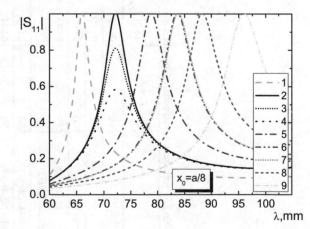

Table 3.1 Resonant wavelengths of the vibrators

$\bar{X}_S$	λ_{res} (mm)	$|\bar{X}_S|^2\big|_{\lambda=\lambda_{res}}$	$\lambda_G^{res} = \dfrac{2\pi}{\sqrt{k_{res}^2 - (\pi/a)^2}}$ (mm)	$\dfrac{\lambda_G^{res}}{4}$ (mm)	$\dfrac{\lambda_G^{res}}{2}$ (mm)	$\dfrac{3\lambda_G^{res}}{4}$ (mm)
$\dfrac{-0.03}{kr}$	66	0.025	80	20	40	60
0	72	0	92	23	46	69
$kr\ln(4.0)\phi_2(s)$	80	0.047	108	27	54	81
$kr\ln(4.0)$	84	0.043	120	30	60	90
$kr\ln(2.7)\phi_1(s)$		0.022				
$kr\ln(4.0)\phi_1(s)$	88	0.039	136	34	68	102
$kr\ln(8.0)\phi_1(s)$	96	0.074	170	42.5	85	127.5

(curves 3.4). The simulation results where obtain when the active impedance $\overline{R}_S = 0.0001$. The relative width of the resonant curve at the half power level, $0.707 = |S_{11}|_{max}$, is within 4–8%.

If the two identical monopole which axes are placed in the plane of the waveguide cross-section symmetrically relative to the longitudinal waveguide axis at a distance $\lambda_G^{res}/4$ from each other, the resonant curve width, $\Delta\lambda/\lambda_{res}$, increases up to 15% in the long-wave part of the waveguide operating range and up to 50% in the short-wave part of the band. Simultaneously the common resonant wavelength of the two vibrator system determined by the condition arg $S_{11} = 0$ decreases as shown in Fig. 3.8.

If the monopoles are placed in the plane $\{x0y\}$ at other distances between each other, the two resonances for reflected and transmitted wave can be observed (Fig. 3.9). The width of the resonant curves of the reflected waves is greater if the monopoles are placed symmetrically ($x_{01} = a - x_{02}$, Fig. 3.9a) than for asymmetric monopole placing ($x_{01} \neq a - x_{02}$, Fig. 3.9b). The same pattern is also observed

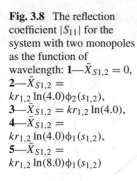

Fig. 3.8 The reflection coefficient $|S_{11}|$ for the system with two monopoles as the function of wavelength: **1**—$\overline{X}_{S1,2} = 0$, **2**—$\overline{X}_{S1,2} = kr_{1,2}\ln(4.0)\phi_2(s_{1,2})$, **3**—$\overline{X}_{S1,2} = kr_{1,2}\ln(4.0)$, **4**—$\overline{X}_{S1,2} = kr_{1,2}\ln(4.0)\phi_1(s_{1,2})$, **5**—$\overline{X}_{S1,2} = kr_{1,2}\ln(8.0)\phi_1(s_{1,2})$

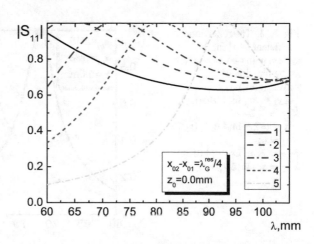

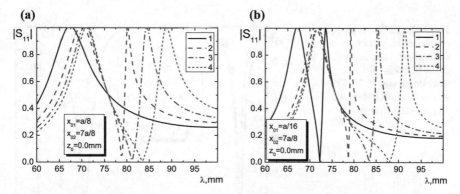

Fig. 3.9 The reflection coefficient $|S_{11}|$ for the system with two monopoles as the function of wavelength ($\overline{X}_{S2} = 0$): **1**—$\overline{X}_{S1} = 0$, **2**—$\overline{X}_{S1} = kr_1\ln(4.0)\phi_2(s_1)$, **3**—$\overline{X}_{S1} = kr_1\ln(4.0)$, **4**—$\overline{X}_{S1} = kr_1\ln(4.0)\phi_1(s_1)$

for the resonance of transmitted wave if the surface impedance is defined by the function $\phi_2(s) = 2[1 - (s/L)]$, while the resonant curves with functions $\phi(s) = 1$ and $\phi_2(s) = 2(s/L)$ wider if the monopoles are placed asymmetrically.

The position of resonances of reflected and transmitted waves can be varied in the operating wavelength range by varying the impedance magnitude, type and also the functions $f(s)$ (Fig. 3.10). If the both monopoles resonate at the same wavelength, only one resonance of the reflected wave is observed (curve 3) (Fig. 3.10). If the resonant wavelengths of the vibrators are at different ends of the operating range, then reflection resonances also occur there, and in the long-wavelength part, in the case of an asymmetric arrangement of monopoles, the curves of the dependences of the modulus of the transmission coefficient on the wavelength are much narrower in comparison with the symmetric arrangement (Fig. 3.11). This type of curves is explained by the different dependence of the values of the impedances of the inductive and capacitive types on the wavelength.

Fig. 3.10 The reflection coefficient $|S_{11}|$ for the system with two monopoles as the function of wavelength: **1**— $-\bar{X}_{S1} = kr_1 \ln(4.0)\phi_1(s_1)$, $\bar{X}_{S2} = 0$; **2**— $-\bar{X}_{S1} = kr_1 \ln(4.0)\phi_2(s_1)$, $\bar{X}_{S2} = 0$; **3**— $-\bar{X}_{S1} = kr_1 \ln(4.0)$, $\bar{X}_{S2} = kr_2 \ln(2.7)\phi_1(s_2)$

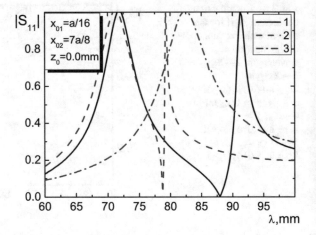

Fig. 3.11 The transmission coefficient $|S_{12}|$ for the system with two monopoles as the function of wavelength $(x_{02} = 7a/8)$: $\bar{X}_{S1} = kr_1 \ln(8.0)\phi_1(s_1)$, $\bar{X}_{S2} = -0.03/(kr_2)$

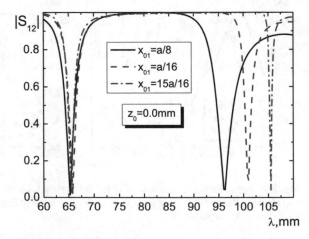

If one monopole is displaced along the longitudinal waveguide axis at distance $n\lambda_G^{reg}/4, n = 0.3\ldots$, the slope of resonance curve is increased as compared to the single monopole and monopole system located in the plane $\{x0y\}$. The slope is also increased if the integer n grows larger (Fig. 3.12). As can be seen, the two resonances of transmitted wave are positioned on both sides of reflected wave resonance.

As can be seen, the simulated and experimental results for the samples of monopoles shown in Fig. 3.13 are in satisfactory agreement with each other (Fig. 3.14).

Thus, positions of the reflected and transmitted wave resonances for the waveguide structure can be varied in the single-mode waveguide range by combining the impedance magnitude, type, and impedance distribution function along the vibrators, and also position of vibrators in the waveguide. The inductive surface impedance, especially if the distribution function $\phi_2(s) = 2(s/L)$ is used, allow us to apply the vibrators and multi-vibrator systems as resonance elements in low-profile waveguides, for example, with dimensions $\{a \times b\} = 58.0$ mm $\times$ 12.5 mm.

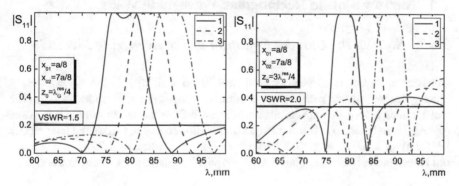

Fig. 3.12 The reflection coefficient $|S_{11}|$ for the system with two monopoles as the function of wavelength: **1**—$\bar{X}_{S1,2} = kr_{1,2}\ln(4.0)\phi_2(s_{1,2})$, **2**—$\bar{X}_{S1,2} = kr_{1,2}\ln(4.0)$, **3**—$\bar{X}_{S1,2} = kr_{1,2}\ln(4.0)\phi_1(s_{1,2})$

(a) **(b)** **(c)**

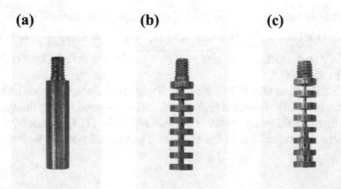

Fig. 3.13 The samples of monopoles used in the experiments

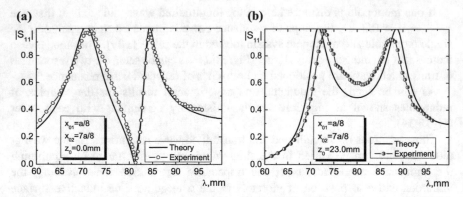

Fig. 3.14 The reflection coefficient $|S_{11}|$ for the system with two monopoles as the function of wavelength: **a** $\bar{X}_{S1} = kr_1 \ln(4.0), \bar{X}_{S2} = 0$; **b** $\bar{X}_{S1} = kr_1 \ln(4.0)\phi_1(s_1), \bar{X}_{S2} = 0$

3.4 Narrow Slots in Rectangular Waveguide Walls

3.4.1 Solving the Current Equation by the Averaging Method

Consider a narrow rectilinear slot cut in a broad wall of a hollow $\varepsilon_2 = \mu_2 = 1$ rectangular waveguide with a cross-section $\{a \times b\}$ (the region index is Wg). The slot radiates a free half-space placed over an infinite perfectly conducting plane (the region index is Hs). The slot length and width, $2L$ and d, satisfy the inequalities $[d/(2L)] \ll 1, [d/\lambda] \ll 1$. The structure analysis, according to (2.22), can be carried out based on the integral equation for the equivalent magnetic current in the slot

$$\left(\frac{d^2}{ds^2} + k^2\right) \int\limits_{-L}^{L} J(s')\big[G_s^{Hs}(s, s') + G_s^{Wg}(s, s')\big]ds' = -i\omega H_{0s}(s), \qquad (3.55)$$

where $H_{0s}(s)$ is the projection of the field of extrinsic sources to the slot axis, s is the local coordinate related to the slot axis, and $G_s^{Hs}(s, s') = 2\frac{e^{-ikR(s,s')}}{R(s,s')}$, $G_s^{Wg}(s, s') = 2\frac{e^{-ikR(s,s')}}{R(s,s')} + g_s^{Wg}(s, s')$, $R(s, s') = \sqrt{(s - s')^2 + (d/4)^2}$ are defined by the formulas (A.2), (2.11) and (2.26b).

The Eq. (3.55) will be solved by the averaging method. First, we made a change of variables according to the formulas (3.6). Then, if the slot is excited by the symmetric and antisymmetric field components $H_{0s}(s) = H_{0s}^s(s) + H_{0s}^a(s)$ mark by indices s and a, the current components in the slot can be obtained with an accuracy up to terms α^2 as

$$J(s) = J_0^s f^s(s) + J_0^a f^a(s) = \alpha \frac{i\omega}{k} \left\{ \int_{-L}^s H_{0s}(s') \sin k(s - s') ds' \right.$$

$$- \frac{\sin k(L + s) \int_{-L}^L H_{0s}^s(s') \sin k(L - s') ds'}{\sin 2kL + \alpha N^s(kd_e, 2kL)}$$

$$\left. - \frac{\sin k(L + s) \int_{-L}^L H_{0s}^a(s') \sin k(L - s') ds'}{\sin 2kL + \alpha N^a(kd_e, 2kL)} \right\}, \tag{3.56}$$

where $\alpha = 1/[8 \ln(d/2L)]$ and $N^s(kd_e, 2kL)$, $N^a(kd_e, 2kL)$ are functions of the slot eigenfield determined by the magnetic Green's functions $G_s^\Sigma = G_s^{Hs} + G_s^{Wg}$ for the coupling volumes equal to

$$N^s(kd_e, 2kL) = \int_{-L}^L [G_s^\Sigma(d_e; s, -L) + G_s^\Sigma(d_e; s, L)] \sin k(L - s) ds,$$

$$N^a(kd_e, 2kL) = \int_{-L}^L [G_s^\Sigma(d_e; s, -L) - G_s^\Sigma(d_e; s, L)] \sin k(L - s) ds, \tag{3.57}$$

where d_e is equivalent slot width, which takes into account the waveguide wall thickness h. The functions d_e and d are related as [17, 18]:

$$d_e = d \frac{K}{E(K)}, \tag{3.58}$$

where $E(K)$ is a complete elliptic integral of the second kind and K is the integral modulus. Convenient relations can be obtained from (3.58) in the two limiting cases [17]:

$$d_e = d \left(1 - \frac{h}{\pi d} \ln \frac{d}{h} \right) \text{ if } \frac{h}{d} \ll 1 \tag{3.59a}$$

$$d_e = d \left(\frac{8}{\pi} e^{-\left(\frac{\pi h}{2d} + 1\right)} \right) \text{ if } \frac{h}{d} \ll 1 \tag{3.59b}$$

The formulas (3.59a) and (3.59b) can be combined with a sufficient accuracy into one expression [18], which is valid in the region $0 \le (h/d) < 1$

$$d_e = d \, e^{-\frac{\pi h}{2d}}. \tag{3.60}$$

First, we will find from (3.56) the current distribution functions for various slot positions relative to the waveguide walls. The functions $f^s(s)$ and $f^a(s)$ can be presented as:

(1)   if the slot axis of is oriented at the angle φ relative to the longitudinal waveguide axis at the distance x_0 between the narrow waveguide wall and slot center ($s = 0$)

$$f^s(s) = \frac{\cos ks \cos k_2 L - \cos kL \cos k_2 s}{\left(\sin \varphi + (k_c/k_g)\cos \varphi\right)^2} e^{ik_c x_0}$$
$$- \frac{\cos ks \cos k_1 L - \cos kL \cos k_1 s}{\left(\sin \varphi - (k_c/k_g)\cos \varphi\right)^2} e^{-ik_c x_0},$$

$$f^a(s) = \frac{\sin ks \sin k_2 L - \sin kL \sin k_2 s}{\left(\sin \varphi + (k_c/k_g)\cos \varphi\right)^2} e^{ik_c x_0}$$
$$+ \frac{\sin ks \sin k_1 L - \sin kL \sin k_1 s}{\left(\sin \varphi - (k_c/k_g)\cos \varphi\right)^2} e^{-ik_c x_0}, \qquad (3.61)$$

where $k_1 = k_c \sin \varphi + k_g \cos \varphi$, $k_2 = k_c \sin \varphi - k_g \cos \varphi$;
(2)   if the slot axis coincides with the longitudinal waveguide axis (longitudinal slot, $\varphi = 0^O$)

$$f^s(s) = \cos ks \cos k_g L - \cos kL \cos k_g s,$$
$$f^a(s) = \sin ks \sin k_g L - \sin kL \sin k_g s; \qquad (3.62)$$

(3)   If the slot axis of the slit is perpendicular to the longitudinal waveguide axis (transverse slot, $\varphi = 90^O$

$$f^s(s) = \cos ks \cos k_c L - \cos kL \cos k_c s,$$
$$f^a(s) = \sin ks \sin k_c L - \sin kL \sin k_c s. \qquad (3.63)$$

If $2L \approx \lambda/2$, $x_0 = a/2$, and $\varphi = 90^O$ (symmetric transverse slot), expressions (3.63) can be satisfactorily approximated by one symmetric function

$$f(s) = f^s(s) = \cos ks - \cos kL. \qquad (3.64)$$

3.4.2   Symmetric Transverse Slot in the Broad Waveguide Wall

Consider the transverse slot cut in the broad wall of rectangular waveguide symmetrically to its longitudinal axis which radiates into free half-space above infinite perfectly conducting plane $\{x0y\}$. The Eq. (3.55) relative to the magnetic current in the slot will be solved by using the generalized method of induced magnetomotive forces (MMF). Then we introduce the Cartesian coordinate system related to the waveguide, as in Fig. 3.6. Assume that the longitudinal slot axis and the local coordinate s coincide with the axis $\{0x\}$.

If the slot is excited by the wave $H_{0s}(s) = H_{0s}^s(s) = H_0 \cos k_c s$, where H_0 is the wave amplitude, and the slot current is approximated by the formula $J(s) = J_0(\cos ks - \cos kL)$ (3.64), the current amplitude according to expression (3.19) can be obtain in the form

$$J_0 = -\frac{i\omega}{k} H_0 \frac{(1/k)f(kL)}{Z^{Hs}(kd_e, kL) + Z^{Wg}(kd_e, kL)} = -\frac{i\omega}{k}\tilde{J}_0,$$

$$f(kL) = \sin kL - kL \cos kL. \tag{3.65}$$

where Si and Cin are the integral sine and cosine

$$Z^{Hs}(kd_e, kL) = (\text{Si}4kL - i\text{Cin}4kL) - 2\cos kL$$

$$\times \left[2(\sin kL - kL \cos kL)\left(\ln\frac{16L}{d_e} - \text{Cin}2kL - i\text{Si}2kL \right) + \sin 2kL e^{-ikL} \right], \tag{3.66}$$

$$Z^{Wg}(kd_e, kL) = \frac{8\pi}{ab} \sum_{m=1,3..}^{\infty} \sum_{n=0,1..}^{\infty} \frac{\varepsilon_n k}{k_z(k^2 - k_x^2)} e^{-k_z\frac{d_e}{4}} [\sin kL \cos k_x L$$

$$- (k/k_x)\cos kL \sin k_x L]^2. \tag{3.67}$$

If the formulas (3.64) and (3.65) are taken into account, the current in the slot is

$$J(s) = -H_0 \frac{i\omega}{k}\tilde{J}_0(\cos ks - \cos kL). \tag{3.68}$$

The energy characteristics of the vibrator-slot structure, the field reflection and transmission coefficients S_{11} and S_{12}, and the power radiation coefficient $|S_\Sigma|^2$ under condition that the amplitude of the exciting H_0-wave is equal to unity, can be determined by the following expressions

$$S_{11} = \frac{8\pi i k_g}{abk^2}\tilde{J}_0 f(kL)e^{2ik_g z}, \tag{3.69}$$

$$S_{12} = 1 + \frac{8\pi i k_g}{abk^2}\tilde{J}_0 f(kL), \tag{3.70}$$

$$|S_\Sigma|^2 = 1 - |S_{11}|^2 - |S_{12}|^2. \tag{3.71}$$

The simulated and experimental results for the slot radiator with the following parameters: $a = 58.0$ mm, $b = 25.0$ mm, $h = 0.5$ mm, $d = 4.0$ mm, $2L = 40.0$ mm are presented in Fig. 3.15.

Fig. 3.15 The slot radiation
coefficient $|S_\Sigma|^2$ as the
function of wavelength

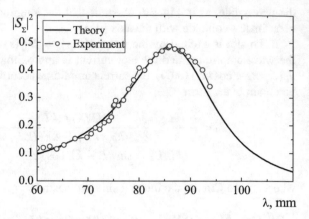

As can be seen from the plot, the theoretical and experimental curves are in satisfactory agreement with one another, confirming the legitimacy of applying the generalized method of induced MMF with the current distribution function found by the averaging method.

If the transverse slot radiates into free half-space above the infinite plane $\{x0z\}$, characterized by constant distributed impedance $\bar{Z}_S = Z_S/Z_0$, the problem solution can also be obtained in a similar way. In this case, the magnetic current in the slot and the energy parameters of the waveguide-slot radiator placed over the impedance plane can be determined by using the above formulas for the perfectly conducting plane if the slot extrinsic conductivity $Z^{Hs}(kd_e, kL)$ is replaced by the sum $Z^{Hs}(kd_e, kL) + \Delta Z^{Hs}(kd_e, kL)$, where $\Delta Z^{Hs}(kd_e, kL)$ is the additional extrinsic slot conductivity defined by the plane impedance.

The explicit expression for $\Delta Z^{Hs}(kd_e, kL)$ can be obtained by using the Green's functions for free space over the impedance plane (A.17). It should be borne in mind that according to the requirement of the impedance condition $[\vec{E}, \vec{y}_0] = -\bar{Z}_S[[\vec{H}, \vec{y}_0], \vec{y}_0]$ on the outer plane surface $x0z$, the slot will be excited not only by the magnetic current $J(s) = J_0(\cos ks - \cos kL)$, as for the perfectly conducting plane ($\bar{Z}_S = 0$), but by the coupled magnetic and electric currents, $\vec{J}^m(x) = \vec{x}_0 J_0(\cos ks - \cos kL)$ and $\vec{J}^e(x) = \vec{z}_0 J_0 \frac{1}{Z_S}(\cos ks - \cos kL)$. In other words, if the plane is perfectly conducting, the slot aperture is assumed to be metalized, while for the impedance surface should be considered to be a closed segment of the impedance surface. That is, in both cases, homogeneity of the slot aperture and the screen surface is guaranteed. This approach was previously used for the electrodynamic analysis of a slot antenna located on an impedance sphere [19].

First, let us present the Green's functions (A.17) in the coordinate system, shown in Fig. 3.6, as the sum of two terms (2.11), using the well-known identity

$$\frac{e^{-ik\sqrt{(x-x')^2+y^2+(z-z')^2}}}{\sqrt{(x-x')^2+y^2+(z-z')^2}} = \frac{1}{2\pi}\int\limits_{-\infty}^{\infty}\int\limits_{-\infty}^{\infty}\frac{e^{-ik_x(x-x')-ik_z(z-z')+y\sqrt{k_x^2+k_z^2-k^2}}}{\sqrt{k_x^2+k_z^2-k^2}}dk_x dk_z.$$

(3.72)

Then, we obtain by applying the method of MMF

$$\Delta Z^{Hs}(kd_e,kL)$$

$$= \frac{2\bar{Z}_S}{\pi i}\int\limits_{-\infty}^{\infty}\int\limits_{-\infty}^{\infty}\left\{\begin{array}{c}\dfrac{\chi^2 k^2(1+\bar{Z}_S^2)-(k^2+\chi^2)(k^2-k_x^2)}{\chi^2 k^2(1+\bar{Z}_S^2)+\bar{Z}_S(k^2+\chi^2)\chi k}\\ \times\left[\dfrac{kk_x\sin kL\cos k_x L-k^2\cos kL\sin k_x L}{(k^2-k_x^2)k_x}\right]^2 e^{-ik_z(d_e/4)}\end{array}\right\}dk_x dk_z,$$

(3.73)

where $\chi^2 = k^2 - k_x^2 - k_z^2$.

Under the condition $\left|\bar{Z}_S^2\right| \ll 1$, the expression (3.73) can be easily converted to

$$\Delta Z^{Hs}(kd_e,kL) = \frac{2\bar{Z}_S}{\pi i}\int\limits_{-\infty}^{\infty}\int\limits_{-\infty}^{\infty}\left\{\begin{array}{c}\frac{\chi^2 k^2-(k^2+\chi^2)(k^2-k_x^2)}{\chi^2 k^2+\bar{Z}_S(k^2+\chi^2)\chi k}\\ \left[\frac{kk_x\sin kL\cos k_x L-k^2\cos kL\sin k_x L}{(k^2-k_x^2)k_x}\right]^2 e^{-ik_z(d_e/4)}\end{array}\right\}dk_x dk_z,$$

(3.74)

As can be seen from formulas (3.73) and (3.74), the additional component of the extrinsic slot conductivity is expressed as a double infinite integral over the frequency parameters k_x and k_z. The additional conductivity component in complex fashion depends on the impedance $\bar{Z}_S$, which can significantly affect the slot radiation coefficient. As expected, the term $\Delta Z^{Hs}(kd_e,kL) = 0$ if the plane is perfectly conducting, $\bar{Z}_S = 0$.

3.4.3 Longitudinal Slot in the Waveguide Broad Wall

Consider a longitudinal slot cut in a broad wall of a rectangular waveguide at a distance x_0 from a narrow wall (in the Cartesian coordinate system shown in Fig. 3.6). The solution of Eq. (3.55) by the generalized method of induced MMF with approximating current functions $f^s(s)$ and $f^a(s)$ define in (3.62) can be written as

$$J(s) = -\frac{i\omega}{2k^2}H_0\cos\frac{\pi x_0}{a}\left[J_0^s f^s(s) + iJ_0^a f^a(s)\right],$$

(3.75)

where

$$J_0^s = \frac{F_k^s}{Z_s^{Hs} + Z_s^{Wg}}, \quad J_0^a = \frac{F_k^a}{Z_a^{Hs} + Z_a^{Wg}},$$

$$F_k^s = 2\cos k_g L_{sl} \frac{\sin k L_{sl} \cos k_g L_{sl} - (k_g/k)\cos k L_{sl} \sin k_g L_{sl}}{1 - (k_g/k)^2}$$
$$- \cos k L_{sl} \frac{\sin 2k_g L_{sl} + 2k_g L_{sl}}{2(k_g/k)},$$

$$F_k^a = 2\sin k_g L_{sl} \frac{\cos k L_{sl} \sin k_g L_{sl} - (k_g/k)\sin k L_{sl} \cos k_g L_{sl}}{1 - (k_g/k)^2}$$
$$- \sin k L_{sl} \frac{\sin 2k_g L_{sl} - 2k_g L_{sl}}{2(k_g/k)},$$

$$Z_s^{Hs} = 2 \left\{ \begin{array}{l} \left[\left(\frac{k_g}{k}\right)\cos k L \sin k_g L - \sin k L \cos k_g L \right] F^s(L) \\ + k\cos k_g L \int_{-L}^{L} \cos ks\, F^s(s)\mathrm{d}s - \frac{k^2+k_g^2}{2k}\cos k L \int_{-L}^{L} \cos k_g s\, F^s(s)\mathrm{d}s \end{array} \right\},$$

$$Z_a^{Hs} = 2 \left\{ \begin{array}{l} \left[\left(\frac{k_g}{k}\right)\sin k L \cos k_g L - \cos k L \sin k_g L \right] F^a(L) \\ + k\sin k_g L \int_{-L}^{L} \sin ks\, F^a(s)\mathrm{d}s - \frac{k^2-k_g^2}{2k}\sin k L \int_{-L}^{L} \sin k_g s\, F^a(s)\mathrm{d}s \end{array} \right\},$$

$$F^{s(a)}(s) = \int_{-L}^{L} f^{s(a)}(s') \frac{e^{-ik\sqrt{(s-s')^2+(d_e/4)^2}}}{\sqrt{(s-s')^2+(d_e/4)^2}}\mathrm{d}s',$$

$$Z_s^{Wg} = \frac{2\pi}{ab} \sum_{m=0}^{\infty}\sum_{n=0}^{\infty} \frac{\varepsilon_m \varepsilon_n}{k^2} \cos k_x x_0 \cos k_x\left(x_0 + \frac{d_e}{4}\right)$$
$$\times \left\{ \left[\cos k_g L \left(\frac{k}{k_z}\sin kL - \cos kL\right)F_e^s \right] \right.$$
$$\left. - \frac{\cos kL}{k_z^2 + k_g^2}\left[(k_z^2 + k^2)\left(\frac{k_g}{k_z}\sin k_g L - \cos k_g L\right)F_e^s + k_c^2 F_k^s \right] \right\},$$

$$F_e^s = \frac{k\cos k_g L}{k_z^2 + k^2}\left[k_z \cos kL\left(1 - e^{-2k_z L}\right) + k\sin kL\left(1 + e^{-2k_z L}\right) \right]$$
$$- \frac{k\cos kL}{k_z^2 + k_g^2}\left[k_z \cos k_g L\left(1 - e^{2k_z L}\right) + k_g \sin k_g L\left(1 + e^{-2k_z L}\right) \right],$$

$$Z_a^{Wg} = \frac{2\pi}{ab} \sum_{m=0}^{\infty}\sum_{n=0}^{\infty} \frac{\varepsilon_m \varepsilon_n}{k^2} \cos k_x x_0 \cos k_x\left(x_0 + \frac{d_e}{4}\right)$$
$$\times \left\{ \left[-\sin k_g L\left(\frac{k}{k_z}\cos kL + \sin kL\right) \right] F_e^a \right\},$$

$$+\frac{\sin kL}{k_z^2 + k_g^2}\left[(k_z^2 + k^2)\left(\frac{k_g}{k_z}\cos k_g L + \sin k_g L\right)F_e^a + k_c^2 F_k^a\right]\Bigg\}$$

$$F_e^a = \frac{k\sin k_g L}{k_z^2 + k^2}\left[k_z \sin kL\left(1 + e^{-2k_z L}\right) - k\cos kL\left(1 - e^{-2k_z L}\right)\right]$$

$$-\frac{k\sin kL}{k_z^2 + k_g^2}\left[k_z \sin k_g L\left(1 + e^{2k_z L}\right) - k_g \cos k_g L\left(1 - e^{-2k_z L}\right)\right],$$

and the axis of the current coordinate s coincides with the axis $\{0z\}$. These relations can be used to simulate the electrodynamic characteristics of longitudinal slot cut in the broad wall of rectangular waveguide which radiates into free half-space above perfectly conducting plane. Under condition $H_0 = 1$, the field reflection S_{11} and transmission S_{12} coefficients in the waveguide and the power radiation coefficient $|S_\Sigma|^2$ can be presented as

$$S_{11} = -\frac{2\pi k_c^2 \cos^2(\pi x_0/a)}{iabk_g k^3}\left(J_0^s F_k^s + J_0^a F_k^a\right)e^{2ik_g z}, \qquad (3.76)$$

$$S_{12} = 1 - \frac{2\pi k_c^2 \cos^2(\pi x_0/a)}{iabk_g k^3}\left(J_0^s F_k^s - J_0^a F_k^a\right), \qquad (3.77)$$

$$|S_\Sigma|^2 = 1 - |S_{11}|^2 - |S_{12}|^2. \qquad (3.78)$$

The plots of the experimental and simulated radiation coefficient $|S_\Sigma|^2$ as the functions of the slot length are shown in Fig. 3.16. The simulation results were obtained by using formulas (3.62), the Galerkin method with the current approximated by trigonometric functions $J(s) = \sum_{p=1}^{P} J_P \sin\frac{p\pi(L+s)}{2L}$ defined in Sect. 2.2.1 and by approximating function consisting of the first two basis functions

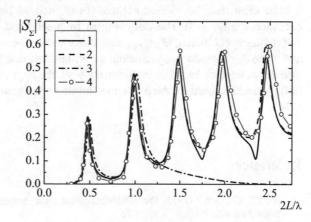

Fig. 3.16 The plots of the radiation coefficient $|S_\Sigma|^2$ as function of the slot length: $a = 23.0\,mm$, $b = 10.0\,mm$, $d = 1.5\,mm$, $\lambda = 30.0\,mm$, $h = 1.0\,mm$, $x_0 = a/4$; 1—simulation by the formula (3.62), 2—simulation by the Galerkin method, 3—simulation by the formula (3.79), 4—experimental data

$$J(s) = J_1 \cos \frac{\pi s}{2L} + J_2 \sin \frac{\pi s}{L}. \tag{3.79}$$

As follows from the plots, the current approximating functions (3.62) and the Galerkin method can be successfully applied up to $(2L/\lambda) = 2.75$, while the approximating function (3.79) can be applied only $(2L/\lambda) = 1.25$.

The problem of the longitudinal slot radiation into free half-space above the impedance plane $\{x0z\}$ characterized by the constant impedance can be solved by replacing in the expressions (3.75–3.78) defining the partial extrinsic slot conductivities $Z_{s(a)}^{Hs}$ for the perfectly conducting plane by the sums $Z_{s(a)}^{Hs}(kd_e, kL) + \Delta Z_{s(a)}^{Hs}(kd_e, kL)$. The additional terms $\Delta Z_{s(a)}^{Hs}(kd_e, kL)$ are determined by the following expressions

$$\Delta Z_{s(a)}^{Hs}(kd_e, kL) = \frac{\bar{Z}_S}{2\pi i} \int\limits_{-L}^{L} f^{s(a)}(s) \int\limits_{-L}^{L} f^{s(a)}(s')$$

$$\times \int\limits_{-\infty}^{\infty} \int\limits_{-\infty}^{\infty} \frac{\chi^2 k^2 (1 + \bar{Z}_S^2) - (k^2 + \chi^2)(k^2 - k_x^2)}{\chi^2 k^2 (1 + \bar{Z}_S^2) + \bar{Z}_S (k^2 + \chi^2)\chi k} e^{-ik_z(s-s') - ik_x(d/4)} dk_x dk_z ds' ds;$$

$$\tag{3.80}$$

$$\Delta Z_{s(a)}^{Hs}(kd_e, kL)\Big|_{|\bar{Z}_S| \ll 1} = \frac{\bar{Z}_S}{2\pi i} \int\limits_{-L}^{L} f^{s(a)}(s) \int\limits_{-L}^{L} f^{s(a)}(s')$$

$$\times \int\limits_{-\infty}^{\infty} \int\limits_{-\infty}^{\infty} \frac{\chi^2 k^2 - (k^2 + \chi^2)(k^2 - k_x^2)}{\chi^2 k^2 + \bar{Z}_S (k^2 + \chi^2)\chi k} e^{-ik_z(s-s') - ik_x(d/4)} dk_x dk_z ds' ds. \tag{3.81}$$

It is clear that the resonant condition for any type of waveguide-slot radiators. As is quite clear must be pointed out that for of, will, as before, is determined by the condition arg $S_{11} = 0$. The slot resonant frequency and the frequency of maximal slot radiation coefficient $|S_\Sigma|_{\max}^2$ may differ due to the influence of waveguide wall and plane thicknesses. In conclusion, we will notice, that if the slot is cut at the angle φ to the waveguide broad or in its narrow wall, the approximated functions (3.61) or (3.62) should be used for the problem solution by the generalized method of induced MMF.

References

1. Wu TT, King RWP (1965) The cylindrical antenna with nonreflecting resistive loading. IEEE Trans Antennas Propag 13:369–373
2. Glushkovsky EA, Isralit AB, Levin BM, Rabinovich EY (1967) Variable surface impedance linear antennas. Antennas 2:154–165 (in Russian)

3. Shen L-C (1967) An experimental study of the antenna with nonreflecting resistive loading. IEEE Trans Antennas Propag 15:606–611
4. Taylor CD (1968) Cylindrical transmitting antenna: tapered resistivity and multiple impedance loadings. IEEE Trans Antennas Propag 16:176–179
5. Rao BLJ, Ferris JE, Zimmerman WE (1969) Broadband characteristics of cylindrical antennas with exponentially tapered capacitive loading. IEEE Trans Antennas Propag 17:145–151
6. Levin BM, Yakovlev AD (1985) Antenna with loads as an impedance vibrator with variable impedance. Radio Eng Electron 30(1):25–33. (in Russian)
7. Nesterenko MV, Katrich VA, Penkin YuM, Dakhov VM, Berdnik SL (2011) Thin impedance vibrators. Theory and applications. Springer Science+Business Media, New York
8. Nesterenko MV (2005) Scattering of electromagnetic waves by thin vibrators with variable surface impedance. Radiophys. Radioastronomy 10(4):408–417. (in Russian)
9. Nesterenko MV, Belogurov EYu, Katrich VA, Kiyko VI (2008) H_{10}-wave scattering by a thin impedance vibrator with a variable radius in a rectangular waveguide. Vestnik Kharkov Univ 806:14–17 (in Russian)
10. Skolnik MI (ed) (1970) Radar Handbook. Mcgraw-Hill Book Company
11. Philatov AN (1974) Asymptotic methods in the theory of differential and integral-differential equations. PHAN, Tashkent (in Russian)
12. Kamke E (1959) Differentialgleichungen Lösungsmethoden und Lösungen. I. Gewöhnliche Differentialgleichungen. 6, Verbesserte Auflage, Leipzig (1959) (in German)
13. King RWP (1956) The theory of linear antennas. Harvard University Press, Cambridge, Massachusetts (1956)
14. Fradin AZ (1977) Antenna-feeder devices. Svyaz', Moskow (in Russian)
15. Eisenberg GZ, Yampolsky VG, Tereshin ON (1977) VHF Antennas. Svyaz', Moskow (in Russian)
16. Wolman VI, Pimenov YuV (1971) Technical electrodynamics. Svyaz', Moskow (in Russian)
17. Garb HL, Levinson IB, Fredberg PSh (1968) Effect of wall thickness in slot problems of electrodynamics. Radio Eng Electron Phys 13:1888–1896
18. Warne LK (1995) Eddy current power dissipation at sharp corners: closely spaced rectangular conductors. J Electromagn Waves Appl 9:1441–1458
19. Penkin YM (1998) Investigation of the conductivity of an impedance spherical slot antenna. Radiophys Radioastronomy 3(3):341–347. (in Russian)

Chapter 4
Combined Radiating Vibrator-Slot Structures in Rectangular Waveguide

4.1 Two-Element Vibrator-Slot Structure

4.1.1 Problem Formulation and Solution the Equations for Currents

Consider a problem of electromagnetic wave scattering by a waveguide vibrator-slot structure presented in Fig. 4.1. A narrow rectilinear transverse slot cut in a broad wall of a hollow ($\varepsilon_1 = \mu_1 = 1$) infinite rectangular waveguide with perfectly conducting walls (the region index is Wg) radiates into half-space above an infinite perfectly conducting plane (the region index is Hs). A thin passive asymmetric vibrator (monopole) with variable surface impedance is placed in the waveguide cross-sectional plane symmetric with respect to its longitudinal axis. The structure parameters are as following: the waveguide cross-section and wall thickness are $\{a \times b\}$ and h, vibrator radius and length are r and $2L_v$, slot width and length are d and $2L_{sl}$, and distance between the vibrator and slot axes is z_0. The thin wire and narrow slot approximations, $[r/(2L_v)] \ll 1$, $[r/\lambda] \ll 1$, $[d/(2L_{sl})] \ll 1$ and $[d/\lambda] \ll 1$, are used.

If the H_{10}—wave propagates in the waveguide from the region $z = -\infty$, the system of integral equations for the electric current in the vibrator and the equivalent magnetic current in the slot can be written according to (2.19) as:

$$
\left(\frac{d^2}{ds_1^2} + k^2\right) \int\limits_{-L_v}^{L_v} J_v(s_1') G_{s_1}^{Wg}(s_1, s_1') ds_1' - ik \int\limits_{-L_{sl}}^{L_{sl}} J_{sl}(s_2') \tilde{G}_{s_2}^{Wg}(s_1, s_2') ds_2'
$$

$$
= -i\omega [E_{0s_1}(s_1) - z_i(s_1) J_v(s_1)], \tag{4.1a}
$$

© The Author(s), under exclusive license to Springer Nature Switzerland AG 2020
M. V. Nesterenko et al., *Combined Vibrator-Slot Structures: Theory and Applications*, Lecture Notes in Electrical Engineering 689,
https://doi.org/10.1007/978-3-030-60177-5_4

Fig. 4.1 The configuration
of the vibrator-slot structure

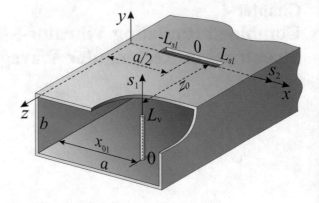

$$\left(\frac{\mathrm{d}^2}{\mathrm{d}s_2^2} + k^2\right) \int\limits_{-L_{sl}}^{L_{sl}} J_{sl}(s_2')[G_{s_2}^{Wg}(s_2, s_2') + G_{s_2}^{Hs}(s_2, s_2')]\mathrm{d}s_2'$$

$$- ik \int\limits_{-L_v}^{L_v} J_v(s_1')\tilde{G}_{s_1}^{Wg}(s_2, s_1')\mathrm{d}s_1' = -i\omega H_{0s_2}(s_2), \qquad (4.1b)$$

where s_1 and s_2 are the local coordinates associated with the vibrator and slot axes, ' $E_{0s_1}(s_1)$ and $H_{0s_2}(s_2)$ are the projections of the fields of extraneous sources on the vibrator and the slot axis, $G_{s_1}^{Wg}(s_1, s_1')$, $G_{s_2}^{Wg}(s_2, s_2')$ and $G_{s_2}^{Hs}(s_2, s_2')$ are the corresponding components of the tensor Green's functions of the rectangular waveguide and half-space above the plane (Appendix A), $-L_v$ is the coordinate of the mirror image of the monopole end relative to the waveguide lower broad wall. The functions $\tilde{G}_{s_1}^{Wg}(s_2, s_1') = \frac{\partial}{\partial z}G_{s_1}^{Wg}[x(s_2), 0, z; x'(s_1'), y'(s_1'), z_0]$ and $\tilde{G}_{s_2}^{Wg}(s_1, s_2') = \frac{\partial}{\partial z}G_{s_2}^{Wg}[x(s_1), y(s_1), z; x'(s_2'), 0, 0]$ are obtained by substitution $z = 0$ into $\tilde{G}_{s_1}^{Wg}$ and $z = z_0$ into $\tilde{G}_{s_2}^{Wg}$ after taking the derivatives.

If the interaction between the monopole and the slot is absent ($z_0 = 0$, $\tilde{G}_{s_2}^{Wg} = \tilde{G}_{s_2}^{Wg} = 0 = \tilde{G}_{s_2}^{Wg} = 0$) the system of coupled Eq. (4.1a) is converted to two independent equations:

$$\left(\frac{\mathrm{d}^2}{\mathrm{d}s_1^2} + k^2\right) \int\limits_{-L_v}^{L_v} J_v(s_1')G_{s_1}^{Wg}(s_1, s_1')\mathrm{d}s_1' = -i\omega[E_{0s_1}(s_1) - z_i(s_1)J_v(s_1)], \quad (4.2a)$$

$$\left(\frac{\mathrm{d}^2}{\mathrm{d}s_2^2} + k^2\right) \int\limits_{-L_{sl}}^{L_{sl}} J_{sl}(s_2')[G_{s_2}^{Wg}(s_2, s_2') + G_{s_2}^{Hs}(s_2, s_2')]\mathrm{d}s_2' = -i\omega H_{0s_2}(s_2). \quad (4.2b)$$

The solution of equations system (4.1a) will be obtained by the generalized method of induced EMMF, using $J_v(s_1) = J_{0v} f_v(s_1)$ and $J_{sl}(s_2) = J_{0sl} f_{sl}(s_2)$ as the approximating expressions for the currents. In the above expressions J_{0v} and J_{0sl} are unknown current amplitudes, and $f_v(s_1)$ and $f_{sl}(s_2)$ are predefined current distribution functions obtained from the solution of Eq. (4.2a) by the asymptotic averaging method. For the arbitrary vibrator-slot structures and coupling electrodynamic volumes, the current distribution functions have symmetric $(f_v^s(s_1), f_{sl}^s(s_2))$ and antisymmetric $(f_v^a(s_1), f_{sl}^a(s_2))$ components relative to the monopole and slot centers. The functions $f_v^{s,a}(s_1)$ and $f_{sl}^{s,a}(s_2)$ can be found based on (3.32) and (3.56) as:

$$f_v^{s,a}(s_1) \sim \left\{ \begin{array}{l} \sin \tilde{k}(L_v - s_1) \int\limits_{-L_v}^{s_1} E_{0s_1}^{s,a}(s_1') \sin \tilde{k}(L_v + s_1')ds_1' \\ + \sin \tilde{k}(L_v + s_1) \int\limits_{s_1}^{L_v} E_{0s_1}^{s,a}(s_1') \sin \tilde{k}(L_v - s_1')ds_1' \end{array} \right\}, \qquad (4.3a)$$

$$f_{sl}^{s,a}(s_2) \sim \left\{ \begin{array}{l} \sin k(L_{sl} - s_2) \int\limits_{-L_{sl}}^{s_2} H_{0s_2}^{s,a}(s_2') \sin k(L_{sl} + s_2')ds_2' \\ + \sin k(L_{sl} + s_2) \int\limits_{s_2}^{L_{sl}} H_{0s_2}^{s,a}(s_2') \sin k(L_{sl} - s_2')ds_2' \end{array} \right\}, \qquad (4.3b)$$

where $E_{0s_1}^{s,a}(s_1)$ and $H_{0s_2}^{s,a}(s_2)$ are projections of symmetric and antisymmetric field component defined by extrinsic sources on the monopole and slot axes. The sign $\sim$ means that only terms depending on the coordinates s_1 and s_2 should be left after integration in the expressions (4.3).

If the vibrator-slot structure is excited by the H_{10}—wave, the approximating function can be written based on the expressions (3.36) and (3.64) as:

$$f_v(s_1) = f_v^s(s_1) = \cos \tilde{k} s_1 - \cos \tilde{k} L_v, \qquad (4.4a)$$

$$f_{sl}(s_2) = f_{sl}^s(s_2) = \cos k s_2 - \cos k L_{sl}, \qquad (4.4b)$$

where $\tilde{k} = k - \frac{i2\pi z_i^{av}}{Z_0 \Omega}$, $z_i^{av} = \frac{1}{2L_v} \int_{-L_v}^{L_v} z_i(s_1)ds_1$ is the internal impedance, averaged over the monopole length, and $\Omega = 2 \ln(2L_v/r)$.

Let us multiply the Eqs. (4.1a) and (4.1b) by the functions $f_v(s_1)$ and $f_{sl}(s_2)$ integrate the resulting equations over the monopole and slot length, respectively. Thus, we obtain the SLAE relative the currents amplitudes J_{0v} and J_{0sl} which solution can be written as:

$$J_{0v}[Z_{11}(kr, \tilde{k}L_v) + F_z(\tilde{k}r, \tilde{k}L_v)] + J_{0sl} Z_{12}(z_0, \tilde{k}L_v, kL_{sl}) = -\frac{i\omega}{2k} E_1(\tilde{k}L_v),$$

$$J_{0sl} Z_{22}^\Sigma(kd_e, kL_{sl}) + J_{0v} Z_{21}(z_0, kL_{sl}, \tilde{k}L_v) = -\frac{i\omega}{2k} H_2(kL_{sl}), \qquad (4.5)$$

where according to (3.46), (3.43), (3.66) and (3.67)

$$Z_{11(22)}(kr_{1(2)}, \tilde{k}_{1(2)}L_{1(2)}) = \frac{4\pi}{ab} \sum_{m=1}^{\infty} \sum_{n=0}^{\infty} \frac{\varepsilon_n(k^2 - k_y^2)\tilde{k}_{1(2)}^2}{kk_z(\tilde{k}_{1(2)}^2 - k_y^2)^2} e^{-k_z r_{1(2)}} \sin^2 k_x x_{01(02)}$$

$$\times [\sin \tilde{k}_{1(2)}L_{1(2)} \cos k_y L_{1(2)}$$
$$- (\tilde{k}_{1(2)}/k_y) \cos \tilde{k}_{1(2)}L_{1(2)} \sin k_y L_{1(2)}]^2,$$

$$F_z(\tilde{k}r, \tilde{k}L_v) = -\frac{i}{r} \int_0^{L_v} f_v^2(s_1) \bar{Z}_S(s_1) ds_1,$$

$$Z_{22}^{Hs}(kd_e, kL_{sl}) = (\text{Si}4kL_{sl} - i\text{Cin}4kL_{sl}) - 2\cos kL_{sl}$$

$$\times \left[2(\sin kL_{sl} - kL_{sl}\cos kL_{sl})\left(\ln \frac{16L_{sl}}{d_e} - \text{Cin}2kL_{sl} - i\text{Si}2kL_{sl}\right) \right.$$
$$\left. + \sin 2kL_{sl} e^{-ikL_{sl}} \right],$$

$$Z_{22}^{Wg}(kd_e, kL_{sl}) = \frac{8\pi}{ab} \sum_{m=1,3..}^{\infty} \sum_{n=0,1..}^{\infty} \frac{\varepsilon_n k}{k_z(k^2 - k_x^2)} e^{-k_z \frac{d_e}{4}}$$

$$\times [\sin kL_{sl} \cos k_x L_{sl} - (k/k_x) \cos kL_{sl} \sin k_x L_{sl}]^2,$$

$$Z_{22}^{\Sigma}(kd_e, kL_{sl}) = Z_{22}^{Wg}(kd_e, kL_{sl}) + Z_{22}^{Hs}(kd_e, kL_{sl}),$$

and

$$Z_{12}(z_0, \tilde{k}L_v, kL_{sl}) = -Z_{21}(z_0, kL_{sl}, \tilde{k}L_v)$$

$$= \frac{4\pi}{ab} \sum_{m=1,3..}^{\infty} \sum_{n=0,1..}^{\infty} \frac{\varepsilon_n k\tilde{k}e^{-k_z z_0}}{i(k^2 - k_x^2)(\tilde{k}^2 - k_y^2)} \sin k_x x_{01}$$

$$\times [\sin \tilde{k}L_v \cos k_y L_v - (\tilde{k}/k_y)\cos \tilde{k}L_v \sin k_y L_v]$$

$$\times [\sin kL_{sl} \cos k_x L_{sl} - (k/k_x)\cos kL_{sl} \sin k_x L_{sl}], \qquad (4.6)$$

$$E_1(\tilde{k}L_v) = 2H_0 \frac{k}{k_g \tilde{k}} \sin \frac{\pi}{a} x_{01} e^{-ik_g z_0} f(\tilde{k}L_v), \quad f(\tilde{k}L_v)$$

$$= \sin \tilde{k}L_v - \tilde{k}L_v \cos \tilde{k}L_v,$$

$$H_2(kL_{sl}) = 2H_0 \frac{1}{k} f(kL_{sl}), \quad f(kL_{sl})$$

$$= \frac{\sin kL_{sl} \cos(\pi L_{sl}/a) - (ka/\pi)\cos kL_{sl} \sin(\pi L_{sl}/a)}{1 - [\pi/(ka)]^2}.$$

The analytical solution of the equation system (4.5) has the following form:

$$
\begin{aligned}
&J_{0v} \\
&= -\frac{i\omega}{2k} \frac{E_1(\tilde{k}L_v)Z_{22}^{\Sigma}(kd_e, kL_{sl}) - H_2(kL_{sl})Z_{12}(z_0, \tilde{k}L_v, kL_{sl})}{[Z_{11}(kr, \tilde{k}L_v) + F_z(\tilde{k}r, \tilde{k}L_v)]Z_{22}^{\Sigma}(kd_e, kL_{sl}) - Z_{21}(z_0, kL_{sl}, \tilde{k}L_v)Z_{12}(z_0, \tilde{k}L_v, kL_{sl})} \\
&= -\frac{i\omega}{2k}\tilde{J}_{0v},
\end{aligned}
$$

$$
\begin{aligned}
&J_{0sl} \\
&= -\frac{i\omega}{2k} \frac{H_2(kL_{sl})[Z_{11}(kr, \tilde{k}L_v) + F_z(\tilde{k}r, \tilde{k}L_v)] - E_1(\tilde{k}L_v)Z_{21}(z_0, kL_{sl}, \tilde{k}L_v)}{[Z_{11}(kr, \tilde{k}L_v) + F_z(\tilde{k}r, \tilde{k}L_v)]Z_{22}^{\Sigma}(kd_e, kL_{sl}) - Z_{21}(z_0, kL_{sl}, \tilde{k}L_v)Z_{12}(z_0, \tilde{k}L_v, kL_{sl})} \\
&= -\frac{i\omega}{2k}\tilde{J}_{0sl}. \quad\quad (4.7)
\end{aligned}
$$

The final expressions for the monopole and slot currents can be written based on the formulas (4.4a) and (4.7) as:

$$
J_v(s_1) = -\frac{i\omega}{2k}\tilde{J}_{0v}(\cos \tilde{k}s_1 - \cos \tilde{k}L_v),
$$

$$
J_{sl}(s_2) = -\frac{i\omega}{2k}\tilde{J}_{0sl}(\cos ks_2 - \cos kL_{sl}). \quad\quad (4.8)
$$

The field reflection and transmitted coefficients, S_{11} and S_{12}, for the structure can be determined by the following expressions:

$$
S_{11} = \frac{4\pi i}{abkk_g}\left\{\frac{2k_g^2}{k}\tilde{J}_{0sl}f(kL_{sl}) - \frac{kk_g}{\tilde{k}}\tilde{J}_{0v}\sin(\pi x_{01}/a)e^{-ik_g z_0}f(\tilde{k}L_v)\right\}e^{2ik_g z}, \quad (4.9)
$$

$$
S_{12} = 1 + \frac{4\pi i}{abkk_g}\left\{\frac{2k_g^2}{k}\tilde{J}_{0sl}f(kL_{sl}) + \frac{kk_g}{\tilde{k}}\tilde{J}_{0v}\sin(\pi x_{01}/a)e^{-ik_g z_0}f(\tilde{k}L_v)\right\}. \quad (4.10)
$$

The power radiation coefficient $|S_{\Sigma}|^2$ of the slot to half-space over the plane is equal to

$$
|S_{\Sigma}|^2 = \frac{P_{\Sigma}}{P_{10}}, \quad\quad (4.11)
$$

where P_{Σ} is the average power radiated through the slot aperture, Umov-Poynting vector flux through the slot, and P_{10} is the input power of the H_{10} waveguide mode. It can be easily verified that

$$
|S_{\Sigma}|^2 = \frac{16\pi k_g}{abk}\left|\tilde{J}_{0sl}\right|^2\left|\text{Im}Z_{22}^{Hs}(kd_e, kL_{sl})\right|, \quad\quad (4.12)
$$

where $\text{Im}Z_{22}^{Hs}(kd_e, kL_{sl})$ is the imaginary part of the extrinsic slot conductivity.

If $\bar{R}_S \neq 0$, the power loss in the impedance vibrator, P_{σ}, can be found by using the energy balance relation

$$|S_{11}|^2 + |S_{12}|^2 + |S_\Sigma|^2 + P_\sigma = 1. \tag{4.13}$$

The equality (4.13) can also be used to verify the numerical algorithms for calculating the characteristics of waveguide-slot structure when the impedance vibrator is lossless and the surface impedance is purely imaginary.

Then, the distribution functions of the imaginary part of the surface impedance, defined in the Sect 3.2.2): $\phi_0(s_1) = 1$, $\phi_1(s_1) = 2[1 - (s_1/L_v)]$ and $\phi_2(s_1) = 2(s_1/L_v)$ can be used. The expressions for $F_{z0,1,2}(\tilde{k}r, \tilde{k}L_v)$ are determined by the formulas (3.49)—(3.51).

4.1.2 Numerical and Experimental Results

The numerical analysis of achievable variations of the energy characteristics of the vibrator-slot structure was carried out based on the developed mathematical model. Results for the vibrator-slot structure and single radiating slot were compared. The plots of the radiation $|S_\Sigma|^2(\lambda)$, reflection $|S_{11}|(\lambda)$, and transmission $|S_{12}|(\lambda)$ coefficients in the single-mode operating range of the waveguide are presented in Figs. 4.2, 4.3, 4.4, 4.5 and 4.6. The common parameters of the structure: $a = 58.0$ mm $b = 25.0$ mm, $h = 0.5$ mm, $r = 2.0$ mm, $L_v = 15.0$ mm, $\bar{R}_s = 0$, $d = 4.0$ mm, $2L_{sl} = 40.0$ mm were selected so that the slot resonant wavelength $\lambda_{sl}^{res} = 86.0$ mm was in the middle of the operating waveguide band, and the monopole resonant wavelength λ_v^{res} was also in this band. The distance z_0 between the vibrator and longitudinal slot axis was varied in fractions of the resonant slot wavelength in the waveguide, $\lambda_G = \lambda_{sl}^{gres} = \dfrac{2\pi}{\sqrt{(2\pi/\lambda_{sl}^{res})^2-(\pi/a)^2}}$, since the system energy parameters defined in (4.9)–(4.13) contain periodic functions of the wavenumber k_g. The simulation results are presented only for vibrators with inductive impedance, $\bar{X}_S > 0$, since in this case, the resonant wavelength of the vibrator λ_v^{res} increases, as compared to that for the perfectly conducting vibrator, $\bar{Z}_S = 0$. Therefore, the gap between the vibrator end and the upper broad wall of the waveguide increases allowing to enlarge the breakdown power for the waveguide device.

Analysis of the plots presented in Figs. 4.2, 4.3, 4.4, 4.5 and 4.6 show that both the perfectly conducting and impedance vibrators can significantly affect the radiation coefficient of the vibrator-slot system. However, this effect turns out to be more significant with impedance vibrators, since the natural wavelengths of the vibrator and slot can be approached to each other. As can be seem from Fig. 4.3, the simple coincidence of the resonances is not optimal from point of view of interaction between the vibrator and-slot. The plots in Figs. 4.4 and 4.5, show that the slot radiation coefficient can closely approach to unity, that is, the limit known from the theory of waveguide-slot radiators in hollow infinite waveguides. The natural resonant wavelengths of the slot and vibrator are slightly spaced in the range, while the common resonance of the vibrator-slot system is shifted to the resonant wavelength of the vibrator. As expected from physical considerations, the mutual interactions between the vibrator and slot

Fig. 4.2 The wavelength plots of energy characteristics of the vibrator-slot structure with $x_{01} = a/8$, $\bar{Z}_S = 0$

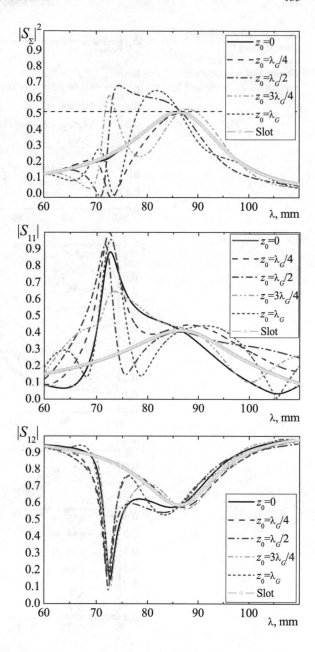

is observed when the distance z_0 is multiple of $\lambda_G/4$. The maximal values of the slot radiation coefficient, close to unity, are observed if z_0 is multiples of $\lambda_G/2$.

If the vibrator is shifted to the waveguide longitudinal axis in the transverse direction when the distance $z_0 = \lambda_G/2$ x_0 when $z_0 = \lambda_G/2$, the maximal radiation coefficient $|S_\Sigma|^2(\lambda)$ slightly increases and its operation band widens considerably

Fig. 4.3 The wavelength plots of energy characteristics of the vibrator-slot structure with $x_{01} = a/8$, $\bar{Z}_S = ikr \ln(5.5)$

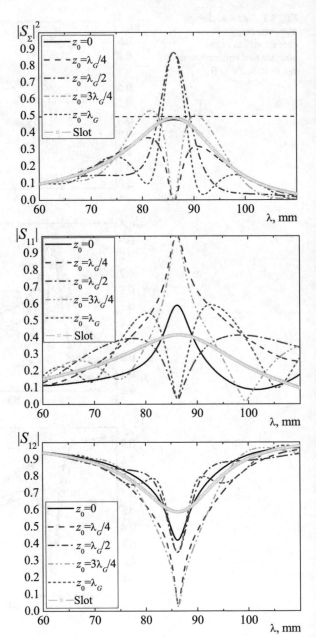

Fig. 4.4 The wavelength plots of energy characteristics of the vibrator slot structure with $x_{01} = a/8$, $\tilde{Z}_S(s_1) = ikr\ln(5.5)\phi_1(s_1)$

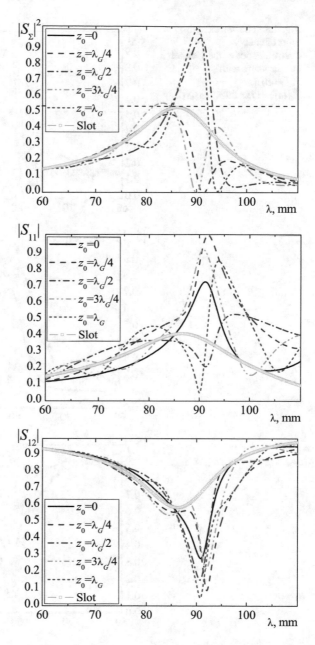

Fig. 4.5 The wavelength plots of energy characteristics of the vibrator slot structure with $x_{01} = a/8$, $\bar{Z}_S(s_1) = ikr \ln(5.5)\phi_2(s_1)$

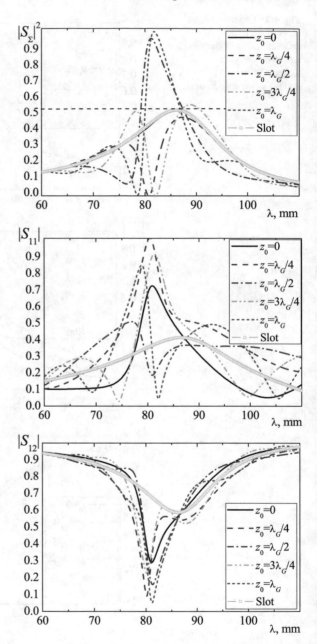

Fig. 4.6 The wavelength
plots of energy
characteristics of the vibrator
slot structure with
$z_0 = \lambda_G/2 = 64.0$ mm,
$\bar{Z}_S = ikr \ln(5.5)$

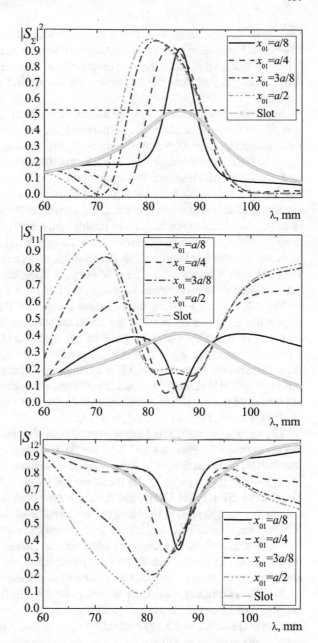

(Fig. 4.6). As seen, if $x_{01} = a/2$, the width of radiation coefficient curve at the level $|S_\Sigma|^2 = 0.6$ increases by 3 times as compared with its value for $x_{01} = a/8$. It should be noted that in this case, the maximum of the radiation coefficient is reached at the wavelength that does not coincide with the natural resonances of the slot and vibrator.

Since the low-profile waveguides are often used in practice to miniaturize waveguide devices, considerable restrictions are imposed on the monopole length. Например, при $b = 12.5$ mm можно выбрать приемлемую длину вибратора $L_v = 10.0$ mm. For example, if $b = 12.5$ mm, the acceptable vibrator length should be $L_v = 10.0$ mm. Therefore, we obtain that $\lambda_{sl}^{res} = 83.0$ mm and $\lambda_{sl}^{gres} = 118.0$ mm. The simulation results have showed that if the other geometric parameters of the vibrator-slot structure are not varied, the short perfectly conducting vibrator is unable to affect the slot characteristics and significantly increase the slot radiation coefficient $|S_\Sigma|^2$ due to the large difference between the natural wavelengths of the slot and vibrator. The electric length of the vibrator can be increased by using the monopole with inductance impedance. The efficiency of using the impedance monopoles in the low-profile waveguide can be estimated from the simulation results presented in Fig. 4.7. As can be seen, the radiation coefficient the monopole with constant impedance increases up to 0.7 if the resonant wavelengths of the vibrator and slot coincide, although the operating band of the structure $|S_\Sigma|^2(\lambda)$ becomes narrower. However, impedance $\bar{Z}_S = ikr \ln(51.0)$ is rather difficult to implement in practice. If the distribution function $\phi_1(s_1)$ is used, the energy characteristics of the radiator is almost identical to that for the vibrator with constant impedance, but at a more acceptable value of the logarithmic coefficient $\ln(17.0)$. As seen, the radiation coefficient $|S_\Sigma|^2$ may reach 1.0 if function $\phi_2(s_1)$ is used.

The simulation results and experimental data obtained were compared by using the setup shown in Figs. 4.8 and 3.13 for the perfectly conducting vibrator and for the vibrator with distributed surface impedance defined as $\bar{Z}_S(s_1) = ikr \ln(4.0)$ and $\bar{Z}_S(s_1) = ikr \ln(4.0)\phi_1(s_1)$. The measurements were carried out using the *S810D Site Master* Broadband Cable and Antenna Analyzer. The simulation and experimental wavelength plots of the energy characteristics of the vibrator-slot structure for different impedance vibrators placed at the fixed distance $z_0 = \lambda_G/2$ are presented in Fig. 4.9. Satisfactory agreement between the simulated and experimental data confirms the physical adequacy of the mathematical model, the correctness of the electrodynamic problem solution and numerical calculations.

The vibrator surface impedance was calculated by the formula (3.52). The variable surface impedance can be carried out by varying the internal radius of the corrugated conductor (Figs. 3.2 and 3.13) defined by the function $r_i(s) = re^{-\ln(r/r_i)\phi_n(s)}$, where r and r_i the external and internal radii of the corrugation.

Thus, the conditions to achieve the maximal slot radiation coefficient, close to 1.0 were established. As is known, from the theory of waveguide-slot emitters, this is unattainable in a hollow infinite rectangular waveguide [1, 2] and besides, the efficiency of impedance monopole application to achieve maximal radiation level in the vibrator-slot systems based on the low-profile rectangular waveguides is shown.

Fig. 4.7 The wavelength plots of energy characteristics of the vibrator slot structure with parameters $a = 58.0$ mm, $b = 12.5$ mm, $2L_{sl} = 40.0$ mm, $L_v = 10.0$ mm, $x_{01} = a/8$, $z_0 = \lambda_{sl}^{gres}/2 = 59.0$ mm: 1—$\bar{Z}_S = 0$; 2—$\bar{Z}_S = ikr\ln(51.0)$; 3—$\bar{Z}_S(s_1) = ikr\ln(17.0)\phi_1(s_1)$; 4—$\bar{Z}_S(s_1) = ikr\ln(17.0)\phi_2(s_1)$; 5—slot without vibrator

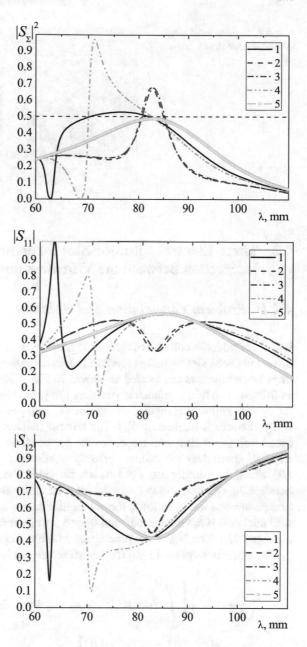

Fig. 4.8 Experimental setup
of the vibrator-slot structure

4.2 Three-Element Vibrator-Slot Structure Without Interaction Between the Vibrators and Slot

4.2.1 Problem Formulation and Solution

Let us consider the problem of electromagnetic wave scattering on the narrow recti-linear transverse slot cut in the broad wall of rectangular waveguide where two passive impedance vibrators are located as shown in Fig. 4.10. The problem is formulated as follows: two thin asymmetric vibrators are placed in the cross-sectional plane of the infinite hollow rectangular waveguide ($\varepsilon_1 = \mu_1 = 1$) with perfectly conducting walls (the area is marked as Wg). The narrow transverse slot cut in the waveguide broad wall symmetric with respect to the longitudinal waveguide axis radiates into free half-space over the infinite perfectly conducting plane (the area is marked as Hs). Since the vibrator and slot axis are placed in the plane $\{x0y\}$, the interaction between the vibrators and slot over the internal waveguide space is absent. The structure parameters are as follows: the waveguide cross-section is $\{a \times b\}$, waveguide wall thickness is h, vibrator radii and length are $r_{1,2}$ and $2L_{1,2}$, slot width and length are d and $2L_3$. The H_{10}- wave propagates in the waveguide from the area $z = -\infty$.

The equations system (2.20) for this structure can be written as:

$$
\left(\frac{d^2}{ds_1^2} + k^2 \right) \left\{ \int_{-L_1}^{L_1} J_1(s_1') G_{s_1}^{Wg}(s_1, s_1') ds_1' + \int_{-L_2}^{L_2} J_2(s_2') G_{s_2}^{Wg}(s_1, s_2') ds_2' \right\}
$$
$$
= -i\omega \left[E_{0s_1}(s_1) - z_{i1}(s_1) J_1(s_1) \right], \tag{4.24a}
$$

Fig. 4.9 The wavelength plots of energy characteristics of the vibrator slot structure with parameters $a = 58.0$ mm, $b = 12.5$ mm, $2L_{sl} = 40.0$ mm, $d = 4.0$ mm, $L_v = 15.0$ mm, $r = 2.0$ mm, $x_{01} = a/8$, $z_0 = 64.0$ mm: **1**—$\bar{Z}_S = 0$; **2**—$\bar{Z}_S = ikr \ln(4.0)$; **3**—$\bar{Z}_S(s_1) = ikr \ln(4.0)\phi_1(s_1)$; **4, 5, 6**—experimental data

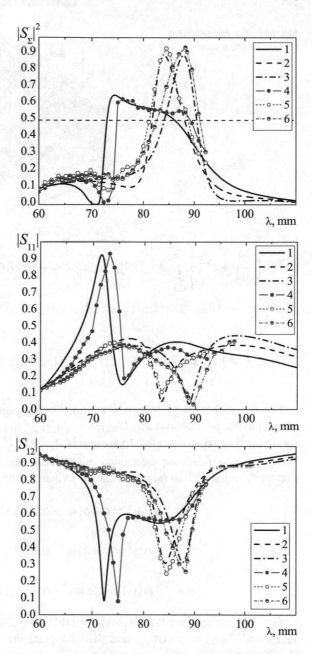

Fig. 4.10 The three-element
vibrator-slot structure

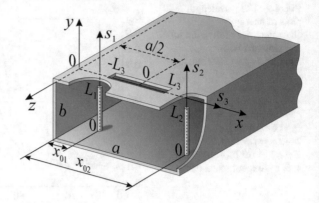

$$\left(\frac{d^2}{ds_2^2} + k^2\right)\left\{\int\limits_{-L_2}^{L_2} J_2(s_2')G_{s_2}^{Wg}(s_2, s_2')ds_2' + \int\limits_{-L_1}^{L_1} J_1(s_1')G_{s_1}^{Wg}(s_2, s_1')ds_1'\right\}$$
$$= -i\omega\left[E_{0s_2}(s_2) - z_{i2}(s_2)J_2(s_2)\right], \tag{4.24b}$$

$$\left(\frac{d^2}{ds_3^2} + k_1^2\right)\int\limits_{-L_3}^{L_3} J_3(s_3')\left[G_{s_3}^{Wg}(s_3, s_3') + G_{s_3}^{Hs}(s_3, s_3')\right]ds_3' = -i\omega H_{0s_3}(s_3). \tag{4.24c}$$

Solution of the Eqs. (4.24) can be obtained by the generalized method of induced
EMMF, using the functions $J_{1(2)}(s_{1(2)}) = J_{1(2)}^0 f_{1(2)}(s_{1(2)})$ and $J_3(s_3) = J_3^0 f_3(s_3)$ as
approximating expressions for the currents ($J_{1(2)}^0$ and J_3^0 are the current amplitudes,
$f_{1(2)}(s_{1(2)})$ and $f_3(s_3)$ are predefined distribution functions). For the vibrator-slot
structure the distribution functions can be written similarly to (4.4) as

$$f_1(s_1) = \cos\tilde{k}_1 s_1 - \cos\tilde{k}_1 L_1, \tag{4.25a}$$

$$f_2(s_2) = \cos\tilde{k}_2 s_2 - \cos\tilde{k}_2 L_2, \tag{4.25b}$$

$$f_3(s_3) = \cos ks_3 - \cos kL_3. \tag{4.25c}$$

First, we multiply the Eqs. (4.24a), (4.24b) and (4.24c) by the functions $f_1(s_1)$,
$f_2(s_2)$ and $f_3(s_3)$ respectively, and then integrate the resulting Eqs. (4.24a) and
(4.24b) over the vibrator lengths and Eq. (4.24c) over the slot length. As a result, we
obtain the SLAE, which solution are the current amplitudes $J_{1,2}^0$ and J_3^0:

$$J_1^0[Z_{11}(kr_1, \tilde{k}_1L_1) + F_1^z(\tilde{k}_1r_1, \tilde{k}_1L_1)] + J_2^0 Z_{12}(\tilde{k}_1L_1, \tilde{k}_2L_2) = -\frac{i\omega}{2k}E_1(\tilde{k}_1L_1),$$

$$J_2^0[Z_{22}(kr_2, \tilde{k}_2L_2) + F_2^z(\tilde{k}_2r_2, \tilde{k}_2L_2)] + J_1^0 Z_{21}(\tilde{k}_2L_2, \tilde{k}_1L_1) = -\frac{i\omega}{2k}E_2(\tilde{k}_2L_2),$$

$$J_3^0[Z_{33}^{Wg}(kd_e, kL_3) + Z_{33}^{Hs}(kd_e, kL_3)] = -\frac{i\omega}{2k}H_3(kL_3). \tag{4.26}$$

where according to (3.43), (3.46), (3.58), (3.66) and (3.67):

$$
Z_{11(22)}(kr_{1(2)}, \tilde{k}_{1(2)}L_{1(2)}) = \frac{4\pi}{ab} \sum_{m=1}^{\infty}\sum_{n=0}^{\infty} \frac{\varepsilon_n(k^2 - k_y^2)\tilde{k}_{1(2)}^2}{kk_z(\tilde{k}_{1(2)}^2 - k_y^2)^2} e^{-k_z r_{1(2)}} \sin^2 k_x x_{01(02)}
$$
$$
\times [\sin\tilde{k}_{1(2)}L_{1(2)} \cos k_y L_{1(2)}
$$
$$
- (\tilde{k}_{1(2)}/k_y)\cos\tilde{k}_{1(2)}L_{1(2)} \sin k_y L_{1(2)}]^2,
$$

$$
Z_{12(21)}(\tilde{k}_{1(2)}L_{1(2)}, \tilde{k}_{2(1)}L_{2(1)}) = \frac{4\pi}{ab}\sum_{m=1}^{\infty}\sum_{n=0}^{\infty}\frac{\varepsilon_n(k^2-k_y^2)\tilde{k}_1\tilde{k}_2 e^{-k_z r_{2(1)}}}{kk_z(\tilde{k}_1^2-k_y^2)(\tilde{k}_2^2-k_y^2)}\sin k_x x_{01}\sin k_x x_{02}
$$
$$
\times[\sin\tilde{k}_1 L_1 \cos k_y L_1 - (\tilde{k}_1/k_y)\cos\tilde{k}_1 L_1 \sin k_y L_1]
$$
$$
\times[\sin\tilde{k}_2 L_2 \cos k_y L_2 - (\tilde{k}_2/k_y)\cos\tilde{k}_2 L_2 \sin k_y L_2],
$$

$$
Z_{33}^{Wg}(kd_e, kL_3) = \frac{8\pi}{ab}\sum_{m=1,3..}^{\infty}\sum_{n=0,1..}^{\infty}\frac{\varepsilon_n k}{k_z(k^2-k_x^2)}e^{-k_z\frac{d_e}{4}}
$$
$$
\times[\sin kL_3 \cos k_x L_3 - (k/k_x)\cos kL_3 \sin k_x L_3]^2,
$$

$$
Z_{33}^{Hs}(kd_e, kL_3) = (\mathrm{Si}4kL_3 - i\mathrm{Cin}4kL_3) - 2\cos kL_3
$$
$$
\times\left[2(\sin kL_3 - kL_3\cos kL_3)\left(\ln\frac{16L_3}{d_e} - \mathrm{Cin}2kL_3 - i\mathrm{Si}2kL_3\right)\right.
$$
$$
+ \sin 2kL_3 e^{-ikL_3}, \tag{4.27}
$$

$$
E_{1(2)}(\tilde{k}_{1(2)}L_{1(2)}) = 2H_0\frac{k}{k_g\tilde{k}_{1(2)}}\sin\frac{\pi}{a}x_{01(02)}f(\tilde{k}_{1(2)}L_{1(2)}),
$$

$$
f(\tilde{k}_{1(2)}L_{1(2)}) = \sin\tilde{k}_{1(2)}L_{1(2)} - \tilde{k}_{1(2)}L_{1(2)}\cos\tilde{k}_{1(2)}L_{1(2)},
$$

$$
H_3(kL_3) = 2H_0\frac{1}{k}f(kL_3), \; f(kL_3)
$$
$$
= \frac{\sin kL_3 \cos(\pi L_3/a) - (ka/\pi)\cos kL_3 \sin(\pi L_3/a)}{1 - [\pi/(ka)]^2},
$$

$$
F_{1(2)}^z(\tilde{k}_{1(2)}r_{1(2)}, \tilde{k}_{1(2)}L_{1(2)}) = -\frac{i}{r_{1(2)}}\int_0^{L_{1(2)}} f_{1(2)}^2(s_{1(2)})\bar{Z}_{S1(2)}(s_{1(2)})\mathrm{d}s_{1(2)}.
$$

The analytical solution of the (4.26) can be presented as

$$
J_1^0 = -\frac{i\omega}{2k}\frac{E_1(Z_{22} + F_2^z) - E_2 Z_{12}}{(Z_{11} + F_1^z)(Z_{22} + F_2^z) - Z_{21}Z_{12}} = -\frac{i\omega}{2k}\tilde{J}_1^0,
$$

$$
J_2^0 = -\frac{i\omega}{2k}\frac{E_2(Z_{11} + F_1^z) - E_1 Z_{21}}{(Z_{11} + F_1^z)(Z_{22} + F_2^z) - Z_{21}Z_{12}} = -\frac{i\omega}{2k}\tilde{J}_2^0,
$$

$$
J_3^0 = -\frac{i\omega}{2k}\frac{H_3(kL_3)}{Z_{33}^{Wg} + Z_{33}^{Hs}} = -\frac{i\omega}{2k}\tilde{J}_3^0. \tag{4.28}
$$

The final expressions for the vibrator and slot currents can be obtained by using the expressions (4.25a) and (4.28) as:

$$
J_{1(2)}(s_{1(2)}) = -\frac{i\omega}{2k}\tilde{J}_{1(2)}^0(\cos \tilde{k}_{1(2)}s_{1(2)} - \cos \tilde{k}_{1(2)}L_{1(2)}),
$$

$$
J_3(s_3) = -\frac{i\omega}{2k}\tilde{J}_3^0(\cos ks_3 - \cos kL_3). \tag{4.29}
$$

The energy characteristics, the field reflection and transmission coefficients, S_{11} and S_{12}, of the structure are determined by the following expressions:

$$
S_{11} = \frac{4\pi i}{abkk_g}\left\{\frac{2k_g^2}{k^2}\frac{f^2(kL_3)}{Z_{33}^{Wg} + Z_{33}^{Hs}} - \frac{k^2}{\tilde{k}_1}\tilde{J}_1^0\sin\left(\frac{\pi x_{01}}{a}\right)f(\tilde{k}_1 L_1)\right.
$$
$$
\left. -\frac{k^2}{\tilde{k}_2}\tilde{J}_2^0\sin\left(\frac{\pi x_{02}}{a}\right)f(\tilde{k}_2 L_2)\right\}e^{2ik_g z}, \tag{4.30}
$$

$$
S_{12} = 1 + \frac{4\pi i}{abkk_g}\left\{\frac{2k_g^2}{k^2}\frac{f^2(kL_3)}{Z_{33}^{Wg} + Z_{33}^{Hs}} + \frac{k^2}{\tilde{k}_1}\tilde{J}_1^0\sin\left(\frac{\pi x_{01}}{a}\right)f(\tilde{k}_1 L_1)\right.
$$
$$
\left. +\frac{k^2}{\tilde{k}_2}\tilde{J}_2^0\sin\left(\frac{\pi x_{02}}{a}\right)f(\tilde{k}_2 L_2)\right\}. \tag{4.31}
$$

The power radiation coefficient was calculated by the formulas (4.11) and (4.12).

4.2.2 Numerical and Experimental Results

The numerical analysis of the energy characteristics of the three-element vibrator-slot structure was carried out based on the developed mathematical model. As in Sect. 4.1.1, the simulation results were obtained by using the functions: $\phi_0(s_{1(2)}) = 1$, $\phi_1(s_{1(2)}) = 2[1 - (s_{1(2)}/L_{1(2)})]$, and $\phi_2(s_{1(2)}) = 2(s_{1(2)}/L_{1(2)})$.

The wavelength plots of the radiation coefficient $|S_\Sigma|^2(\lambda)$ and the modules of reflection and transmission coefficients, $|S_{11}|(\lambda)$ and $|S_{12}|(\lambda)$, in the single-mode operation range of the waveguide are presented in Figs. 4.11, 4.12, 4.13 and 4.14. The structure parameters were as follows: $a = 58.0\,\text{mm}$, $b = 25.0\,\text{mm}$, $h = 0.5\,\text{mm}$, $r_{1,2} = 2.0\,\text{mm}$, $L_{1,2} = 15.0\,\text{mm}$, $\bar{R}_{S1(2)} = 0$, $d = 4.0\,\text{mm}$, $2L_3 = 40.0\,\text{mm}$. As expected, the curves $|S_\Sigma|^2(\lambda)$ for the structure with various distances x_{01}, x_{02}, and single slots practically coincide (Fig. 4.11). That is, the radiation coefficient of the

Fig. 4.11 The wavelength plots of energy characteristics of the vibrator slot-structure with parameters $x_{01} = a/8$, $\bar{Z}_{S1} = 0$, $\bar{Z}_{S2} = ikr_2 \ln(4.0)$

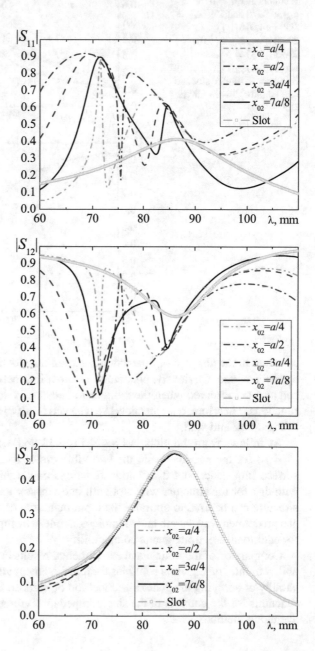

Fig. 4.12 The wavelength plots of energy characteristics of the vibrator- slot structure with symmetric placing of the vibrators at $x_{01} = a/8$, $x_{02} = 7a/8$ and $\bar{Z}_{S1}(s_1) = ikr_1 \ln(4.0)\phi_1(s_1)$: **1**—$\bar{Z}_{S2} = ikr_2 \ln(4.0)$; **2**—$\bar{Z}_{S2}(s_2) = ikr_2 \ln(4.0)\phi_2(s_2)$; **3**–$\bar{Z}_{S2} = 0$; **4**—the slot and one vibrator; **5**—the single slot

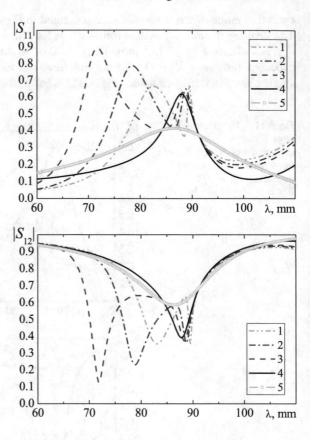

structure in the absence of interaction between the slot and vibrator are determined by the slot size. Obviously, polarization decoupling between the slot and vibrators can also be observed when the vibrator impedance is defined by other distribution functions. Therefore, only the plots $|S_{11}|(\lambda)$ and $|S_{12}|(\lambda)$ in the waveguide are shown in Figs. 4.12 and 4.13.

As follows from the plots in Figs. 4.11, 4.12 and 4.13, the curves $|S_{11}|(\lambda)$ and $|S_{12}|(\lambda)$ for the structure with the two vibrators of fixed length with the different surface impedance and distribution functions can significantly vary as compared with that for the structure with slot with one vibrator and single slots. Besides, the structure can be used to optimize the input matching of the waveguide-slot radiator at a predefined wavelength and to implement one-way input or output signal filtering for additional electromagnetic compatibility.

Comparison of theoretical and experimental results in Fig. 4.14 indicates the reliability of integral equation solutions for multi-element vibrator-slot structure and the validity of applying the generalized method of induced EMMF with approximating functions for the currents in the single impedance vibrator and slot obtained by the averaging method.

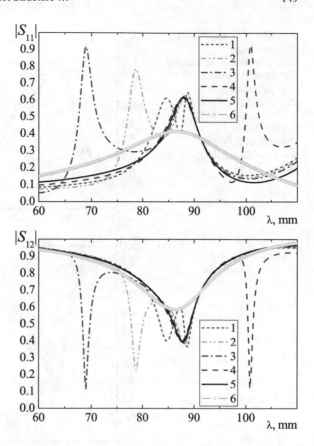

Fig. 4.13 The wavelength plots of energy characteristics of the vibrator-slot structure with asymmetric placing of the vibrators at $x_{01} = a/8$, $x_{02} = 15a/16$ and $\bar{Z}_{S1}(s_1) = ikr_1 \ln(4.0)\phi_1(s_1)$: $1 - \bar{Z}_{S2} = ikr_2 \ln(4.0)$; $2 - \bar{Z}_{S2}(s_2) = ikr_2 \ln(4.0)\phi_2(s_2)$; $3 - \bar{Z}_{S2} = 0$; $4 - \bar{Z}_{S2}(s_2) = ikr_2 \ln(8.0)\phi_1(s_2)$; 5—the slot and one vibrator; 6—the single slot

Thus, the approach to solving the electrodynamic problem tested earlier in Sect. 4.1 is extended to multi-element vibrator-slot structures. The vibrator-slot structure with the transverse slot and two vibrators in waveguide cross-sectional plane (Fig. 4.10) which allows to use the simplified of equation system was analyzed. The interaction between the vibrators and slot is absent, since they are decoupled by polarization. The possibility to control the reflection and transmission coefficients over a wide range by using passive vibrators with varying surface impedance and various distribution functions was proved.

Fig. 4.14 The wavelength plots of energy characteristics of the vibrator slot structure with parameters at $x_{01} = a/8$, $\bar{Z}_{S1} = 0$, $x_{02} = 7a/8$: **1**–$\bar{Z}_{S2} = ikr_2 \ln(4.0)$; **2**—$\bar{Z}_{S2}(s_2) = ikr_2 \ln(4.0)\phi_1(s_2)$; **3**, **4**—experimental data; **5**—single slot

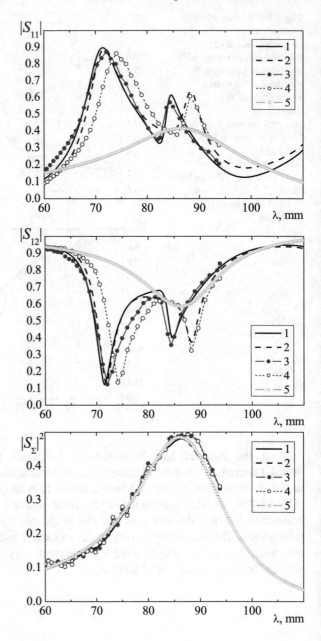

4.3 Three-Element Vibrator-Slot Structures with Interaction Between the Vibrators and Slot

4.3.1 Problem Formulation and Solution

The three-element vibrator-slot structure is shown in Fig. 4.15

According to (2.19), the equation system relative to electric currents on the vibrators and equivalent magnetic current in the slot intended for analysis of the structure, can be presented as:

$$
\left(\frac{d^2}{ds_1^2} + k^2\right)\left\{\int_{-L_1}^{L_1} J_1(s_1')G_{s_1}^{Wg}(s_1, s_1')ds_1' + \int_{-L_2}^{L_2} J_2(s_2')G_{s_2}^{Wg}(s_1, s_2')ds_2'\right\}
$$
$$
- ik\int_{-L_3}^{L_3} J_3(s_3')\tilde{G}_{s_3}^{Wg}(s_1, s_3')ds_3' = -i\omega\left[E_{0s_1}(s_1) - z_{i1}(s_1)J_1(s_1)\right],
$$

$$
\left(\frac{d^2}{ds_2^2} + k^2\right)\left\{\int_{-L_2}^{L_2} J_2(s_2')G_{s_2}^{Wg}(s_2, s_2')ds_2' + \int_{-L_1}^{L_1} J_1(s_1')G_{s_1}^{Wg}(s_2, s_1')ds_1'\right\}
$$
$$
= -i\omega\left[E_{0s_2}(s_2) - z_{i2}(s_2)J_2(s_2)\right], \tag{4.32}
$$

$$
\left(\frac{d^2}{ds_3^2} + k_1^2\right)\int_{-L_3}^{L_3} J_3(s_3')\left[G_{s_3}^{Wg}(s_3, s_3') + G_{s_3}^{Hs}(s_3, s_3')\right]ds_3'
$$
$$
- ik\int_{-L_1}^{L_1} J_1(s_1')\tilde{G}_{s_1}^{Wg}(s_3, s_1')ds_1' = -i\omega H_{0s_3}(s_3),
$$

Fig. 4.15 The three-element vibrator-slot structure

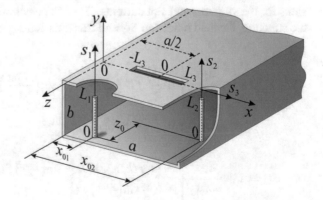

where z_0 is the distance between the first vibrator and slot axes, $\tilde{G}_{s_1}^{Wg}(s_3, s_1') = \frac{\partial}{\partial z} G_{s_1}^{Wg}[x(s_3), 0, z; x'(s_1'), y'(s_1'), z_0]$, $\tilde{G}_{s_3}^{Wg}(s_1, s_3') = \frac{\partial}{\partial z} G_{s_3}^{Wg}[x(s_1), y(s_1), z; x'(s_3'), 0, 0]$ after substitution $z = 0$ in $\tilde{G}_{s_1}^{Wg}$ and $z = z_0$ into $\tilde{G}_{s_3}^{Wg}$ in after the first derivation. The remaining notations in (4.32) coincide with that in the Sect. 4.2.

After application of the generalized method of induced EMMF with current distribution functions (4.25a), we obtain the SLAE relative to unknown current amplitudes:

$$
\begin{cases}
J_1^0 Z_{11}^\Sigma + J_2^0 Z_{12} + J_3^0 Z_{13} = -\frac{i\omega}{2k} \int_{-L_1}^{L_1} f_1(s_1) E_{0s_1}(s_1) ds_1, \\
J_2^0 Z_{22}^\Sigma + J_1^0 Z_{21} = -\frac{i\omega}{2k} \int_{-L_2}^{L_2} f_2(s_2) E_{0s_2}(s_2) ds_2, \\
J_3^0 Z_{33}^\Sigma + J_1^0 Z_{31} = -\frac{i\omega}{2k} \int_{-L_3}^{L_3} f_3(s_3) H_{0s_3}(s_3) ds_3,
\end{cases}
\tag{4.33}
$$

where $Z_{11(22)}^\Sigma = Z_{11(22)} + F_{1(2)}^z$, $Z_{33}^\Sigma = Z_{33}^{Hs} + Z_{33}^{Wg}$,

$$
\begin{aligned}
Z_{12} = Z_{21} = \frac{4\pi}{ab} \sum_{m=1}^\infty \sum_{n=0}^\infty & \left\{ \frac{\varepsilon_n(k^2 - k_y^2)\tilde{k}_1\tilde{k}_2 e^{-k_z z_0}}{kk_z(\tilde{k}_1^2 - k_y^2)(\tilde{k}_2^2 - k_y^2)} \sin k_x x_{01} \right. \\
& \times \sin k_x x_{02} \left[\sin\tilde{k}_1 L_1 \cos k_y L_1 - (\tilde{k}_1/k_y)\cos\tilde{k}_1 L_1 \sin k_y L_1 \right] \\
& \left. \times \left[\sin\tilde{k}_2 L_2 \cos k_y L_2 - (\tilde{k}_2/k_y)\cos\tilde{k}_2 L_2 \sin k_y L_2 \right] \right\},
\end{aligned}
$$

$$
\begin{aligned}
Z_{13} = -Z_{31} = \frac{4\pi}{ab} \sum_{m=1}^\infty \sum_{n=0}^\infty & \left\{ \frac{\varepsilon_n k\tilde{k}_1 e^{-k_z z_0}}{i(\tilde{k}_1^2 - k_y^2)(k^2 - k_x^2)} \right. \\
& \times \sin k_x x_{01} \sin \frac{k_x a}{2} \left[\sin\tilde{k}_1 L_1 \cos k_y L_1 - \frac{\tilde{k}_1}{k_y}\cos\tilde{k}_1 L_1 \sin k_y L_1 \right] \\
& \left. \times \left[\sin k L_3 \cos k_x L_3 - (k/k_x)\cos k L_3 \sin k_x L_3 \right] \right\}.
\end{aligned}
$$

The analytical solution of the equation system (4.33) allows to obtain the expressions for the vibrators and slot currents. Then the electrodynamic characteristics of the structure, the field reflection S_{11} and transmission S_{12} coefficients can be written as:

$$
\begin{aligned}
S_{11} = \frac{4\pi i}{abkk_g} & \left\{ J_3 \frac{2k_g^2}{k^2} f(k L_3) - J_1 \frac{k_g}{\tilde{k}_1} \sin\left(\frac{\pi x_{01}}{a}\right) f(\tilde{k}_1 L_1) e^{-ik_g z_0} \right. \\
& \left. - J_2 \frac{k_g}{\tilde{k}_2} \sin\left(\frac{\pi x_{02}}{a}\right) f(\tilde{k}_2 L_2) \right\} e^{2ik_g z},
\end{aligned}
\tag{4.34}
$$

$$
S_{12} = 1 + \frac{4\pi i}{abkk_g} \left\{ \begin{array}{l} J_3 \frac{2k_g^2}{k^2} f(k L_3) + J_1 \frac{k_g}{\tilde{k}_1} \sin\left(\frac{\pi x_{01}}{a}\right) f(\tilde{k}_1 L_1) e^{ik_g z_0} \\ + J_2 \frac{k_g}{\tilde{k}_2} \sin\left(\frac{\pi x_{02}}{a}\right) f(\tilde{k}_2 L_2) \end{array} \right\},
\tag{4.35}
$$

where

$$
J_1 = \frac{1}{\left(Z_{11}^{\Sigma} Z_{22}^{\Sigma} Z_{33}^{\Sigma} - Z_{21} Z_{12} Z_{33}^{\Sigma} - Z_{31} Z_{13} Z_{22}^{\Sigma}\right)} \left[\frac{k^2}{k_g \tilde{k}_1} \sin \frac{\pi x_{01}}{a} f_1(\tilde{k}_1 L_1) e^{-i k_g z_0} Z_{22}^{\Sigma} Z_{33}^{\Sigma} \right.
$$
$$
\left. - \frac{k^2}{k_g \tilde{k}_2} \sin \frac{\pi x_{02}}{a} f_2(\tilde{k}_2 L_2) Z_{12} Z_{33}^{\Sigma} - f_3(k L_3) Z_{13} Z_{22}^{\Sigma} \right],
$$

$$
J_2 = \frac{1}{\left(Z_{11}^{\Sigma} Z_{22}^{\Sigma} Z_{33}^{\Sigma} - Z_{21} Z_{12} Z_{33}^{\Sigma} - Z_{31} Z_{13} Z_{22}^{\Sigma}\right)} \left[\frac{k^2}{k_g \tilde{k}_2} \sin \frac{\pi x_{02}}{a} f_2(\tilde{k}_2 L_2) (Z_{11}^{\Sigma} Z_{33}^{\Sigma} - Z_{31} Z_{13}) \right.
$$
$$
\left. - \frac{k^2}{k_g \tilde{k}_1} \sin \frac{\pi x_{01}}{a} f_1(\tilde{k}_1 L_1) e^{-i k_g z_0} Z_{21} Z_{33}^{\Sigma} + f_3(k L_3) Z_{13} Z_{21} \right],
$$

$$
J_3 = \frac{1}{\left(Z_{11}^{\Sigma} Z_{22}^{\Sigma} Z_{33}^{\Sigma} - Z_{21} Z_{12} Z_{33}^{\Sigma} - Z_{31} Z_{13} Z_{22}^{\Sigma}\right)} \left[\begin{array}{l} f_3(k L_3)(Z_{11}^{\Sigma} Z_{22}^{\Sigma} - Z_{21} Z_{12}) \\[4pt] + \dfrac{k^2}{k_g \tilde{k}_2} \sin \dfrac{\pi x_{02}}{a} f_2(\tilde{k}_2 L_2) Z_{12} Z_{31} \end{array} \right.
$$
$$
\left. - \frac{k^2}{k_g \tilde{k}_1} \sin \frac{\pi x_{01}}{a} f_1(\tilde{k}_1 L_1) e^{-i k_g z_0} Z_{31} Z_{22}^{\Sigma} \right],
$$

$$
f_{1(2)}(\tilde{k}_{1(2)} L_{1(2)}) = \sin \tilde{k}_{1(2)} L_{1(2)} - \tilde{k}_{1(2)} L_{1(2)} \cos \tilde{k}_{1(2)} L_{1(2)},
$$

$$
f_3(k L_3) = \frac{\sin k L_3 \cos(\pi L_3/a) - (ka/\pi) \cos k L_3 \sin(\pi L_3/a)}{1 - [\pi/(ka)]^2}.
$$

The power radiation coefficient can be calculated by the formulas (4.11) and (4.12).

4.3.2   Numerical and Experimental Results

As noted in Sect. 4.1, the greatest mutual interaction between the first vibrator and slot is observed if the z_0 distance is multiples of $\lambda_G/4$ ($\lambda_G = 2\pi/\sqrt{\left(2\pi/\lambda_{sl}^{res}\right)^2 - (\pi/a)^2}$ is the slot resonant wavelength in the waveguide, λ_{sl}^{res} is the slot resonant wavelength in free space over the plane). The curves in Figs. 4.16, 4.17 and 4.18 were obtained with the following parameters: $a = 58.0$ mm, $b = 25.0$ mm, $h = 0.5$ mm, $r_{1,2} = 2.0$ mm, $L_{1,2} = 15.0$ mm, $\bar{R}_{S1(2)} = 0$, $d = 4.0$ mm, $2L_3 = 40.0$ mm, $x_{01} = a/8$, $x_{02} = 7a/8$. As can be seen from Fig. 4.16, if the vibrator surface impedances have equal values, $\bar{Z}_{S1} = \bar{Z}_{S2}$, the acceptable level of reflection coefficient when the radiation coefficient is high. The maximal value of the radiation coefficient with satisfactory matching of the waveguide transmission line can be achieved if surface impedance of one vibrator varies along its length (Figs.°4.17 and 4.18). The surface impedance of the vibrators was calculated by the formula (3.52). In this case, the variable impedance was implemented by varying the internal radius of the corrugated conductor as (Figs. 3.2 and 3.13) according to the law, where and

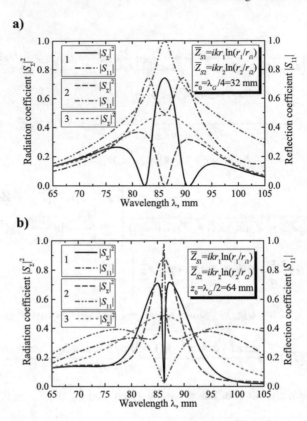

Fig. 4.16 The electrodynamic characteristics versus wavelength at $\bar{Z}_{S1} = \bar{Z}_{S2}, r_{1(2)}/r_{i1(2)} = 5.5$: 1—slot *plus* two vibrators; 2—slot *plus* one vibrator; 3—solitary slot. **a** $z_0 = \lambda_G/4 = 32$ mm. **b** $z_0 = \lambda_G/2 = 64$ mm

$r_i(s) = re^{-\ln(r/r_i)\phi_n(s)}$, where r and r_i are the external and internal corrugation radii. As seen from Fig. 4.18, the results of mathematical modeling are in satisfactory agreement with the experimental data.

In conclusion, we note that the three-element vibrator-slot structures considered in Sects. 4.2 and 4.3 can also be used in low-profile rectangular waveguides (Sect. 4.1) to reduce the probability of electrical breakdown between the end of the monopole and the upper wall of the waveguide when operating at high powers.

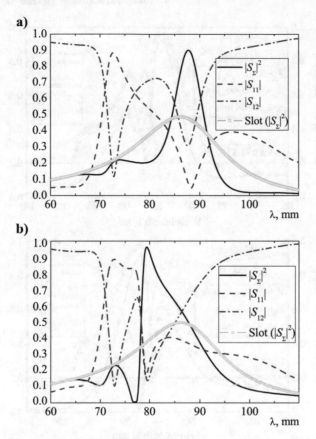

Fig. 4.17 The electrodynamic characteristic of the vibrator-slot structure with parameters at $z_0 = \lambda_G/2 = 64.0$ mm, $\bar{Z}_{S1} \neq \bar{Z}_{S2}$ ($\bar{Z}_{S2} = 0$), $r_{i1}(s_1) = r_1 \exp[-\ln(4.0)\phi_{1(2)}(s_1)]$: **a** $\bar{Z}_{S1}(s_1) - ikr_1 \ln(4.0)\phi_1(s_1)$; **b** $\bar{Z}_{S1}(s_1) = ikr_1 \ln(4.0)\phi_2(s_1)$

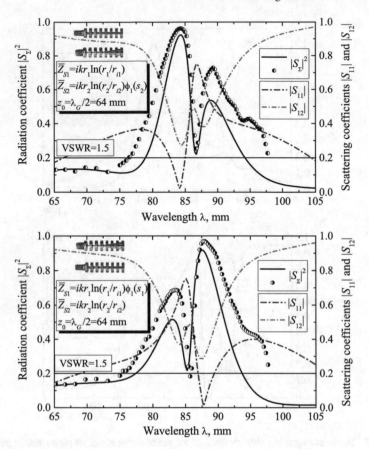

Fig. 4.18 The electrodynamic characteristics versus wavelength at $\bar{Z}_{S1} \neq \bar{Z}_{S2}$, $r_{1(2)}/r_{i1(2)} = 4.0$, $r_{i1(2)}(s_{1(2)}) = r_{1(2)} \exp[-\ln(4.0)\phi_1(s_{1(2)})]$ (circles, experimental data)

References

1. Nesterenko MV, Katrich VA, Penkin YM, Berdnik SL (2008) Analytical and hybrid methods in theory of slot-hole coupling of electrodynamic volumes. Springer Science + Business Media, New York
2. Nesterenko MV, Berdnik SL, Katrich V.A., Penkin Y.M.: Electromagnetic waves excitation by thin impedance vibrators and narrow slots in electrodynamic volumes. In: Chapter in book advanced electromagnetic waves, Bashir SO (ed.), InTech, Rijeka, Chapter 6, pp 147–175 (2015)

Chapter 5
T-Junctions of Rectangular Waveguides with Vibrator-Slot Structures in Coupling Areas

5.1 E-Plane T-Junction of Equal-Size Waveguides with the Two-Element Vibrator-Slot Structure

At present, a variety of planar E- and H- waveguide junctions are widely used in the antenna-waveguide technique at microwave and extremely high frequencies [1–6]. Electrodynamic characteristics of waveguide junctions can be varied by using inductive and capacitive diaphragms [7, 8] and resonant metal rods that do not block the cross section of the waveguide [8, 9]. In the known publications devoted to this problem, the monopoles in coupling region are assumed to be perfectly conducting and arranged symmetrically relative to the waveguide walls. In this subsection, the coupling problem through the resonant slot cut in a common wall between a main infinite and side semi-infinite rectangular waveguides will be solved by the generalized method of induced EMMF. In the T-junction the asymmetric impedance vibrator is placed in the coupling area of the main waveguide, and the end wall of the side waveguide is characterized by impedance coating which, in the general case, may be a magnetodielectric. The energy characteristics of the E-plane T-junction operating in the single-mode regime of waveguides were applied to study using impedance coatings as non-mechanical control elements.

5.1.1 Problem Formulation and Solution

Consider a T-junction shown in Fig. 5.1 consisting of a main infinite hollow rectangular waveguide with perfectly conducting walls marked as $Wg1$ and a side semi-infinite rectangular waveguide with an impedance end wall marked as $Wg2$. Let us introduce rectangular coordinate systems $\{x, y, z\}$ and $\{x', y', z'\}$ related to the

© The Author(s), under exclusive license to Springer Nature Switzerland AG 2020 157
M. V. Nesterenko et al., *Combined Vibrator-Slot Structures: Theory and Applications*, Lecture Notes in Electrical Engineering 689,
https://doi.org/10.1007/978-3-030-60177-5_5

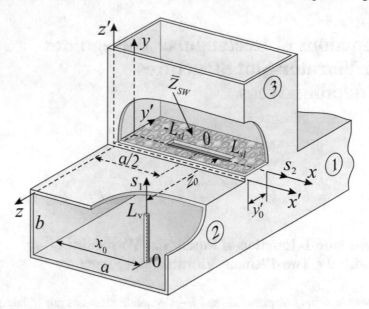

Fig. 5.1 The T-junction geometry

main and side waveguides. A H_{10}- wave propagates in the main waveguide from the region $z = -\infty$. A narrow transverse coupling slot cut in the broad wall of the main waveguide symmetrically its longitudinal axis radiates into the side waveguide. A thin asymmetric vibrator (monopole) with variable surface impedance is located in the cross-sectional plane of the main waveguide parallel to its narrow walls. The parameter of the T-junction are as follows: the cross-sections of both waveguides are $\{a \times b\}$, the thickness of the waveguide walls is h ($h/\lambda \ll 1$), the slot width and length are d and $2L_{sl}$, the vibrator radius and length are r and L_v, the vibrator displacement the in the waveguide cross sectional plane is x_0, the distance between the slot and broad wall of side waveguide is $y_0' = y_0$, and the distance between the vibrator axis and the slot is z_0.

The system of integral equations defining the electric current $J_v(s_1)$ on the vibrator and the equivalent magnetic current in the slot $J_{sl}(s_2)$ can be written based on the expression (2.19) as:

$$\left(\frac{d^2}{ds_1^2} + k^2\right) \int\limits_{-L_v}^{L_v} J_v(s_1') G_{s_1}^{Wg1}(s_1, s_1') ds_1'$$

$$- ik \int\limits_{-L_{sl}}^{L_{sl}} J_{sl}(s_2') \tilde{G}_{s_2}^{Wg1}(s_1, s_2') ds_2' = -i\omega \left[E_{0s_1}(s_1) - z_i(s_1) J_v(s_1) \right], \qquad (5.1a)$$

$$\left(\frac{d^2}{ds_2^2} + k^2\right) \int\limits_{-L_{sl}}^{L_{sl}} J_{sl}(s_2') \left[\begin{array}{c} G_{s_2}^{Wg1}(s_2, s_2') + \\ + G_{s_2}^{Wg2}(s_2, s_2') \end{array}\right] ds_2'$$

$$- ik \int\limits_{-L_v}^{L_v} J_v(s_1') \tilde{G}_{s_1}^{Wg1}(s_2, s_1') ds_1' = -i\omega H_{0s_2}(s_2), \tag{5.1b}$$

where s_1 and s_2 are the local coordinates associated with the vibrator and slot axis, $z_i(s_1)$ is internal linear impedance of the vibrator [Ohm/m], $E_{0s_1}(s_1)$ and $H_{0s_2}(s_2)$ are the projections of the extraneous sources fields on the vibrator and slot axes; $G_{s_1}^{Wg1}(s_1, s_1')$ and $G_{s_2}^{Wg1,2}(s_2, s_2')$ are the corresponding components of the electric and magnetic tensor Green's functions of the main and side waveguides (see Appendix A), L_v is the end coordinate of the mirror vibrator image relatively broad waveguide wall $(J_v(\pm L_v) = 0, J_{sl}(\pm L_{sl}) = 0)$,; $k = 2\pi/\lambda$, and the tensors $\tilde{G}_{s_1}^{Wg1}(s_2, s_1') = \frac{\partial}{\partial z} G_{s_1}^{Wg1}[x(s_2), 0, z; x'(s_1'), y'(s_1'), z_0]$ and $\tilde{G}_{s_2}^{Wg1}(s_1, s_2') = \frac{\partial}{\partial z} G_{s_2}^{Wg1}[x(s_1), y(s_1), z; x'(s_2'), 0, 0]$ are the Green's tensors obtained by substitution $z = 0$ in $\tilde{G}_{s_1}^{Wg1}$ and $z = z_0$ in $\tilde{G}_{s_2}^{Wg1}$ after differentiation.

As in the Sect. 4.1, the solution of the equation system (5.1a) will be obtained by the generalized method of induced EMMF, by using the expressions $J_v(s_1) = J_{0v} f_v(s_1)$ and $J_{sl}(s_2) = J_{0sl} f_{sl}(s_2)$ for the vibrator and slot currents in which J_{0v} and J_{0sl} unknown current amplitudes, $f_v(s_1)$ and $f_{sl}(s_2)$ are predefined function in the form (4.4):

$$f_v(s_1) = \cos \tilde{k} s_1 - \cos \tilde{k} L_v, \tag{5.2a}$$

$$f_{sl}(s_2) = \cos k s_2 - \cos k L_{sl}, \tag{5.2b}$$

where $\tilde{k} = k - \frac{i 2\pi z_i^{av}}{Z_0 \Omega}$, $z_i^{av} = \frac{1}{2L_v} \int_{-L_v}^{L_v} z_i(s_1) ds_1$ is the internal impedance, averaged over the monopole length, and $\Omega = 2ln(2L_v/r)$.

Let us multiply the Eqs. (5.1a) and (5.1b) by the functions $f_v(s_1)$ and $f_{sl}(s_2)$ and integrate the resulting equations along the vibrator and slot length. Thus, the system of linear algebraic equations can be obtained in the form

$$J_{0v}(Z_{11} + F_z) + J_{0sl} Z_{12} = -\frac{i\omega}{2k} E_1,$$
$$J_{0sl}(Z_{22}^{Wg1} + Z_{22}^{Wg2}) + J_{0v} Z_{21} = -\frac{i\omega}{2k} H_2, \tag{5.3}$$

where, as in the Sect. 4.1,

$$Z_{11} = \frac{4\pi}{ab} \sum_{m=1}^{\infty} \sum_{n=0}^{\infty} \left[\frac{\varepsilon_n (k^2 - k_y^2) \tilde{k}^2}{k k_z (\tilde{k}^2 - k_y^2)^2} e^{-k_z r} \sin^2 k_x x_0 \right.$$

$$\left. \left[\sin \tilde{k} L_v \cos k_y L_v - (\tilde{k}/k_y) \cos \tilde{k} L_v \sin k_y L_v \right]^2 \right],$$

$$Z_{12} = Z_{21} = \frac{4\pi}{ab} \sum_{m=1,3\dots}^{\infty} \sum_{n=0}^{\infty} \left[\frac{\varepsilon_n k \tilde{k} e^{-k_z z_0}}{i(k^2 - k_x^2)(\tilde{k}^2 - k_y^2)} \sin k_x x_0 \right.$$

$$\times \left[\sin \tilde{k} L_v \cos k_y L_v - (\tilde{k}/k_y) \cos \tilde{k} L_v \sin k_y L_v \right]$$

$$\times \left[\sin k L_{sl} \cos k_x L_{sl} - (k/k_x) \cos k L_{sl} \sin k_x L_{sl} \right],$$

$$Z_{22}^{Wg1}$$

$$= \frac{8\pi}{ab} \sum_{m=1,3\dots}^{\infty} \sum_{n=0}^{\infty} \left[\frac{\varepsilon_n k e^{-k_z(d_e/4)}}{k_z(k^2 - k_x^2)} \left[\sin k L_{sl} \cos k_x L_{sl} - (k/k_x) \cos k L_{sl} \sin k_x L_{sl} \right]^2 \right],$$

$$(5.4)$$

$$E_1 = 2H_0 \frac{k}{k_g \tilde{k}} \sin \frac{\pi}{a} x_0 e^{-ik_g z_0} f(\tilde{k} L_v), \ f(\tilde{k} L_v)$$

$$= \sin \tilde{k} L_v - \tilde{k} L_v \cos \tilde{k} L_v, \ H_2 = 2H_0(1/k) f(k L_{sl}),$$

$$f(k L_{sl}) = \frac{\sin k L_{sl} \cos \frac{\pi L_{sl}}{a} - \frac{ka}{\pi} \cos k L_{sl} \sin \frac{\pi L_{sl}}{a}}{1 - [\pi/(ka)]^2},$$

$$F_z = -\frac{i}{r} \int_0^{L_v} f_v^2(s_1) \bar{Z}_{SV}(s_1) ds_1,$$

and

$$Z_{22}^{Wg2} = \frac{16\pi}{ab} \sum_{m=1,3\dots}^{\infty} \sum_{n=0}^{\infty} \left[\frac{\varepsilon_n k F(\bar{Z}_{SW})}{k_z(k^2 - k_x^2)} \cos k_y y_0 \cos k_y \left(y_0 + \frac{d_e}{4} \right) \right.$$

$$\times \left[\sin k L_{sl} \cos k_x L_{sl} - (k/k_x) \cos k L_{sl} \sin k_x L_{sl} \right]^2],$$

$$Z_{22}^{\Sigma} = Z_{22}^{Wg1} + Z_{22}^{Wg2}, \qquad\qquad (5.5)$$

$$F(\bar{Z}_{SW}) = \frac{k k_z (1 + \bar{Z}_{SW}^2)}{(ik + k_z \bar{Z}_{SW})(k \bar{Z}_{SW} - ik_z)} \left(1 - i \frac{k k_z \bar{Z}_{SW}}{k^2 - k_x^2} \right),$$

$$\varepsilon_n = \begin{cases} 1, n = 0 \\ 2, n \neq 0 \end{cases}, k_x = \frac{m\pi}{a}, k_y = \frac{m\pi}{b}, \text{ m and n are integers, } k_z = \sqrt{k_x^2 + k_y^2 - k^2},$$

$\bar{Z}_{SV}(s_1) = \bar{R}_{SV} + i \bar{X}_{SV} \phi(s_1) = 2\pi r_i(s_1), \left| \bar{Z}_{SV}(s_1) \right|^2 \ll 1$ is complex surface impedance normalized to Z_0, $\bar{Z}_{SV}(s_1) = 2\pi r z_i(s_1)$, $\phi(s_1)$ is predefined function, $\bar{Z}_{SW}$ is the normalized surface impedance of the end wall of the side T-junction arm 3 ($|\bar{Z}_{SW}|^2 \ll 1$), d_e is the equivalent slot width taking into account the real waveguide wall thickness (see (3.59), (3.60)), H_0 is the H_{10}-wave amplitude. The solution of the equations system (5.3) has the form

$$J_{0v} = -\frac{i\omega}{2k} \frac{E_1 Z_{22}^{\Sigma} - H_2 Z_{12}}{(Z_{11} + F_z)Z_{22}^{\Sigma} - Z_{21}Z_{12}} = -\frac{i\omega}{2k}\tilde{J}_{0v},$$

$$J_{0sl} = -\frac{i\omega}{2k} \frac{H_2(Z_{11} + F_z) - E_1 Z_{21}}{(Z_{11} + F_z)Z_{22}^{\Sigma} - Z_{21}Z_{12}} = -\frac{i\omega}{2k}\tilde{J}_{0sl}. \tag{5.6}$$

The vibrator and slot currents can be obtained based on the formulas (5.2a) and (5.6) in the form:

$$J_v(s_1) = -\frac{i\omega}{2k}\tilde{J}_{0v}(\cos \tilde{k}s_1 - \cos \tilde{k}L_v),$$

$$J_{sl}(s_2) = -\frac{i\omega}{2k}\tilde{J}_{0sl}(\cos ks_2 - \cos kL_{sl}). \tag{5.7}$$

The field reflection and transmission coefficients, S_{11} and S_{12}, in the main waveguide can be determined by the expressions:

$$S_{11} = \frac{4\pi i}{abk}\left\{ \frac{2k_g}{k}\tilde{J}_{0sl} f(kL_{sl}) - \frac{k}{\tilde{k}}\tilde{J}_{0v} \sin\frac{\pi x_0}{a}e^{-ik_g z_0} f(\tilde{k}L_v)\right\}e^{2ik_g z}, \tag{5.8}$$

$$S_{12} = 1 + \frac{4\pi i}{abk}\left\{ \frac{2k_g}{k}\tilde{J}_{0sl} f(kL_{sl}) + \frac{k}{\tilde{k}}\tilde{J}_{0v} \sin\frac{\pi x_0}{a}e^{ik_g z_0} f(\tilde{k}L_v)\right\}. \tag{5.9}$$

The T-junction energy characteristics, the power reflection and transmission coefficients in the main waveguide, P_{11} and P_{12}, and the transmission coefficient, P_{13}, in the side waveguide can be found from the following relations:

$$P_{11} = |S_{11}|^2, \quad P_{12} = |S_{12}|^2, \quad P_{13} = 1 - P_{11} - P_{12}. \tag{5.10}$$

The power losses P_{ov} and $P_{\sigma W}$ in the vibrator and coating of the side waveguide end wall are automatically taken into account in the coefficients P_{11} and P_{12}. The quantitative losses assessment can be determined from the energy balance equations defined by the operating mode of the T-junction. These equations are as follows: $\sum_{q=1}^{2} P_{1q} + P_{ov} = 1$ at the alternative frequency and $\sum_{q=1}^{3} P_{1q} + P_{\sigma W} = 1$ at the main frequency. The coefficient P_{13} can be determined by the expression

$$P_{13} = \left| \frac{16\pi k_g f^2(kL_{sl})(1 + \bar{Z}_{SW}^2)}{abk^3 Z_{22}^{\Sigma}[1 + (k_g/k)\bar{Z}_{SW}]} \right|^2. \tag{5.11}$$

5.1.2   Numerical and Experimental Results

The wavelength energy characteristics $P_{1q} = f(\lambda)$ of the slot without the vibrator for the T-junction with the perfectly conducting waveguide end wall are presented in Fig. 5.2. The simulations were carried out with following T-junction parameters: the cross-sections of the waveguides are $\{58.0 \times 25.0\}$ mm^2, $2L_{sl} = 40$ mm, $d = 40$ mm, and $h = 0.5$ mm. The levels P_{11} corresponding to coefficient of voltage standing wave ratio (VSWR) equal to 2 are marked in this and following Figs. As can be seen, the powers of the transmitted waves are equal to each other at the slot resonant wavelength $\lambda_{sl}^{res} = 83.5$ mm and, in total, make up about 90% of the incident wave power, and only about 10% of power is reflected.

If the perfectly conducting monopole with parameters $L_v = 18.0$ mm, $r = 2.0$ mm and $\lambda_v^{res} = 83.5$ mm is placed in the coupling region of the T-junction (in the plane $\{x0y\}$, $z_0 = 0.0$ mm), the coefficient P_{11} increases, the coefficient P_{12} decreases, while the coefficient P_{13} stays constant (Fig. 5.3a).

If the displacement z_0 between the vibrator and slot longitudinal axes is the multiples of $\lambda_G^{Sres}/8$ ($\lambda_G^{Sres} = \dfrac{2\pi}{\sqrt{(2\pi/\lambda_{sl}^{res})^2-(\pi/a)^2}}$ is the slot resonant wavelength in waveguide, λ_{sl}^{res} is the slot resonant length in the free half-space), the coefficient P_{13} increases up to about 90%, while the coefficients P_{12} and P_{11} decrease at the slot and monopole resonant wavelengths is placed at $z_0 = \lambda_G^{Sres}/2 = 60.0$ mm (Fig. 5.3b).

The T-junction can also be used for power division between the T-junction arms in a predefined ratio at some wavelength from operating band within 60.0−77.0 mm with almost complete matching in the arm 1 if $z_0 = 3\lambda_G^{Sres}/8 = 45.0$ mm (Fig. 5.3c).

If the monopole is shifted in the plane of the waveguide cross section, the width of the transmission coefficient curve at the half power level from $\Delta\lambda \approx 7.5$ mm (Fig. 5.3b) to $\Delta\lambda \approx 12.5$ mm (Fig. 5.4). The simulation results have showed that variation of the slot position in the end wall of the side arm has practically no effect on the T-junction energy characteristics.

Fig. 5.2 The wavelength plots of the energy characteristics of the single slot

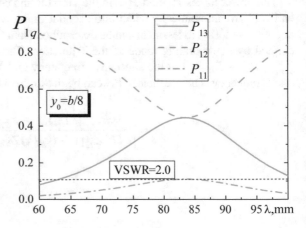

Fig. 5.3 The plots of the wavelength energy characteristics of the T-junction with parameters $L_v = 18.0\,\text{mm}$, $r = 2.0\,\text{mm}$, $\bar{Z}_{SV} = 0$; **a**—$z_0 = 0.0\,\text{mm}$, **b**— $z_0 = \lambda_G^{Sres}/2 = 60.0\,\text{mm}$, **c**—$z_0 = 3\lambda_G^{Sres}/8 = 45.0\,\text{mm}$

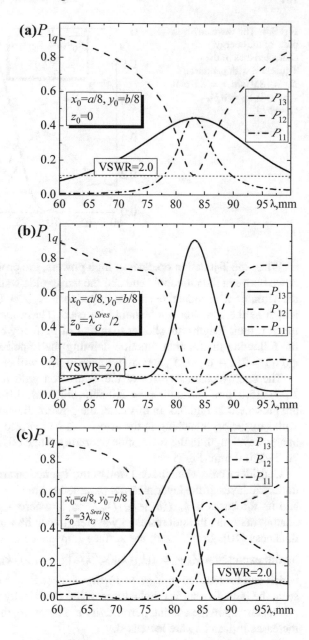

To transfer a predetermined amount of power at a specific wavelength to the arm 3, the monopoles with losses distributed impedance (resistance) can be used. In this case, the coefficients $P_{12} \approx 0$ and $P_{11} \approx 0$, and the losses in the vibrator material are nonzero (Fig. 5.5).

Fig. 5.4 The wavelength
plots of the energy
characteristics of the
T-junction with parameters:
$L_v = 18.0$ mm, $r = 2.0$ mm,
$\bar{Z}_{SV} = 0$, $x_0 = a/4$,
$y_0 = b/4$

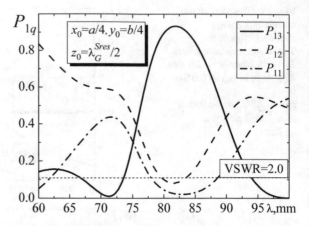

When the T-junction operates at high powers, the probability of electrical break-down between the monopole end and the waveguide wall increases [1]. In this case, the vibrator with inductive surface impedance $\bar{X}_{SV} > 0$ varying over the vibrator length as the function $\phi(s_1)$ should be used. These monopoles can be resonantly tuned if their lengths are shorter than that of perfectly conducting vibrator. Below, the following distribution function defining the impedance variation: $\phi_0(s_1) = 1$, $\phi_1(s_1) = 2[1 - (s_1/L_v)]$, and $\phi_2(s_1) = 2(s_1/L_v)$ will be used.

The energy characteristics of the T-junction with monopoles of equal lengths $L_v = 15.0$ mm and radii $r = 2.0$ mm, but with different resonant wavelengths are presented in Fig. 5.6. In this case, the powers division of the transmitted waves with satisfactory matching in the arm 1 can be realized in any predetermined ratio at any wavelength in the waveguide operating range 60.0−91.5 mm (Fig. 5.6b) and 76.5−110.0 mm (Fig. 5.6c).

Consider a case when the end wall of the T-junction arm 3 is coated by a magneto-dielectric layer. If the layer material parameters are $\varepsilon' = \varepsilon' - \varepsilon''$ and $\mu = \mu' - i\mu''$, and its thickness is h_d $((h + h_d)/\lambda_{\varepsilon\mu} \ll 1$, where $\lambda_{\varepsilon\mu}$ is the wavelength in the coating material, the slot resonant wavelength can be varied depending on the layer thickness [10]. In this case, the surface impedance of the end wall is determined by the expression $\bar{Z}_{SW} = i(k_d/k_W)\sqrt{\mu/\varepsilon}\,\mathrm{tg}\,(k_W h_d)$, $k_W = \sqrt{k_d^2 - (\pi/a)^2}$ (1.71), which can be reduced to $\bar{Z}_{SW} \approx ik\mu h_d$ if inequality $k_W h_d \ll 1$ holds. As can be seen, the inductive surface impedance of the electrically thin layer ($\bar{X}_{SW} > 0$), does not depend upon the coating permittivity ε. Therefore, the resonant wavelength λ_{sl}^{res} increases if μ and h_d are increased.

The wavelength plots of energy characteristics $P_{1q} = f(\lambda)$ for the T-junction with end wall coated by the magneto-dielectric with $\mu = 4.7$ and $P_{\sigma w} = 0$ [11] are presented in Fig. 5.7 for the various coating thickness. The resonant wavelength λ_{sl}^{res} and λ_G^{Sres} as function of the layer thickness h_d are summarized in the Table 5.1.

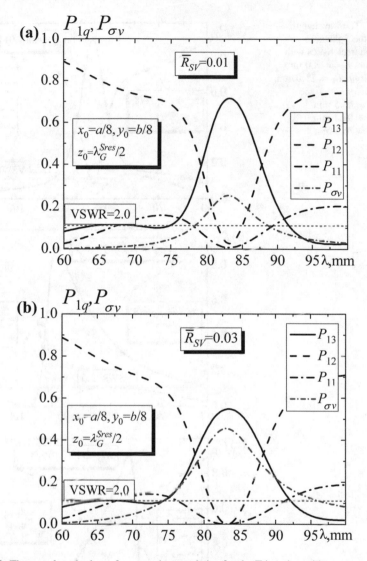

Fig. 5.5 The wavelength plots of energy characteristics for the T-junction with parameters: $L_v =$ 18.0 mm, $r = 2.0$ mm, $\bar{Z}_{SV} = \bar{R}_{SV}$: **a**—$\bar{R}_{SV} = 0.01$; **b**—$\bar{R}_{SV} = 0.03$

Thus, the energy characteristics of the T-junction can be varied over a wide wavelength range by varying the magnitude, distribution functions, and the distance z_0 even if geometrical dimensions of the vibrator and slot are constant (Fig. 5.7).

The simulation and experimental results obtained for the T-junctions in the following configurations: with the perfectly conducting monopole made of solid brass cylinder with radius $r = 2.0$ mm (Fig. 3.13a) and the impedance monopole, the corrugated brass cylinder with external radius $r = 2.0$ mm, internal radius r

Fig. 5.6 The wavelength plots of the T-junction energy characteristics with parameters $L_v = 18.0$ mm, $\bar{Z}_{SV} = \bar{R}_{SV}$, $L_v = 15.0$ mm, $\bar{Z}_{SV} = i\bar{X}_S$:
a—$\lambda_v^{res} = 83.5$ mm,
b—$\lambda_v^{res} = 91.5$ mm,
c—$\lambda_v^{res} = 76.5$ mm

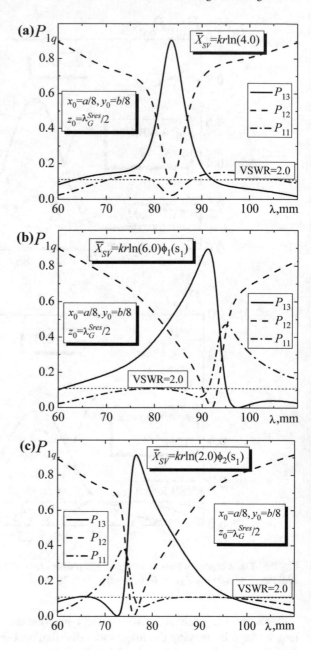

Fig. 5.7 The wavelength plots of the energy characteristics for the T-junction with parameters $L_v = 15.0$ mm, $\bar{Z}_{SV} = i\bar{X}_{SV}$, $\bar{Z}_{SW} = i\bar{X}_{SW}$

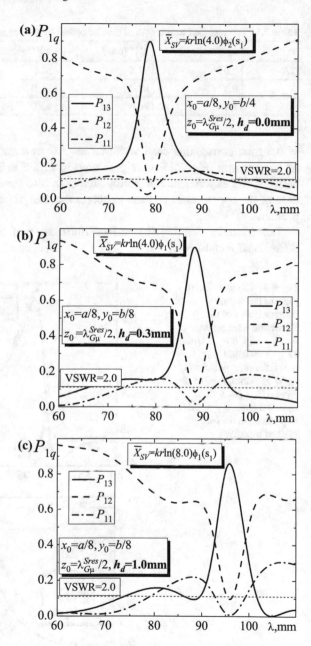

Table 5.1 The resonant wavelength of the slot coated by the magneto-dielectric with $\mu = 4.7$

#	h_d (mm)	λ_{sl}^{res} (mm)	λ_G^{Sres} (mm)	Shown in
1	0.0	78.5	54.0	Fig. 5.7a
2	0.3	88.0	68.0	Fig. 5.7b
3	1.0	96.0	85.0	Fig. 5.7c

$= 0.5$ mm, corrugation crest width $L_1 = 1.0$ mm and corrugation trough width $L_2 = 1.0$ mm (Fig. 3.13b) are presented in Fig. 5.8. The satisfactory agreement between the experimental and simulated data also confirms the physical correctness of the approximations used to obtain the analytical solution of the diffraction problem.

The formulas defining the surface impedance of thin vibrators are given in Sects. 1.3.2.6 and 1.3.2.7.

Fig. 5.8 The wavelength plots of the energy characteristics of the T-junction with parameters $a = 5.8$ mm, $b = 25.0$ mm, $2L_{sl} = 40.0$ mm, $h = 0.5$ mm, $d = 0.5$ mm, $L_v = 15.0$ mm, $r = 0.5$ mm, $x_0 = a/8$, $y_0 = b/2$, $z_0 = 0.5$ mm, $\bar{Z}_{SV} = i\bar{X}_S$. Experimental data are marked by circles

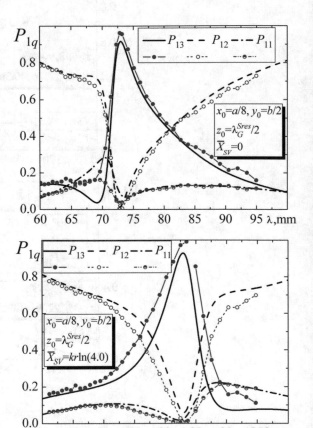

5.2   *E*-Plane T-Junction with Two Waveguides of Different Cross-Sections and Two-Element Vibrator Slot-Structure

5.2.1   *Features of the Problem Formulation*

The T-junction with waveguide of different cross-section is presented in Fig. 5.9. If the impedance distributions along the vibrator are represented by the contrast and cosine function, $\phi_0(s_1) = 1$ and $\phi_3(s_1) = \frac{\pi}{2}\cos\frac{\pi s_1}{2L_v}$ In this case, the following function can be obtained base on the relation (5.4):

$$F_z^{\phi_0} = -\frac{i L_v (\bar{R}_{SV} + i\bar{X}_{SV})}{2r}\left[(2 + \cos 2\tilde{k}L_v) - 3\frac{\sin 2\tilde{k}L_v}{2\tilde{k}L_v}\right] = F_z^c(\bar{R}_{SV} + i\bar{X}_{SV})\Phi_z^c$$

(5.12)

for the constant distribution function and

$$F_z^{\phi_3} = F_z^c\left\{\bar{R}_{SV}\Phi_z^c + i\bar{X}_{SV}\left[\frac{\pi^2\cos 2\tilde{k}L_v}{\pi^2 - (4\tilde{k}L_v)^2} - 2\cos\tilde{k}L_v\frac{\pi^2 + (2\tilde{k}L_v)^2}{\pi^2 - (2\tilde{k}L_v)^2} + 1\right]\right\}$$

(5.13)

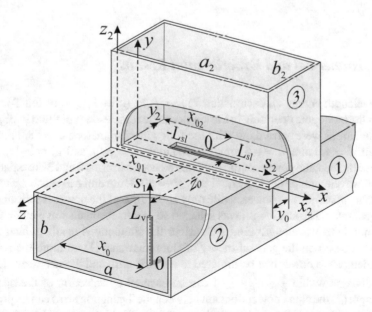

Fig. 5.9 The T-junction of two waveguides with different cross-sections and the two-element vibrator-slot structure

for the cosine distribution function.

The number of propagating wave types in the coupling waveguide section ($Wg2$) depends on the size of its cross-section $\{a_2 \times b_2\}$. Therefore, the general scattering matrix of the T-junction and the power reflection and transmission coefficients in the main waveguide, $P_{11} = |S_{11}|^2$, $P_{12} = |S_{12}|^2$ (coefficients $|S_{11}|$ and $|S_{12}|$ are determined by expressions (5.8) and (5.9)) contain a number of the power transmission coefficients P_{13}^{mn} of propagating waves in the coupled waveguide. The coefficients P_{13}^{mn} can be found by the formula $P_{13}^{mn} = P_{13} \frac{P_{sl}^{mn}}{P_{\Sigma sl}^{mn}}$, where $P_{13} = 1 - P_{11} - P_{12} - P_{\sigma v}$, $\frac{P_{sl}^{mn}}{P_{\Sigma sl}^{mn}}$ is the ratio of the partial wave power to the total power of the propagating waves in the arm $Wg2$ normalized to the power of H_{10}-wave propagating from the generator, $P_{\sigma v}$ is the power loss in the monopole, determined from the energy balance equation $P_{11v} + P_{12v} + P_{\sigma v} = 1$ valid if the slot is metalized.

If the side arm of the T-junction is placed asymmetric relative to the main waveguide under conditions $y_0 = b_2/2$, the expressions for propagating waves of the magnetic type in the side arm

$$P_{sl}^{m0} = \frac{a_2 b_2 k_g}{ab k_g^{m0}} \left| \frac{16\pi k_g^{m0} \sin \frac{\pi x_{01}}{a} f(kL_{sl}) \sin \frac{m\pi x_{02}}{a_2} f^{m0}(kL_{sl})}{a_2 b_2 k^3 Z_{22}^{\Sigma}} \right|^2, \qquad (5.14)$$

where $k_g^{m0} = \sqrt{k^2 - \left(\frac{m\pi}{a}\right)^2}$, $f^{m0}(kL_{sl}) = \frac{\sin kL_{sl} \cos \frac{m\pi L_{sl}}{a_2} - \frac{ka_2}{m\pi} \cos kL_{sl} \sin \frac{m\pi L_{sl}}{a_2}}{1 - [m\pi/(ka_2)]^2}$, $P_{\Sigma sl}^{m0} = \sum_{m=1}^{M} P_{sl}^{m0}$.

5.2.2   Numerical and Experimental Results

The wavelength energy characteristics $P_{1q} = f(\lambda)$ ($q = 1, 2, 3$) of the T-junction without the monopole operating in the single-mode regime are presented in Fig. 5.10. The simulations were carried out with the following parameters: $a = 58.0$ mm, $b = 25.0$ mm, $d = 4.0$ mm, $h = 0.5$ mm, $r = 2.0$ mm, $x_0 = a/8$, and $y_0 = b_2/2$.

As can be seen, the powers of the transmitted waves in arms 2 and 3 are equal at the resonant wavelength of the slot equal to $\lambda_{sl}^{res} \approx 80.0$ mm and can reach up to about 90% of the incident wave power, while only about 10% of the power is reflected back to the generator. The slot length was selected so that its resonant wavelength belong to the middle of wavelength range. Consider the simulations results which can be interpreted based on the general theory of slot radiations: (1) no more than 50% of the incident wave power can be directed to the side arm, and the necessary level of power division within $\frac{1}{6} \leq \frac{P_{13}^{10}}{P_{12}} \leq 1$ can be ensured by selection of the operating wavelength (2) the equal power division between the T-junction arms can be provided only at the resonant slot wavelength that guarantee the highest possible level of the reflection coefficient P_{11} in the main waveguide; (3) the resonant waveguide coupling determines the relative (narrow band operation as compared with bridge-type junction

Fig. 5.10 The wavelength plots of the energy characteristics of the single slot: $b_2 = b$, $x_{01} = x_{02} = a/2$

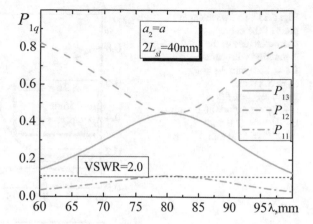

in any mode of power division between the arms. The first two restrictions can be overcome by placing the impedance monopole into the main waveguide in the slot region.

If the monopole with the constant impedance is placed at the distance z_0 from the longitudinal slot axis equal to $\lambda_G^{Sres}/2$, the wave power transmitted to the arm 3 increases up to 90%, while the coefficients P_{12} and P_{11} are simultaneously decreased at the resonant wavelength of the slot and monopole $\lambda_{sl}^{res} = \lambda_v^{res} \approx 80.0$ mm (Fig. 5.11a). The power of the transmitted waves can be also divided in a predetermined ratio at any wavelength in the operating range within 60.0–80.0 mm with satisfactory matching in the arm 1.

In Figs. 5.11b, c presents the energy characteristics of monopoles having the same length $L_v = 15.0$ mm, like the monopole in Fig. 5.11a, but different resonant wavelengths: b – $\lambda_v^{res} \approx 72.0$ mm (perfectly conducting monopole), c – $\lambda_v^{res} \approx 88.0$ (monopole with variable impedance). In these cases, it is possible to realize the division of the powers of the transmitted waves in a given ratio at a certain wavelength for the entire operating range of the waveguide (within 73.0 -100.0 mm, Fig. 5.11b, and 60.0–85.0 mm, Fig. 5.11c) with agreement satisfactory for practice in shoulder 1 VSWR ~ 2.0). The simulation results for the T-junction with variable surface impedance were compared with experimental data obtained by using a standard meter with monopole presented in Fig. 3.13c are shown in Fig. 5.12. The satisfactory agreement between the experimental and simulated data confirms the physical correctness of the approximations used to obtain the analytical solution of the diffraction problem.

Since the most of the incident power can be transferred from the main waveguide to the side arm, additional modal channels for power splitting in the side arm under condition of the multimode operations can be realized. The modes of equal division of incident power between the modes in the T-junction side arm and between all physical channels of power division are of particular interest for practical applications. The simulations have showed that such operational modes can be realized. If the broad wall of T-junction $x_{01} = a/2$, $x_{02} = a/4$ is increased two times, the two types of

Fig. 5.11 The wavelength plots of the energy characteristics of the T-junction with parameters $L_v = 15.0$ mm, $b_2 = b$, $z_0 = 5.8$ mm, $x_{01} = x_{02} = a/2$:
a—$\lambda_{sl}^{res} = \lambda_v^{res} \approx 80.0$ mm;
b—$\lambda_v^{res} \approx 72.0$ mm;
c—$\lambda_v^{res} \approx 88.0$ mm

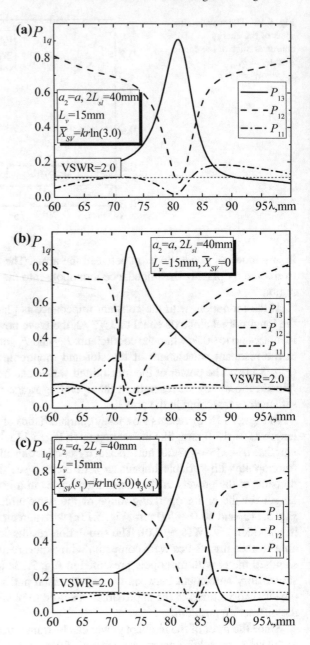

Fig. 5.12 The wavelength plots of the energy characteristics of the T-junction with parameters $b_2 = b$, $z_0 = 54$ mm, $x_{01} = x_{02} = a/2$. Experimental data are marked by circles

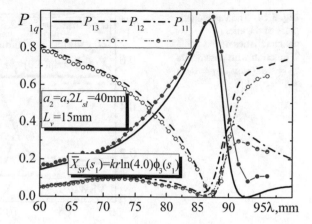

waves H_{10} and H_{20} are effectively excited (Fig. 5.13). In this case, the amplitude of the H_{30}-wave is negligible.

Fig. 5.13 The wavelength plots of the energy characteristics of the T-junction with parameters $b_2 = b$, $x_{01} = a/2$, $x_{02} = a_2/4$: **a**—$z_0 = 68.0$ mm; **b**—$z_0 = 85.0$ mm

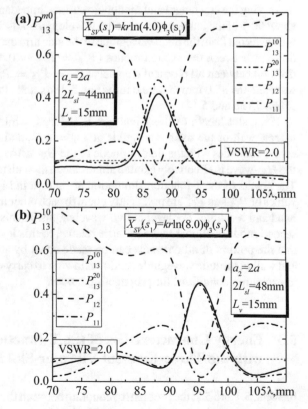

Fig. 5.14 The wavelength plots of the energy characteristics of the T-junction with parameters $z_0 = 46.0$ mm, $x_{01} = a/2$, $x_{02} = a_2/4$

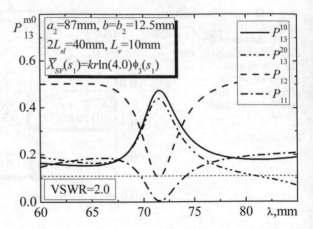

As can be seen, the approximately equal power division, $P_{13}^{10} \approx P_{13}^{20}$, between the transmitted waves H_{10} and H_{20} in the wavelength intervals $\lambda \in [85.0; 92.5]$ mm (Fig. 5.13a) and $\lambda \in [93.0; 98.5]$ mm (Fig. 5.13b) can be ensured with VSWR ≤ 2.0 by varying the slot length and monopole surface impedance. With optimal matching, about 90% of the incident power at the wavelengths $\lambda \approx 87.5$ mm and $\lambda \approx 95.0$ mm is transferred from the main waveguide to the side arm under condition of equal power division between the excited modes ($P_{13}^{10} = P_{13}^{20} \approx 0.45$). The mode equal power division between all physical channels ($P_{12} = P_{13}^{10} = P_{13}^{20} \approx 0.3$) can be realized at wavelengths 85.0 mm and 90.0 mm (Fig. 5.13a), and also at 92.5 mm and 97.5 mm (Figs. 5.13b and 5.14).

If the slot length of the T-junction is reduced, while its center is shifted to the narrow wall of the main waveguide at $x_{01} = a/3$, and hence $x_{02} = a_2/6$, the slot center coincides with the field antinode of the H_{30}-wave. In this case, in the side arm, the H_{30}-wave is effectively excited simultaneously with the H_{10}-wave and H_{20}-wave. The results of the T-junction simulation are shown in Fig. 5.15. As can be seen, the values of P_{13}^{10} and P_{13}^{30} are practically equal to each other in the middle of the operating band and reach the P_{12} level at two wavelengths. In this case, the power P_{13}^{20} of the second mode is almost twice as large relative to this level. However, one may state that the powers of all channels can be made equal by shifting the slot center in the end wall of the side waveguide section allowing to carry out the necessary correction to excitation conditions for propagating modes.

5.3 Energy Characteristics of the T-Junction with the Three-Element Vibrator-Slot Structure

In Sects. 5.1 and 5.2, the problem of coupling between the main infinite and side semi-infinite rectangular waveguides through the resonant slot was investigated. In the

Fig. 5.15 The wavelength plots of the energy characteristics of the T-junction with parameters $b_2 = b$, $z_0 = 46$ mm, $x_{01} = a/3$, $x_{02} = a_2/6$

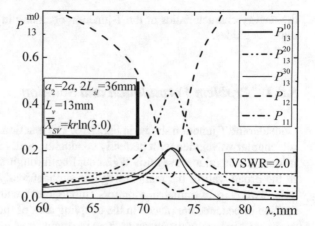

main waveguide the asymmetric impedance vibrator was located in the coupling area of the main waveguide, and the end wall of the side waveguide was characterized by the impedance coating which, in general case, can be coated by a magnetodielectric. The simulation results have reviled that characteristic of the E-plane T-junctions can be controlled by the impedance coatings. The slot resonant length can be varied by the impedance coating of the end wall of the side waveguide, while the power division between the output arms of the T-junction in a given proportion can be realized by using the impedance monopole. It should be emphasized that the proposed T-junction allows to redirect almost all of the input power of the main waveguide to the side waveguide.

However, all aspects of the possible application of the vibrator-slot coupling in the T-junction with the two vibrators in the main waveguide have not been studied [12–16]. Thus, the optimal variation range of the end wall impedance coating as function of the slot resonant length has not been defined. As it was estimated earlier, the surface impedance of a natural magnetodielectric layer with material parameters $\varepsilon = \varepsilon' - i\varepsilon''$, $\mu = \mu' - i\mu''$ and thickness h_d is determined by the expression $\bar{Z}_{SW} = i\frac{k_d}{k_W}\sqrt{\frac{\mu}{\varepsilon}}\mathrm{tg}(k_W h_d)$, $k_d = k\sqrt{\varepsilon\mu}$ and $k_W = \sqrt{k_d^2 - (\pi/a)^2}$ which, under condition $|k_W h_d| \ll 1$ is reduced to $\bar{Z}_{SW} \approx ik\mu h_d$. As can be seen, the surface impedance of electrically thin magnetodielectric layer is inductive and does not depend on the material permittivity. As a result, the slot resonant length can be increased if the permittivity μ or layer thickness h_d is increased (Fig. 5.7). On the other hand, artificial metamaterials are currently known [17–19], which are characterized by negative real parts of material parameters ($\varepsilon' < 0$ and/or $\mu' < 0$) which can provide the capacitive impedance of the coating layer.

In this subsection, a mathematical model defining the diffraction fields of the E-plane T-junction of two equal-sized rectangular waveguides with the three-element vibrator-slot coupling structure, characterized by the variable surface impedance along the monopole axes and the constant impedance of the magnetodielectric layer distributed at the end wall of the side waveguide. The model will be used to define the energy characteristics of the T-junction operating in the single-mode regime of the waveguides.

5.3.1 Problem Formulation and Solution

Consider the T-junction shown in Fig. 5.16. The junction consists of a hollow infinite rectangular waveguide with perfectly conducting walls marked as *Wg1* and a semi-infinite rectangular waveguide *Wg2* coupling through a narrow transverse slot cut in the broad wall of the main waveguide symmetrically relative to its longitudinal axis. Two thin asymmetric vibrators (monopoles) with variable along their axes surface impedance are placed in the coupling area of the T-junction. The H_{10}-wave propagates in the main waveguide from the region $z = -\infty$. The cross-section of the both waveguides is $\{a \times b\}$, vibrators radii and lengths are $r_{1(2)}$ and $L_{1(2)}$, slot width and length are d and $2L_3$. The vibrators and slot satisfy the thin wire and narrow slot approximations $r_{1(2)}/L_{1(2)} \ll 1$, $r_{1(2)}/\lambda \ll 1$, $d/(2L_3) \ll 1$, and $d/\lambda \ll 1$. The vibrators are displaced in the waveguide cross-section plane at the distance $x_{01(02)}$ and the slot is shifted in the end wall of the side waveguide at the distance $y_0' = y_0$. The distance between the first vibrator and slot axes is equal z_0. The second vibrator is located directly below the slot axis, and it is decoupled from the slot by polarizations (Fig. 5.16).

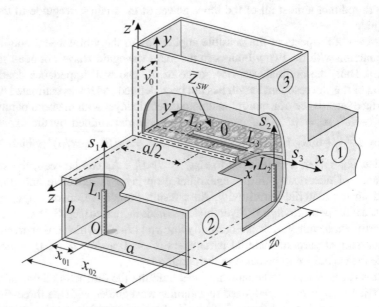

Fig. 5.16 The T-junction with the three-element vibrator-slot structure

The system of integral equations relative to electric currents on the vibrators $J_{1(2)}(s_{1(2)})$ and the equivalent magnetic current in the slot $J_3(s_3)$ can be written based on (2.19) as

$$\left(\frac{d^2}{ds_1^2} + k^2\right)\left\{\int_{-L_1}^{L_1} J_1(s_1')G_{s_1}^{Wg1}(s_1,s_1')ds_1' + \int_{-L_2}^{L_2} J_2(s_2')G_{s_2}^{Wg1}(s_1,s_2')ds_2'\right\}$$

$$-ik\int_{-L_3}^{L_3} J_3(s_3')\tilde{G}_{s_3}^{Wg1}(s_1,s_3')ds_3' = -i\omega\left[E_{0s_1}(s_1) - z_{i1}(s_1)J_1(s_1)\right], \tag{5.15a}$$

$$\left(\frac{d^2}{ds_2^2} + k^2\right)\left\{\int_{-L_2}^{L_2} J_2(s_2')G_{s_2}^{Wg1}(s_2,s_2')ds_2' + \int_{-L_1}^{L_1} J_1(s_1')G_{s_1}^{Wg1}(s_2,s_1')ds_1'\right\}$$

$$= -i\omega\left[E_{0s_2}(s_2) - z_{i2}(s_2)J_2(s_2)\right], \tag{5.15б}$$

$$\left(\frac{d^2}{ds_3^2} + k_1^2\right)\int_{-L_{sl}}^{L_{sl}} J_3(s_3')\left[G_{s_3}^{Wg1}(s_3,s_3') + G_{s_3}^{Wg2}(s_3,s_3')\right]ds_3'$$

$$-ik\int_{-L_1}^{L_1} J_1(s_1')\tilde{G}_{s_1}^{Wg1}(s_3,s_1')ds_1' = -i\omega H_{0s_3}(s_3). \tag{5.15в}$$

In the equation system (5.15a), $s_{1(2)}$ and s_3 are the local coordinates associated with the vibrator and slot axes, $z_{i1(i2)}(s_{1(2)})$ are internal linear impedance of vibrators ([Ohm / m]), $E_{0s_{1(2)}}(s_{1(2)})$ and $H_{0s_3}(s_3)$ are projections of the extraneous sources fields on the vibrator and slot axis, $G_{s_1}^{Wg1}(s_1,s_1')$, $G_{s_2}^{Wg1}(s_2,s_2')$, $G_{s_2}^{Wg1}(s_1,s_2')$, $G_{s_1}^{Wg1}(s_2,s_1')$ and $G_{s_3}^{Wg1,2}(s_3,s_3')$ are the components of the electric and magnetic tensor Green's functions of the infinite rectangular waveguide and the semi-infinite waveguide whose end wall is characterized by the normalized impedance $\bar{Z}_{SW}$, $-L_{1(2)}$ are the coordinates of the mirror image of the vibrators ends relative to the waveguide broad wall, $J_{1(2)}(\pm L_{1(2)}) = 0$, $J_3(\pm L_3) = 0$, $\tilde{G}_{s_1}^{Wg1}(s_3,s_1') = \frac{\partial}{\partial z}G_{s_1}^{Wg1}[x(s_3),0,z;x'(s_1'),y'(s_1'),z_0]$ and $\tilde{G}_{s_3}^{Wg1}(s_1,s_3') = \frac{\partial}{\partial z}G_{s_3}^{Wg1}[x(s_1),y(s_1),z;x'(s_3'),0,0]$ are the Green's tensors obtained by substitution $z = 0$ in $\tilde{G}_{s_1}^{Wg1}$ and $z = z_0$ in $\tilde{G}_{s_3}^{Wg1}$ after differentiation.

The solution to the equation system (5.15a), can be obtained by the generalized method of induced EMMF using approximating expressions for the currents in the vibrators $J_{1(2)}(s_{1(2)}) = J_{1(2)}^0 f_{1(2)}(s_{1(2)})$ and slot $J_3(s_3) = J_3^0 f_3(s_3)$, where $J_{1(2)3}^0$ are unknown current amplitudes. The functions $f_{1(2)}(s_{1(2)})$ and $J_3^0 f_3(s_3)$ can be written as

$$f_{1(2)}(s_{1(2)}) = \cos\tilde{k}_{1(2)}s_{1(2)} - \cos\tilde{k}_{1(2)}L_{1(2)}, \tag{5.16a}$$

$$f_3(s_3) = \cos ks_3 - \cos kL_3, \tag{5.16b}$$

where $\tilde{k}_{1(2)} = k - \frac{i2\pi z_{i1(i2)}^{av}}{Z_0 \Omega_{1(2)}}$, $\Omega_{1(2)} = 2\ln(2L_{1(2)}/r_{1(2)})$, $z_{i1(i2)}^{av} = \frac{1}{2L_{1(2)}} \int_{-L_{1(2)}}^{L_{1(2)}} z_{i1(i2)}(s_{1(2)})ds_{1(2)}$ - are the internal impedances averaged over the vibrator lengths. The system of integral Eqs. (5.15a) can be reduce to a system of linear algebraic equations for unknown current amplitudes by using the standard procedure:

$$J_1^0 Z_{11}^{\Sigma} + J_2^0 Z_{12} + J_3^0 Z_{13} = -\frac{i\omega}{2k} \int_{-L_1}^{L_1} f_1(s_1) E_{0s_1}(s_1)ds_1,$$

$$J_2^0 Z_{22}^{\Sigma} + J_1^0 Z_{21} = -\frac{i\omega}{2k} \int_{-L_2}^{L_2} f_2(s_2) E_{0s_2}(s_2)ds_2,$$ (5.17)

$$J_3^0 Z_{33}^{\Sigma} + J_1^0 Z_{31} = -\frac{i\omega}{2k} \int_{-L_{sl}}^{L_{sl}} f_3(s_3) H_{0s_3}(s_3)ds_3,$$

where the matrix coefficients are equal to:

$$Z_{11(22)} = \frac{4\pi}{ab} \sum_{m=1}^{\infty} \sum_{n=0}^{\infty} \left[\frac{\varepsilon_n (k^2 - k_y^2)\tilde{k}_{1(2)}^2}{kk_z(\tilde{k}_{1(2)}^2 - k_y^2)^2} e^{-k_z r_{1(2)}} \sin^2 k_x x_{01(2)} \right.$$
$$\left. \times \left[\sin \tilde{k}_{1(2)} L_{1(2)} \cos k_y L_{1(2)} - (\tilde{k}_{1(2)}/k_y) \cos \tilde{k} L_{1(2)} \sin k_y L_{1(2)} \right]^2 \right],$$

$$F_{z1(2)} = -\frac{i}{r_{1(2)}} \int_0^{L_{1(2)}} f_{1(2)}^2(s_{1(2)}) \bar{Z}_{SV1(V2)}(s_{1(2)})ds_{1(2)},$$

$$Z_{12(21)} = \frac{4\pi}{ab} \sum_{m=1}^{\infty} \sum_{n=0}^{\infty} \frac{\varepsilon_n (k^2 - k_y^2)\tilde{k}_1 \tilde{k}_2 e^{-k_z r_{2(1)}}}{kk_z(\tilde{k}_1^2 - k_y^2)(\tilde{k}_2^2 - k_y^2)} \sin k_x x_{01} \sin k_x x_{02}$$
$$\times [\sin \tilde{k}_1 L_1 \cos k_y L_1 - (\tilde{k}_1/k_y) \cos \tilde{k}_1 L_1 \sin k_y L_1]$$
$$\times [\sin \tilde{k}_2 L_2 \cos k_y L_2 - (\tilde{k}_2/k_y) \cos \tilde{k}_2 L_2 \sin k_y L_2],$$

$$Z_{13} = Z_{31} = \frac{4\pi}{ab} \sum_{m=1,3...}^{\infty} \sum_{n=0}^{\infty} \left[\frac{\varepsilon_n k \tilde{k}_1 e^{-k_z z_0}}{i(k^2 - k_x^2)(\tilde{k}_1^2 - k_y^2)} \right.$$
$$\times \sin k_x x_{01} \left[\sin \tilde{k} L_1 \cos k_y L_1 - (\tilde{k}_1/k_y) \cos \tilde{k} L_1 \sin k_y L_1 \right]$$
$$\left. \times [\sin k L_3 \cos k_x L_3 - (k/k_x) \cos k L_3 \sin k_x L_3]],$$

$$Z_{33}^{Wg1} = \frac{8\pi}{ab} \sum_{m=1,3..}^{\infty} \sum_{n=0}^{\infty} \left[\frac{\varepsilon_n k e^{-k_z(d_e/4)}}{k_z(k^2 - k_x^2)} \right.$$
$$\left. \times [\sin k L_3 \cos k_x L_3 - (k/k_x) \cos k L_3 \sin k_x L_3]^2 \right]$$

$$Z_{33}^{Wg2} = \frac{16\pi}{ab} \sum_{m=1,3...}^{\infty} \sum_{n=0}^{\infty} \left[\frac{\varepsilon_n k F(\bar{Z}_{SW})}{k_z(k^2 - k_x^2)} \cos k_y y_0 \cos k_y \left(y_0 + \frac{d_e}{4} \right) \right.$$

$$\times \left. [\sin kL_3 \cos k_x L_3 - (k/k_x) \cos kL_3 \sin k_x L_3]^2 \right],$$

$$F(\bar{Z}_{SW}) = \frac{kk_z(1 + \bar{Z}_{SW}^2)}{(ik + k_z \bar{Z}_{SW})(k\bar{Z}_{SW} - ik_z)} \left(1 - i\frac{kk_z \bar{Z}_{SW}}{k^2 - k_x^2} \right),$$

$$Z_{11(22)}^{\Sigma} = Z_{11(22)} + F_{z1(2)}, \quad Z_{33}^{\Sigma} = Z_{33}^{Wg1} + Z_{33}^{Wg2},$$

where $\varepsilon_n = \begin{cases} 1, & n = 0 \\ 2, & n \neq 0 \end{cases}$; $k_x = \frac{m\pi}{a}$; $k_y = \frac{n\pi}{b}$; m and n are integers; $k_z = \sqrt{k_x^2 + k_y^2 - k^2}$; $d_e = d exp(-\pi h/2d)$ is the equivalent slot width.

When the unknown current amplitudes are obtained as solution of the equation system (5.17), the field reflection and transmission coefficients in the main waveguide, S_{11} and S_{12}, can be found by using the known electrodynamic relations as:

$$S_{11} = \frac{4\pi i}{abkk_g} \left\{ J_3 \frac{2k_g^2}{k^2} f(kL_3) - J_1 \frac{k_g}{\tilde{k}_1} \sin\left(\frac{\pi x_{01}}{a}\right) f(\tilde{k}_1 L_1) e^{-ik_g z_0} \right.$$
$$\left. - J_2 \frac{k_g}{\tilde{k}_2} \sin\left(\frac{\pi x_{02}}{a}\right) f(\tilde{k}_2 L_2) \right\} e^{2ik_g z}, \tag{5.18a}$$

$$S_{12} = 1 + \frac{4\pi i}{abkk_g} \left\{ J_3 \frac{2k_g^2}{k^2} f(kL_3) \right.$$

$$\left. + J_1 \frac{k_g}{\tilde{k}_1} \sin\left(\frac{\pi x_{01}}{a}\right) f(\tilde{k}_1 L_1) e^{ik_g z_0} + J_2 \frac{k_g}{\tilde{k}_2} \sin\left(\frac{\pi x_{02}}{a}\right) f(\tilde{k}_2 L_2) \right\} \tag{5.18b}$$

where $k_g = \sqrt{k^2 - (\pi/a)^2}$,

$$J_1 = \tilde{J}_0 \left[\frac{k^2}{k_g \tilde{k}_1} \sin\frac{\pi x_{01}}{a} f_1(\tilde{k}_1 L_1) e^{-ik_g z_0} Z_{22}^{\Sigma} Z_{33}^{\Sigma} \right.$$

$$\left. - \frac{k^2}{k_g \tilde{k}_2} \sin\frac{\pi x_{02}}{a} f_2(\tilde{k}_2 L_2) Z_{12} Z_{33}^{\Sigma} - f_3(kL_3) Z_{13} Z_{22}^{\Sigma} \right],$$

$$J_2 = \tilde{J}_0 \left[\begin{array}{l} \frac{k^2}{k_g \tilde{k}_2} \sin\frac{\pi x_{02}}{a} f_2(\tilde{k}_2 L_2)(Z_{11}^{\Sigma} Z_{33}^{\Sigma} - Z_{31} Z_{13}) \\ \\ - \frac{k^2}{k_g \tilde{k}_1} \sin\frac{\pi x_{01}}{a} f_1(\tilde{k}_1 L_1) e^{-ik_g z_0} Z_{21} Z_{33}^{\Sigma} + f_3(kL_3) Z_{13} Z_{21} \end{array} \right],$$

$$J_3 = \tilde{J}_0 \left[\begin{array}{c} f_3(kL_3)(Z^{\Sigma}_{11}Z^{\Sigma}_{22} - Z_{21}Z_{12}) + \dfrac{k^2}{k_g\tilde{k}_2}\sin\dfrac{\pi x_{02}}{a}f_2(\tilde{k}_2 L_2)Z_{12}Z_{31} \\[3mm] - \dfrac{k^2}{k_g\tilde{k}_1}\sin\dfrac{\pi x_{01}}{a}f_1(\tilde{k}_1 L_1)e^{-ik_g z_0}Z_{31}Z^{\Sigma}_{22} \end{array} \right],$$

$$\tilde{J}_0 = 1/\left(Z^{\Sigma}_{11}Z^{\Sigma}_{22}Z^{\Sigma}_{33} - Z_{21}Z_{12}Z^{\Sigma}_{33} - Z_{31}Z_{13}Z^{\Sigma}_{22} \right),$$

$$f_{1(2)}(\tilde{k}_{1(2)}L_{1(2)}) = \sin\tilde{k}_{1(2)}L_{1(2)} - \tilde{k}_{1(2)}L_{1(2)}\cos\tilde{k}_{1(2)}L_{1(2)},$$

$$f_3(kL_3) = \frac{\sin kL_3 \cos(\pi L_3/a) - (ka/\pi)\cos kL_3 \sin(\pi L_3/a)}{1 - [\pi/(ka)]^2}.$$

The T-junction energy characteristics, the power reflection and transmission coefficients in the main waveguide and the transmission coefficient to the side arm, P_{11}, P_{12} and P_{13} can be found as in Sect. 5.1:

$$P_{11} = |S_{11}|^2, \qquad P_{12} = |S_{12}|^2, \qquad P_{13} = 1 - P_{11} - P_{12}. \tag{5.19}$$

The power losses in the vibrators $P_{\sigma V1(V2)}$ and in the end wall coating, $P_{\sigma W}$, are automatically taken into account in the coefficients P_{11} and P_{12}. The losses can be determined by using the energy balance equations $\sum_{q=1}^{2} P_{1q} + P_{\sigma V1} + P_{\sigma V2} = 1$ for the two auxiliary problems ($q = 1, 2, 3$): 1) for the metallized slot $\sum_{q=1}^{2} P_{1q} + P_{\sigma V1} + P_{\sigma V2} = 1$; 2) for the junction without the vibrators $\sum_{q=1}^{3} P_{1q} + P_{\sigma W} = 1$ with the coefficient P_{13} determined by the expression $P_{13} = \left| \dfrac{16\pi k f_3^2(kL)(1+\bar{Z}^2_{SW})}{abk^3 Z^{\Sigma}_{33}[1+(k_g/\bar{Z}_{SW})]} \right|^2$

5.3.2 Numerical and Experimental Results

At the first stage of simulation, the mathematical model was verified by comparing the simulated and experimental results for the T-junction with one impedance vibrator. The results obtained at this stage coincide with that presented in the Sect. 5.1. At the second stage, simulated and experimental plots of the energy parameters for the T-junction with the three-element vibrator-slot structure and the perfectly conducting end of the side arm wall ($\bar{Z}_{SW} = 0$) are presented in Fig. 5.17. In what follows, $\bar{Z}_{SV1(V2)}(s_{1(2)}) = \bar{R}_{SV1(V2)} + i\bar{X}_{SV1(V2)}\phi(s_{1(2)})$ is the vibrators complex distributed surface impedance, $\bar{Z}_{SV1(V2)}(s_{1(2)}) = 2\pi r_{1(2)}z_{i1(i2)}\phi(s_{1(2)})$, $\phi(s_{1(2)})$ is the predefined function, $|\bar{Z}_{SV1(V2)}(s_{1(2)})|^2 \ll 1$, $\bar{Z}_{SW} = \bar{R}_{SW} + i\bar{X}_{SW}$ is normalized surface impedance of the side waveguide end wall, $|\bar{Z}_{SW}|^2 \ll 1$. The results for T-junction in which the first the vibrator impedance is $\bar{X}_{SV1} = kr_1 \ln(4.0)$ (Fig. 3.13b), while the second vibrator is perfectly conductive ($\bar{X}_{SV2} = 0$, Fig. 3.13a) are presented in

Fig. 5.17 Simulation and
experimental results:
a—$\bar{X}_{SV1} = kr_1 \ln(4.0)$,
$\bar{Z}_{SV2} = 0$;
b—$\bar{X}_{SV1} = kr_1 \ln(4.0)$,
$\bar{X}_{SV2}(s_2) =$
$kr_2 \ln(4.0)\phi(s_2)$. The
experimental results are
marked by circles

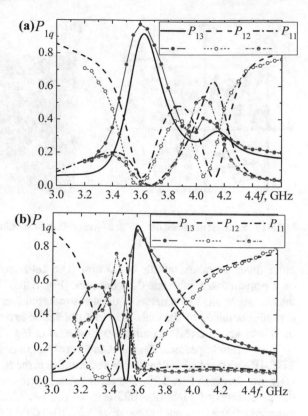

Fig. 5.17a. The results for the T-junction whose second vibrator is characterized by variable impedance defined by the function $\phi(s_2) = 2[1 - (s_2/L_2)]$, (Fig. 3.13a) are shown in Fig. 5.17b. As can be seen, the simulated and experimental results are in rather satisfactory agreement.

As one would expect, the two-resonance mode of power division between the T-junction arms can be realized if the two vibrators are tuned to different resonant frequencies. The operating waveband of the T-junction can be widened by selecting the geometrical and electrophysical parameters of the vibrators, which ensure a close position of the resonant frequencies.

The surface impedance of metamaterial layers on perfectly conducting plane can be calculated by using formula (1.72), $\bar{Z}_{SW} = \bar{R}_{SW} + i\bar{X}_{SW} = \pm i\frac{k_d}{k_w}\sqrt{\frac{\mu}{\varepsilon}}\mathrm{tg}(k_w h_d)$, where the plus or minus signs are used if the permittivity $\varepsilon' > 0$ or $\varepsilon' < 0$. The formula for the surface impedance of electrically thin layer $\bar{Z}_{SW} \approx ik\mu h_d$ does not depend on the permittivity sings. If the permeability $\mu' < 0$, the surface impedance is capacitive $\bar{X}_{SW} < 0$, while the impedance of a natural magnetodielectric is always inductive ($\bar{X}_{SW} > 0$). The simulation was carried out for an isotropic metamaterial LR-5I [18]. A cell of this material consists of four three-turn spirals made of nichrome wire with parameters: the wire diameter is 0.4 mm, the pitch of helix is 1.0 mm, the

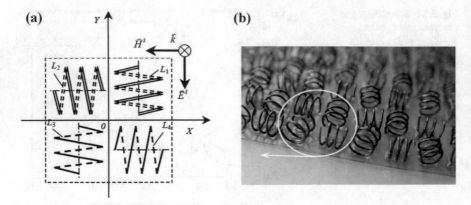

Fig. 5.18 Isotropic metamaterial LR-5I: **a**—unit cell; **b**—experimental sample of the metamaterial

outer diameter of the spirals is 5.0 mm. The cells are arranged in a special way on a polyurethane substrate 0.2 mm thick (Fig. 5.18). The experimental values of the electrophysical parameters of the meta-material and the curves plotted based on formulas obtained by the authors to interpolate the experimental data which will be used later in numerical modeling are presented in Fig. 5.19.

The surface impedance $\bar{Z}_{SW}$ as a function of frequency for the materials LR-5I and TDK-IR-A095 calculated by the above formulas in the frequency range $f = 2.7-4.0$ GHz are shown in Fig. 5.20. The layer thickness for both material $h_d = 5.2$ mm, the parameters of TDK-IR-A095 is $\varepsilon = 6.2 - i0.32$, $\mu = 0.60 - i0.32$, were measured in the frequency range $f = 2.5-10.0$ GHz [20]. As can be seen from the graphs, in contrast to the traditional magnetodielectric, for this metamaterial there is a frequency region in which the imaginary part of the surface impedance takes negative values. In the rest part of the wavelength range, the diffraction properties of the metamaterial LR-5I are close to that of the lossy metallized screen characterized by inductive impedance.

Consider the T-junction with only one vibrator to verify possibility of the resonant slot tuning by changing the impedance $\bar{Z}_{SW}$, which can be inductive or capacitive. The length L_1 of the perfectly conducting vibrator and its displacement $z_0 = \lambda_G^{Sres}/2$ along the axis {0 z} for various impedance of the waveguide end wall are based on the condition of transferring maximum possible power to the side arm. In the formula for z_0, $\lambda_G^{Sres} = \dfrac{2\pi}{\sqrt{(2\pi/\lambda_{sl}^{res})^2 - (\pi/a)^2}}$ is the slot resonant wavelength in the waveguide, λ_{sl}^{res} is the slot resonant length in the free half-space. The simulation results are summarized in Table 5.2.

The influence of the resonant slot length upon the energy characteristics can be observed by analyzing the plots in Fig. 5.21a for the T-junction without vibrators. The thickness of the coating layers was chosen equal to $h_d = 5.2$ mm that technology proposed in [18] can be used.

As expected, impedance coating of the end wall provides the resonant mode of power transfer to the side arm both in the short- and long-wavelengths of the main

Fig. 5.19 Permittivity and permeability of the metamaterial LR-5I

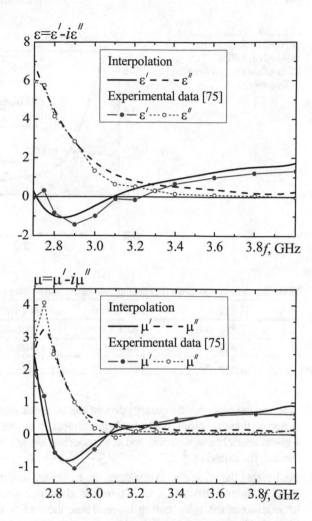

waveguide operating range if the end wall is coated by the TDK-IR-A095 magnetodielectric or by the metamaterial LR-5I. As can be seen from Fig. 5.21, the losses in the coating materials can significantly decrease the power transmitted to the side waveguide. For these materials, the losses are comparable, since the imaginary parts of the magnetic permeability of the materials are close in the operating wavelength range.

As quite clear, significant losses in the control elements of microwave devices prevent their practical use. Therefore, properties of the coating layer material for the end wall should be taken into account. Thus, to obtain coating with inductive surface impedance, the permeability of material layers should have the imaginary part close to zero. As far as the metamaterials are concerned the losses are fundamentally obligatory. Since the metamaterials are dispersing composites, the well-known Kramers–Kronig relations [21] impose rather stringent restrictions on the relationship

Fig. 5.20 Surface
impedance $\bar{Z}_{SW}$ for the
materials LR-5I и
TDK-IR-A095 as a function
of frequency

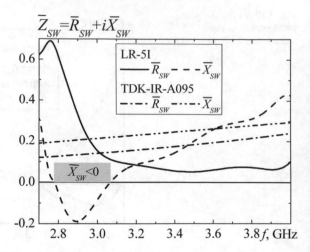

Table 5.2 Vibrator parameters providing maximum power level in the side shoulder

#	$\bar{Z}_{SW}$	λ_{sl}^{res} (mm)	z_0 (mm)	L_1 (mm)
1	$\bar{Z}_{SW} = 0$	80.0	54.0	17.3
2	$\bar{X}_{SW} > 0;$	90.0	90.0	21.5
3	$\bar{X}_{SW} < 0.$	91.5	74.0	20.0

between the real and imaginary parts of the complex permittivity and permeability. Based on these relations, the lower limit of electric and magnetic losses, below which a metamaterial cannot be realized at a frequency ω_0 was established in [190]. Nevertheless, the criterion $\frac{2}{\pi} \int_0^\infty \omega^3 \frac{\varepsilon''(\omega)\mu'(\omega) + \mu''(\omega)\varepsilon'(\omega)}{\omega^2 - \omega_0^2} d\omega \leq -1$ obtained in [22], allows minimizing the μ'' value. In this case, the losses are determined only by the metamaterial complex permittivity, which, as was shown above, does not affect the surface impedance of the thin coating layer. Thus, the end of the side waveguide should be coated by layers characterized by magnetic permeability with an imaginary part $\mu'' \to 0$. Therefore, the losses in the coating materials TDK-IR-A095 and LR-5I cause significant decrease of the power transmitted to the side waveguide. Thus, the monopole with length $\lambda_{sl}^{res}/4$ positioned at the distance $z_0 = \lambda_G^{Sres}/2$ from the slot axis allows transmitting at resonant frequencies to the arm 3no more than 80% of the input power.

Let us define conditions required for the power transfer from the main waveguide to the side waveguide with the end wall covered by the metamaterial layer. In this case, the resonant tuning in the operating frequency range can be realized based on condition arg $S_{11}^3 = 0$ (arg S_{11}^3 is reflection coefficient of the single slot) [23, 24]: (1) one conventional resonance ($2L_3 = 37.0$ mm); (2) one specific resonance ($2L_3 = 43.0$ mm); (3) two resonances ($2L_3 = 40.0$ mm). The frequency plots of arg S_{11}^3 and arg S_{11}^1 are shown in Fig. 5.22a. The energy characteristics of the T-junctions are shown in Fig. 5.22b. The frequency curves P_{11}, P_{12} and P_{13} for the first

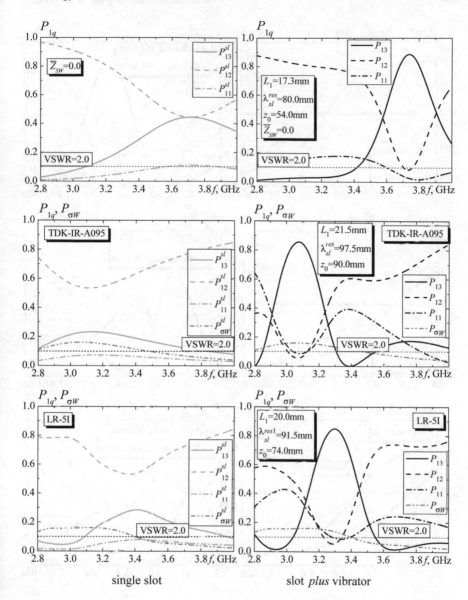

Fig. 5.21 Energy characteristics of the T-junction with two element vibrator-slot structure. The levels corresponding to VSWR = 2.0 are shown in the plots

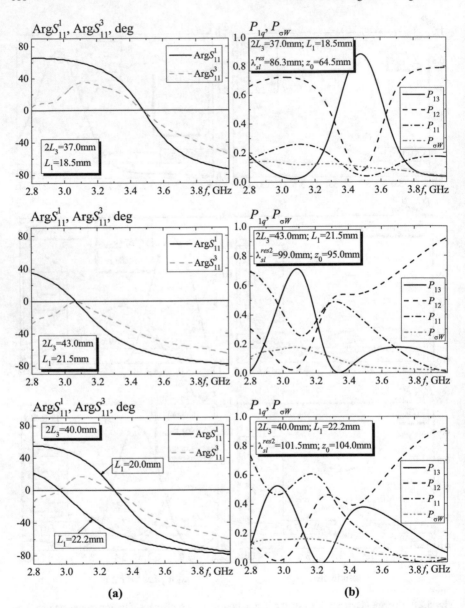

Fig. 5.22 The T-junction with the two-element vibrator-slot structure and end wall coated by the metamaterial ($h_d = 5.2$ mm): **a**—resonant conditions; **b**—the energy characteristics

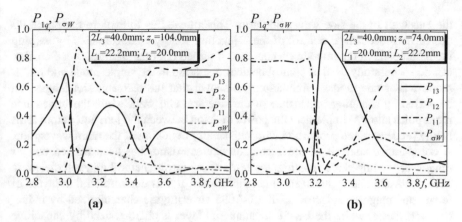

Fig. 5.23 The T-junction with the three-element vibrator-slot structure and end wall coated by the metamaterial ($h_d = 5.2$ mm): **a**—resonant conditions; **b**—the energy characteristics

variant, are similarly to those presented in Fig. 5.21b, while they differ significantly for the second and third options. Thus, two pronounced resonances at the curve P_{13} can be observed for the third option: at the frequency $f \approx 2.95$ GHz $P_{13} \gg P_{12}$, while at the frequency $f \approx 3.45$ GHz $P_{13} \approx P_{12}$.

If the slot length $2L_3 = 40.0$ mm and the monopole lengths correspond to the slot resonances: $2L_{1(2)} = 22.2$ mm at the frequency $f = 2.96$ GHz and $2L_{2(1)} = 20.0$ mm at $f = 3.28$ GHz, the T-junction characteristics, obtained in this case, cannot be realized when the waveguide end wall is perfectly conducting or is coated by the natural magnetodielectric layer (Fig. 5.23). As can be seen from the plots, the following power transfer modes can be observed: three-resonant transfer to the side waveguide (Fig. 5.23a) and superposition of P_{13} resonances at close frequencies (Fig. 5.23b).

The simulation results presented in Figs. 5.21, 5.22 and 5.23 were obtained for perfectly conducting vibrators for simple controlling their resonant frequency. However, the electric length of vibrators can be easily varied by using impedance vibrators. This possibility can be used to reduce probability of electrical breakdown between the monopole end and waveguide wall of low-profile main waveguide.

Thus, the approximate analytical solution of the internal diffraction problem for the E-plane T-junction with the slot cut in the impedance end wall of the side waveguide and one or two semi-infinite impedance vibrators was studied in the approximations of electrically thin linear radiators, The actual wall waveguide thickness was also taken into account. The obtained solution was verified by comparing the simulation and experimental data. The solution allows calculating the energy characteristics of the T-junction with an efficiency much greater than that can be obtained by using commercial software.

The simulation has been carried out taking into account variable surface impedance distributed over the monopoles and constant impedance distributed at the end wall of the side waveguide arm. This allowed us to explore the possibilities of non-mechanical control of the T-junction energy characteristics by varying the parameters of the impedance coatings. It turned out that the impedance coating can be successfully used as control devices operating in the single-mode and multimode waveguide modes. Simulation has shown that the resonant slot length can be varied by changing impedance of the side arm end wall, while the impedance monopoles allow to implement the power division between the junction output arms in any predetermined proportion with satisfactory matching of the main waveguide. A comparative analysis of the T-junction characteristics has been carried out for three types of impedance coating: lossless natural ferrite, natural magnetodielectric TDK-IR-A095 and artificial metamaterial LR-5I. It was determined that layers of ferrite and magnetodielectric TDK-IR-A095 covering are characterized by inductive impedance, while the LR-5I metamaterial layer is characterized by capacitive surface impedance. If the metamaterial LR-5I is used, some frequency region exists where the imaginary part of the surface impedance is negative, while its diffraction properties in the rest of the range are close to that of a metallized screen, characterized by inductive impedance. Thus, such frequency-dependent surface impedance can provide a three-resonance mode of power transmission from the main to side waveguide. It should be noted that losses in the metamaterial can require minimizing the imaginary part of its permeability when thin-layer of metamaterials are applied.

The proposed T-junction differs from the devices currently used in practice, since it allows to redirect almost all the supplied power from the main to side waveguide. However, the resonant nature of the vibrator-slot coupling structure is manifested in the relative narrow-band T-junction operating modes. This drawback has left the T-junctions of this type beyond an area interesting for researchers. However, at present, due to the demand for multifunctional devices attention to such structures has begun to increase. Particular attention is paid to structures with multimode functioning of power division which allow controlling division ratio. It is shown that incident power division between the T-junction arms and the propagating wave types over a wide range can be achieved by varying the geometric dimensions of the T-junction constituent and parameters of the monopole surface impedance.

The presented results can be used to develop antenna-waveguide devices, part of which are T-junctions, including different-sized waveguides and structures with lossy structural elements. The term multidimensional waveguides means that the electric dimensions of the waveguide cross sections can be different, therefore, geometrically identical waveguides with different uniform fillings can also be studied in the framework of this model.

References

1. Southworth JK (1950) Principles and application of a waveguide transmission. New York
2. Lewin L (1975) Theory of waveguides. Techniques for the Solution of Waveguide Problems, Newnes-Butterworths, London

3. Arndt F, Ahrens I, Papziner U, Wiechmann U, Wilkeit R (1987) Optimized E-plane T-junction series power dividers. IEEE Trans Microwave Theory Tech 35:1052–1059
4. Yao H-W, Abdelmonem AE, Liang J-F, Liang X-P, Zaki KA, Martin A (1993) Wide-band waveguide and ridge waveguide T-junctions for diplexer applications. IEEE Trans Microwave Theory Tech 41:2166–2173
5. Kirilenko AA, Senkevich SL, Tkachenko VI, Tysik BG (1994) Waveguide diplexer and multiplexer design. IEEE Trans Microwave Theory Tech 42:1393–1396
6. Widarta A, Kuwano S, Kokubun K (1995) Simple and accurate solutions of the scattering coefficients of E-plane junctions in rectangular waveguides. IEEE Trans Microwave Theory Tech 43:2716–2718
7. Abdelmonem A, Yao H-W, Zaki KA (1994) Slit coupled E-plane rectangular T-junctions using single port mode matching technique. IEEE Trans Microwave Theory Tech 42:903–907
8. Blas AAS, Mira F, Boria VE, Gimeno B, Bressan M, Arcioni P (2007) On the fast and rigorous analysis of compensated waveguide junctions using off-centered partial-height metallic posts. IEEE Trans Microwave Theory Tech 55:168–175
9. Wu K-L, Wang HA (2001) A rigorous modal analysis of H-plane waveguide T-junction loaded with a partial-height post for wide-band applications. IEEE Trans Microwave Theory Tech 49:893–901
10. Nesterenko MV, Katrich VA, Penkin YM (2005) H_{10}-wave diffraction by the stepwise junction of two rectangular waveguides with an impedance slot diaphragm. Telecommun Radio Eng 63(7):569–588
11. Bretones AR, Martín RG, García IS (1995) Time-domain analysis of magnetic-coated wire antennas. IEEE Trans Antennas Propag 43:591–596
12. Penkin YuM, Katrich VA, Nesterenko MV, Berdnik SL, Dakhov VM (2019) Electromagnetic fields excited in volumes with spherical boundaries. Springer Nature Swizerland AG, Cham, Swizerland
13. Nesterenko MV, Katrich VA, Penkin DYu, Berdnik SL, Kijko VI (2012) Electromagnetic waves scattering and radiation by vibrator-slot structure in a rectangular waveguide. Prog Electromagn Res M 24:69–84
14. Penkin DYu, Berdnik SL, Katrich VA, Nesterenko MV, Kijko VI (2013) Electromagnetic fields excitation by a multiclement vibrator-slot structures in coupled electrodynamics volumes. Prog Electromagn Res B 49:235–252
15. Berdnik SL, Katrich VA, Kijko VI, Nesterenko MV (2015) Scattering of electromagnetic waves by a system of vibrators with variable impedance in a rectangular waveguide. Radiophys Radioastronomy 20(1):64–75. (in Russian)
16. Berdnik SL, Katrich VA, Nesterenko MV, Penkin YM, Penkin DY (2015) Radiation and scattering of electromagnetic waves by a multi-element vibrator-slot structure in a rectangular waveguide. IEEE Trans Antennas Propag AP-63(9):4256–4259
17. Lagarkov AN, Kisel VN, Semenenko VN (2008) Wide-angle absorption by the use of a metamaterial plate. Prog Electromagn Res Lett 1:35–44
18. Lagarkov AN, Semenenko VN, Basharin AA, Balabukha NP (2008) Abnormal radiation pattern of metamaterial waveguide. PIERS Online 4(6):641–644
19. Vendik IB, Vendik OG (2013) Metamaterials and their application in microwave technology (Overview). J Tech Phys 83(1):3–28. (in Russian)
20. Yoshitomi K, Sharobim HR (1994) Radiation from a rectangular waveguide with a lossy flange. IEEE Trans Antennas Propag 42:1398–1403
21. Akhiezer AI, Akhiezer IA (1985) Electromagnetism and electromagnetic waves. Vysshaya Shkola, Moskow (in Russian)
22. Stockman MI (2007) Criterion for negative refraction with low optical losses from a fundamental principle of causality. Phys Rev Lett 98(17):177404/1–177404/4
23. Wolman VI, Pimenov YV (1971) Technical electrodynamics. Svyaz', Moskow (in Russian)
24. Nesterenko MV, Katrich VA, Penkin YM, Berdnik SL (2008) Analytical and hybrid methods in theory of slot-hole coupling of electrodynamic volumes. Springer Science+Business Media, New York

Chapter 6
Waveguide Radiation of the Combined Vibrator-Slot Structures

At the present time, thin vibrators are often used to control electrodynamic characteristics of slot radiators (Chaps. 4, 5). The vibrators can be located in a half-space over an infinite perfectly conducting plane with a hole cut for coupling with another electrodynamic volume such as half-space above the plane, waveguide, resonator, etc. [1–9]. The vibrator can also be located in various waveguide transmission lines and resonators [10–22]. A special place among combined vibrator-slot structures is occupied by Clavin elements, named after their inventor [23, 24], which consists of a narrow radiating slot and two identical passive vibrators (monopoles) on both sides of slot at equal distances from its center [25–28]. The Clavin elements are characterized by identical radiation patterns (RPs) in the E- and H-planes. These elements can be used as standalone radiators, exciters for mirror antennas, part of multi-element phased arrays, etc. It should be noted that in publications on the subject only perfectly conducting vibrators were considered. To widen possibility to control the characteristics of vibrator-slot radiators, the monopoles with distributed surface impedance, including that varying along the vibrator axes, can be used [29]. An extended list of the authors' publications on the study of the electrodynamic characteristics of the combined vibrator-slot structures considered in this chapter is presented in Bibliography.

6.1 Problem Formulation and Initial Equations in the General Case for Longitudinal Slot Element

Consider a microwave device consisting of a rectangular waveguide section with a longitudinal slot cut in its wide wall with two asymmetric impedance vibrators (monopoles) located near the slot outside the waveguide and a monopole located inside the waveguide in the plane $\{x0y\}$ parallel to its narrow wall (Fig. 6.1). A H_{10}-wave propagates in the waveguide with cross-section $\{a \times b\}$ (the region is marked

© The Author(s), under exclusive license to Springer Nature Switzerland AG 2020   191
M. V. Nesterenko et al., *Combined Vibrator-Slot Structures: Theory and Applications*, Lecture Notes in Electrical Engineering 689,
https://doi.org/10.1007/978-3-030-60177-5_6

Fig. 6.1 The geometry of
the problem and accepted
notations

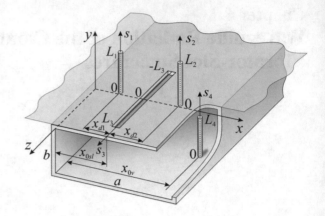

by index Wg) from the direction $z = -\infty$. The parameters of the waveguide filling
material are ε_1, μ_1. The slot radiates into half-space, marked as Hs, located above
an infinite perfectly conducting plane. The material parameters of the half-space are
ε_2, μ_2. The slot length and width are $2L_3$ and d, its slot center is located at the point
$z = 0$, the slot axis is displaced relative the waveguide narrow wall at the distance
x_{0sl}. The outside monopoles with lengths and radii L_1, L_2 and r_1, r_2, are placed at
distances x_{d1} and x_{d2} from the longitudinal slot axis. The length and radius of the
internal monopole are L_4 and r_4.

    If dimensions of the slot and vibrators satisfy the thin radiator approximations:
$\frac{r_m}{2L_m} \ll 1, \frac{r_m}{\lambda_{1,2}} \ll 1, m = 1, 2, 4, \frac{d}{2L_3} \ll 1, \frac{d}{\lambda_{1,2}} \ll 1$ ($\lambda_{1,2}$ are wavelengths in the
corresponding media) the electric currents in the vibrators and equivalent magnetic
current in the slot satisfy the boundary conditions $J_m(\pm L_m) = 0, J_3(\pm L_3) = 0$
($-L_m$ are the end coordinates the monopole mirror images in the plane and waveguide
bottom). The system of Eqs. (2.1) can be transformed, taking into account (2.2) and
(2.19), to the following system of integral equations

$$\left(\frac{d^2}{ds_1^2} + k_2^2\right)\left\{\int_{-L_1}^{L_1} J_1(s_1')G_{s_1}^{HsE}(s_1, s_1')ds_1' + \int_{-L_2}^{L_2} J_2(s_2')G_{s_2}^{HsE}(s_1, s_2')ds_2'\right\}$$

$$- ik\vec{e}_{s_1}\mathrm{rot}\int_{-L_3}^{L_3} J_3(s_3')G_{s_3}^{HsM}(s_1, s_3')ds_3' = i\omega\varepsilon_2 z_{i1}(s_1)J_1(s_1),$$

$$\left(\frac{d^2}{ds_2^2} + k_2^2\right)\left\{\int_{-L_2}^{L_2} J_2(s_2')G_{s_2}^{HsE}(s_2, s_2')ds_2' + \int_{-L_1}^{L_1} J_1(s_1')G_{s_1}^{HsE}(s_2, s_1')ds_1'\right\}$$

$$- ik\vec{e}_{s_2}\mathrm{rot}\int_{-L_3}^{L_3} J_3(s_3')G_{s_3}^{HsM}(s_2, s_3')ds_3' = i\omega\varepsilon_2 z_{i2}(s_2)J_2(s_2), \qquad (6.1)$$

$$\frac{1}{\mu_1}\left(\frac{d^2}{ds_3^2}+k_1^2\right)\int_{-L_3}^{L_3}J_3(s_3')G_{s_3}^{WgM}(s_3,s_3')ds_3'+\frac{1}{\mu_2}\left(\frac{d^2}{ds_3^2}+k_2^2\right)\int_{-L_3}^{L_3}J_3(s_3')G_{s_3}^{HsM}(s_3,s_3')ds_3'$$

$$+ik\vec{e}_{s_3}\mathrm{rot}\left\{\begin{array}{l}\int_{-L_1}^{L_1}J_1(s_1')G_{s_1}^{HsE}(s_3,s_1')ds_1'+\int_{-L_2}^{L_2}J_2(s_2')G_{s_2}^{HsE}(s_3,s_2')ds_2'+\\+\int_{-L_4}^{L_4}J_4(s_4')G_{s_4}^{WgE}(s_3,s_4')ds_4'\end{array}\right\}=-i\omega H_{0s_3}(s_3),$$

$$\left(\frac{d^2}{ds_4^2}+k_1^2\right)\left\{\int_{-L_4}^{L_4}J_4(s_4')G_{s_4}^{WgE}(s_4,s_4')ds_4'\right\}$$

$$-ik\vec{e}_{s_4}\mathrm{rot}\int_{-L_3}^{L_3}J_3(s_3')G_{s_3}^{HsM}(s_4,s_3')ds_3'=-i\omega\varepsilon_1\left[E_{0s_4}(s_4)-z_{i4}(s_4)J_4(s_4)\right].$$

Here $z_{im}(s_m)$ are the intrinsic linear impedances of the monopoles ([Ohm/m]), $H_{0s_3}(s_3)$ and $E_{0s_4}(s_4)$ are the projections of the extraneous force fields on the slot and monopole axes, $G_{s_{1,2}}^{HsE}(s_{1,2,3},s_{1,2}')$, $G_{s_4}^{WgE}(s_3,s_4')$ and $G_{s_3}^{HsM,WgM}(s_{1,2,3,4},s_3')$ are the components of the electric (E) and magnetic (M) tensor Green's functions for the vector potential of the corresponding electrodynamic volumes (see Appendix A), $k=2\pi/\lambda$ (λ is the wavelength in free space), $k_{1,2}=k\sqrt{\varepsilon_{1,2}\mu_{1,2}}=2\pi/\lambda_{1,2}$, $\vec{e}_{s_m}$ and $\vec{e}_{s_3}$ are the unit vectors directed along the vibrator and the slot axes, s_m and s_3 are the local coordinates associated with the vibrators and the slot axes.

At $\varepsilon_{1,2}=\mu_{1,2}=1$ and $H_{0s_3}(s_3)=H_0\cos\frac{\pi x_{0sl}}{a}e^{-ik_gs_3}=H_0^s(s_3)+H_0^a(s_3)$, $E_{0s_4}(s_4)=H_0\frac{k}{k_g}\sin\frac{\pi x_{0v}}{a}$ (H_0 is the amplitude of H_{10}-wave), the slot current can be presented as $J_3(s_3)=J_3^s(s_3)+J_3^a(s_3)$ where $J_3^s(s_3)$ and $J_3^a(s_3)$ are symmetric and antisymmetric components of the slot current relative to its center. Hence, the Eq. (6.1) can be transformed to:

$$\left(\frac{d^2}{ds_1^2}+k^2\right)\left\{\int_{-L_1}^{L_1}J_1(s_1')G_{s_1}^{HsE}(s_1,s_1')ds_1'+\int_{-L_2}^{L_2}J_2(s_2')G_{s_2}^{HsE}(s_1,s_2')ds_2'\right\}$$

$$+ik\int_{-L_3}^{L_3}J_3^s(s_3')\tilde{G}_{s_3}^{HsM}(s_1,s_3')ds_3'=i\omega z_{i1}(s_1)J_1(s_1),\qquad(6.2a)$$

$$\left(\frac{d^2}{ds_2^2}+k^2\right)\left\{\int_{-L_2}^{L_2}J_2(s_2')G_{s_2}^{HsE}(s_2,s_2')ds_2'+\int_{-L_1}^{L_1}J_1(s_1')G_{s_1}^{HsE}(s_2,s_1')ds_1'\right\}$$

$$+ik\int_{-L_3}^{L_3}J_3^s(s_3')\tilde{G}_{s_3}^{HsM}(s_2,s_3')ds_3'=i\omega z_{i2}(s_2)J_2(s_2),\qquad(6.2b)$$

$$\left(\frac{d^2}{ds_3^2} + k^2\right) \int_{-L_3}^{L_3} J_3^s(s_3')\left[G_{s_3}^{WgM}(s_3, s_3') + G_{s_3}^{HsM}(s_3, s_3')\right]ds_3'$$

$$- ik \left\{ \begin{array}{l} \int_{-L_1}^{L_1} J_1(s_1')\tilde{G}_{s_1}^{HsE}(s_3, s_1')ds_1' + \int_{-L_2}^{L_2} J_2(s_2')\tilde{G}_{s_2}^{HsE}(s_3, s_2')ds_2' \\ + \int_{-L_4}^{L_4} J_4(s_4')\tilde{G}_{s_4}^{WgE}(s_3, s_4')ds_4' \end{array} \right\} = -i\omega H_0^s(s_3),$$

$$\hspace{11cm}(6.2c)$$

$$\left(\frac{d^2}{ds_3^2} + k^2\right) \int_{-L_3}^{L_3} J_3^a(s_3')\left[G_{s_3}^{WgM}(s_3, s_3') + G_{s_3}^{HsM}(s_3, s_3')\right]ds_3' = -i\omega H_0^a(s_3), \quad (6.2d)$$

$$\left(\frac{d^2}{ds_4^2} + k^2\right) \int_{-L_4}^{L_4} J_4(s_4')G_{s_4}^{WgE}(s_4, s_4')ds_4' - ik \int_{-L_3}^{L_3} J_3^s(s_3')\tilde{G}_{s_3}^{WgM}(s_4, s_3')ds_3'$$

$$= -i\omega[E_0(s_4) - z_{i4}(s_4)J_4(s_4)]. \hspace{5cm}(6.2e)$$

The solution to the of equation system (6.2) can be found by the generalized method of induced EMMF, using the functions $J_m(s_m) = J_{0m}f_m(s_m)$ and $J_3^{s,a}(s_3) = J_{03}^{s,a}f_3^{s,a}(s_3)$ as approximating expressions for the currents. Here, J_{0m} and $J_{03}^{s,a}$ are current amplitudes, $f_m(s_m)$ and $f_{03}^{s,a}(s_3)$ are predefined current distribution functions, which can be obtained by solving the equations for currents in a standing alone vibrator and slot by the averaging method. The following expressions for the distribution function can be obtained based on (4.3):

$$f_1(s_1) = \cos \tilde{k}_1 s_1 - \cos \tilde{k}_1 L_1, \hspace{4cm}(6.3a)$$

$$f_2(s_2) = \cos \tilde{k}_2 s_2 - \cos \tilde{k}_2 L_2, \hspace{4cm}(6.3b)$$

$$f_3^s(s_3) = \cos k s_3 \cos k_g L_3 - \cos k L_3 \cos k_g s_3, \hspace{2.5cm}(6.3c)$$

$$f_3^a(s_3) = \sin k s_3 \sin k_g L_3 - \sin k L_3 \sin k_g s_3, \hspace{2.7cm}(6.3d)$$

$$f_4(s_4) = \cos \tilde{k}_4 s_4 - \cos \tilde{k}_4 L_4, \hspace{4cm}(6.3e)$$

where $k_g = \frac{2\pi}{\lambda_g} = \sqrt{k^2 - k_c^2}$, $k_c = \frac{2\pi}{\lambda_c} = \frac{\pi}{a}$, λ_g is the wavelength in the waveguide, λ_c is the H_{10}-wave critical wavelength, $\tilde{k}_m = k + \frac{i\alpha 2\pi z_{im}^{av}}{Z_0}$, $z_{im}^{av} = \frac{1}{2L_m}\int_{-L_m}^{L_m} z_{im}(s_m)ds_m$ are the intrinsic impedances average over the vibrator length, and $Z_0 = 120\pi$ Ohm.

Let us first multiply each of Eqs. (6.2a), (6.2b), and (6.2c) by the function $f_1(s_1)$, $f_2(s_2)$, $f_3^s(s_3)$, $f_3^a(s_3)$, and $f_4(s_4)$ respectable, and then integrate the multiplication result of Eqs. (6.2a), (6.2b) and (6.2e) over the vibrator lengths, and Eqs. (6.2c) and (6.2d) over the slot length. Thus, the following system of linear algebraic equations can be obtained:

$$\begin{cases} J_{01}(Z_{11} + F_1^{\overline{Z}}) + J_{02}Z_{12} + J_{03}^s Z_{13} = 0, \\ J_{02}(Z_{22} + F_2^{\overline{Z}}) + J_{01}Z_{21} + J_{03}^s Z_{23} = 0, \\ J_{03}^s(Z_{33}^{sWg} + Z_{33}^{sHs}) + J_{01}Z_{31} + J_{02}Z_{32} + J_{04}Z_{34} = -\frac{i\omega}{2k}H_3^s, \\ J_{03}^a(Z_{33}^{aWg} + Z_{33}^{aHs}) = -\frac{i\omega}{2k}H_3^a, \\ J_{04}(Z_{44} + F_4^{\overline{Z}}) + J_{03}^s Z_{43} = -\frac{i\omega}{2k}E_4, \end{cases} \quad (6.4)$$

where $H_3^s = H_0 \cos\frac{\pi x_{0sl}}{a} \int_{-L_3}^{L_3} \cos k_g s_3 f_3^s(s_3) ds_3$, $H_3^a = -i H_0 \cos\frac{\pi x_{0sl}}{a} \int_{-L_3}^{L_3} \sin k_g s_3 f_3^u(s_3) ds_3$, $E_4 = H_0 \frac{k}{k_g} \sin\frac{\pi x_{0v}}{a} \int_{-L_4}^{L_4} f_4(s_4) ds_4$, Z_{mn} $(m, n = 1, 2, 3, 4)$ and $F_m^{\overline{Z}}$ are dimensionless coefficients. When the amplitudes of the currents J_{0m} and $J_{03}^{s,a}$ are found as solution of equation system (6.4), the electrodynamic characteristics of this vibrator-slot structure can be easily obtained.

6.2 Clavin Element for the General Case

The following relations for the Clavin element are valid: $2L_1 = 2L_2 = 2L_v, r_1 = r_2 = r, \alpha_1 = \alpha_2 = \alpha, 2L_4 = 2L_V, r_4 = r_V, \alpha_4 = \alpha_V, \overline{Z}_{S1}(s_1) = \overline{Z}_{S2}(s_2) = \overline{Z}_S(s_v) = 2\pi r_v z_{iv}(s_v)/Z_0, \overline{Z}_{S4}(s_4) = \overline{Z}_{SV}(s_V) = 2\pi r_V z_{iV}(s_V)/Z_0, \tilde{k}_1 = \tilde{k}_2 = \tilde{k} = k + i(\alpha/r)\overline{Z}_S^{av}, \tilde{k}_4 = \tilde{k}_V = k + i(\alpha_V/r_V)\overline{Z}_{SV}^{av}, x_{d1} = x_{d2} = x_d, F_1^{\overline{Z}} = F_2^{\overline{Z}} = F_v^{\overline{Z}}, F_4^{\overline{Z}} = F_V^{\overline{Z}}, H_3^{s,a} = H_{sl}^{s,a}, (Z_{33}^{s,aWg} + Z_{33}^{s,aHs}) = Z_{sl}^{s,a\Sigma}, f_1(s_1) = f_2(s_2) = f_v(s_v), f_4(s_4) = f_V(s_V), f_3^{s,a}(s_3) = f_{sl}^{s,a}(s_{sl}), Z_{11} + F_1^{\overline{Z}} = Z_{22} + F_2^{\overline{Z}} = Z_v + F_v^{\overline{Z}} = Z_v^{\Sigma}, Z_{44} + F_4^{\overline{Z}} = Z_V^{\Sigma}, Z_{12} = Z_{21} = Z_{vv}, Z_{13} = Z_{23} = -2Z_{31} = -2Z_{32} = Z_c, Z_{34} = -Z_{43} = Z_{Wg}.$

In this case, the SLAE (6.4) can be transformed to the following equation system:

$$\begin{cases} J_{0v}Z_v^{\Sigma} + J_{0v}Z_{vv} + J_{0sl}^s Z_c = 0, \\ J_{0sl}^s Z_{sl}^{s\Sigma} - J_{0v}Z_c + J_{0v}Z_{Wg} = -\frac{i\omega}{2k}H_{sl}^s, \\ J_{0sl}^a Z_{sl}^{a\Sigma} = -\frac{i\omega}{2k}H_{sl}^a, \\ J_{0V}Z_V^{\Sigma} - J_{0sl}^s Z_{Wg} = -\frac{i\omega}{2k}E_V. \end{cases} \quad (6.5)$$

The solution of equations system (6.5) can be written as:

$$J_{0v} = \frac{i\omega}{2k}HE\frac{Z_c}{\tilde{Z}_{sl}\tilde{Z}_v + Z_c^2}, \quad J_{0V} = -\frac{i\omega}{2k}\left[E_V\frac{1}{Z_V^{\Sigma}} + HE\frac{\tilde{Z}_v Z_{Wg}}{(\tilde{Z}_{sl}\tilde{Z}_v + Z_c^2)Z_V^{\Sigma}}\right],$$

$$J_{0sl}^s = -\frac{i\omega}{2k} H E \frac{\tilde{Z}_v}{\tilde{Z}_{sl}\tilde{Z}_v + Z_c^2}, \quad J_{0sl}^a = -\frac{i\omega}{2k} H_{sl}^a \frac{1}{Z_{sl}^{a\Sigma}} \tag{6.6}$$

where $\tilde{Z}_v = Z_v^\Sigma + Z_{vv}$, $\tilde{Z}_{sl} = Z_{sl}^{s\Sigma} + Z_{Wg}^2/Z_V^\Sigma$, $HE = H_{sl}^s - E_V Z_{Wg}/Z_V^\Sigma$. The components of the tensor Green's functions for the Clavin element are:

$$G_{S_{v(vv)}}^{HsE}(s_v, s_v') = \frac{e^{-ik\sqrt{(s_v-s_v')^2+r^2(4x_d^2)}}}{\sqrt{(s_v - s_v')^2 + r^2(4x_d^2)}}, \quad G_{Ssl}^{HsM}(s_{sl}, s_{sl}') = 2\frac{e^{-ik\sqrt{(s_{sl}-s_{sl}')^2+(d_e/4)^2}}}{\sqrt{(s_{sl} - s_{sl}')^2 + (d_e/4)^2}},$$

$$G_{Ssl}^{WgM}(s_{sl}, s_{sl}') = \frac{2\pi}{ab} \sum_{m=0}^{\infty} \sum_{n=0}^{\infty} \frac{\varepsilon_m \varepsilon_n}{k_z} e^{-k_z|s_{sl}-s_{sl}'|} \cos k_x x_0 \cos k_x \left(x_0 + \frac{d_e}{4}\right),$$

$$\tilde{G}_{S_v}^{HsE}(s_{sl}, s_v') = \frac{\partial}{\partial x_{Hs}} G_{S_v}^{HsE}[x_{Hs}, 0, z_{Hs}(s_{sl}); x_d, y_{Hs}'(s_v'), 0] \text{ at } x_{Hs} = 0,$$

$$\tilde{G}_{Ssl}^{HsM}(s_v, s_{sl}') = \frac{\partial}{\partial x_{Hs}} G_{Ssl}^{HsM}[x_{Hs}, y_{Hs}(s_v), 0; 0, 0, z_{Hs}'(s_{sl}')] \text{ at } x_{Hs} = x_d,$$

$$\tilde{G}_{Sv}^{WgE}(s_{sl}, s_V') = \frac{\partial}{\partial x_{Wg}} G_{Sv}^{WgE}[x_{Wg}, 0, z_{Wg}(s_{sl}); \cos k_x x_{0V}, y_{Wg}'(s_V'), 0] \text{ at } x_{Wg} = x_{0sl},$$

$$\tilde{G}_{Ssl}^{WgM}(s_V, s_{sl}') = \frac{\partial}{\partial x_{Wg}} G_{Ssl}^{WgM}[x_{Wg}, y_{Wg}(s_V), 0; \cos k_x x_{0sl}, 0, z_{Wg}'(s_{sl}')] \text{ at } x_{Wg} = x_{0V}.$$

where $\varepsilon_n = \begin{cases} 1, & n = 0 \\ 2, & n \neq 0 \end{cases}$, $k_x = \frac{m\pi}{a}$, $k_y = \frac{n\pi}{b}$, $k_z = \sqrt{k_x^2 + k_y^2 - k^2}$, m and n are integers; d_e is the effective slot width [(3.59), (3.60)], which takes into account the thickness of the waveguide wall. All quantities in the expressions for the current amplitudes (6.6) are:

$$Z_{v(vv)} = \frac{1}{2k} \int_{-L_v}^{L_v} f_v(s) \left[\int_{-L_v}^{L_v} f_v(s') \left(\frac{d^2}{ds^2} + k^2\right) G_{S_{v(vv)}}^{HsE}(s, s') ds' \right] ds$$

$$= \left(\frac{\tilde{k}}{k}\right) \sin \tilde{k} L_v F_{v(vv)}(L_v) - \frac{k}{2} \cos \tilde{k} L_v \int_{-L_v}^{L_v} F_{v(vv)}(s) ds$$

$$+ \frac{k^2 - \tilde{k}^2}{2k} \int_{-L_v}^{L_v} \cos \tilde{k} s \, F_{v(vv)}(s) ds,$$

$$F_{v(vv)}(s) = \int_{-L_v}^{L_v} f_v(s') \frac{e^{-ik\sqrt{(s-s')^2+r^2(4x_d^2)}}}{\sqrt{(s - s')^2 + r^2(4x_d^2)}} ds',$$

$$F_v^{\overline{Z}} = -\frac{i}{r} \int_0^{L_v} f_v^2(s) \overline{Z}_S(s) ds, \quad F_V^{\overline{Z}} = -\frac{i}{r_V} \int_0^{L_V} f_V^2(s) \overline{Z}_{SV}(s) ds, \quad (6.7)$$

where $\overline{Z}_S(s) = \overline{R}_S + i\overline{X}_S \phi(s)$, $\overline{Z}_{SV}(s) = \overline{R}_{SV} + i\overline{X}_{SV}\phi_V(s)$, $\phi(s)$ and $\phi_V(s)$ are predefined functions,

$$Z_c = ix_d \int_{-L_v}^{L_v} f(s) \left\{ \int_{-L_{sl}}^{L_{sl}} f_{sl}^s(s') \left[\frac{e^{-ik\sqrt{s^2+s'^2+x_d^2}}}{(s^2+s'^2+x_d^2)^{3/2}} \left(ik\sqrt{s^2+s'^2+x_d^2} + 1 \right) \right] ds' \right\} ds,$$

$$Z_{sl}^{s(a)Hs\{Wg\}} = \frac{1}{2k} \int_{-L_{sl}}^{L_{sl}} f_{sl}^{s(a)}(s) \left[\int_{-L_{sl}}^{L_{sl}} f_{sl}^{s(a)}(s') \left(\frac{d^2}{ds^2} + k^2 \right) G_{S_{sl}}^{Hs\{Wg\}M}(s, s') ds' \right] ds,$$

$$Z_{sl}^{sHs} = 2 \left\{ \begin{array}{l} \left[\left(\frac{k_g}{k}\right) \cos kL_{sl} \sin k_g L_{sl} - \sin kL_{sl} \cos k_g L_{sl} \right] F_{sl}^s(L_{sl}) \\ + k \cos k_g L_{sl} \int_{-L_{sl}}^{L_{sl}} \cos ks\, F_{sl}^s(s) ds - \frac{k^2+k_g^2}{2k} \cos kL_{sl} \int_{-L_{sl}}^{L_{sl}} \cos k_g s\, F_{sl}^s(s) ds \end{array} \right\},$$

$$Z_{sl}^{aHs} = 2 \left\{ \begin{array}{l} \left[\left(\frac{k_g}{k}\right) \sin kL_{sl} \cos k_g L_{sl} - \cos kL_{sl} \sin k_g L_{sl} \right] F_{sl}^a(L_{sl}) \\ + k \sin k_g L_{sl} \int_{-L_{sl}}^{L_{sl}} \sin ks\, F_{sl}^a(s) ds - \frac{k^2-k_g^2}{2k} \sin kL_{sl} \int_{-L_{sl}}^{L_{sl}} \sin k_g s\, F_{sl}^a(s) ds \end{array} \right\},$$

$$F_{sl}^{s(a)}(s) = \int_{-L_{sl}}^{L_{sl}} f_{sl}^{s(a)}(s') \frac{e^{-ik\sqrt{(s-s')^2+(d_e/4)^2}}}{\sqrt{(s-s')^2+(d_e/4)^2}} ds',$$

$$Z_{sl}^{sWg} = \frac{2\pi}{ab} \sum_{m=0}^{\infty} \sum_{n=0}^{\infty} \frac{\varepsilon_m \varepsilon_n}{k^2} \cos k_x x_0 \cos k_x \left(x_0 + \frac{d_e}{4} \right)$$

$$\times \left\{ \begin{array}{l} \left[\cos k_g L_{sl} \left(\frac{k}{k_z} \sin kL_{sl} - \cos kL_{sl} \right) \right] F_e^s \\ - \frac{\cos kL_{sl}}{k_z^2+k_g^2} \left[(k_z^2+k^2) \left(\frac{k_g}{k_z} \sin k_g L_{sl} - \cos k_g L_{sl} \right) F_e^s + k_c^2 F_k^s \right] \end{array} \right\},$$

$$F_e^s = \frac{k \cos k_g L_{sl}}{k_z^2 + k^2} \left[k_z \cos kL_{sl} \left(1 - e^{-2k_z L_{sl}} \right) + k \sin kL_{sl} \left(1 + e^{-2k_z L_{sl}} \right) \right]$$

$$- \frac{k \cos kL_{sl}}{k_z^2 + k_g^2} \left[k_z \cos k_g L_{sl} \left(1 - e^{-2k_z L_{sl}} \right) + k_g \sin k_g L_{sl} \left(1 + e^{-2k_z L_{sl}} \right) \right],$$

$$F_k^s = 2 \cos k_g L_{sl} \frac{\sin kL_{sl} \cos k_g L_{sl} - (k_g/k) \cos kL_{sl} \sin k_g L_{sl}}{1 - (k_g/k)^2}$$

$$- \cos kL_{sl} \frac{\sin 2k_g L_{sl} + 2k_g L_{sl}}{2(k_g/k)}$$

$$Z_{sl}^{aWg} = \frac{2\pi}{ab} \sum_{m=0}^{\infty} \sum_{n=0}^{\infty} \frac{\varepsilon_m \varepsilon_n}{k^2} \cos k_x x_0 \cos k_x \left(x_0 + \frac{d_e}{4} \right)$$

$$\times \left\{ \begin{array}{l} \left[-\sin k_g L_{sl} \left(\frac{k}{k_z} \cos k L_{sl} + \sin k L_{sl} \right) \right] F_e^a \\ + \frac{\sin k L_{sl}}{k_z^2 + k_g^2} \left[(k_z^2 + k^2) \left(\frac{k_g}{k_z} \cos k_g L_{sl} + \sin k_g L_{sl} \right) F_e^a + k_c^2 F_k^a \right] \end{array} \right\},$$

$$F_e^a = \frac{k \sin k_g L_{sl}}{k_z^2 + k^2} \left[k_z \sin k L_{sl} \left(1 + e^{-2k_z L_{sl}} \right) - k \cos k L_{sl} \left(1 - e^{-2k_z L_{sl}} \right) \right]$$

$$- \frac{k \sin k L_{sl}}{k_z^2 + k_g^2} \left[k_z \sin k_g L_{sl} \left(1 + e^{-2k_z L_{sl}} \right) - k_g \cos k_g L_{sl} \left(1 - e^{-2k_z L_{sl}} \right) \right],$$

$$F_k^a = 2 \sin k_g L_{sl} \frac{\cos k L_{sl} \sin k_g L_{sl} - (k_g/k) \sin k L_{sl} \cos k_g L_{sl}}{1 - (k_g/k)^2}$$

$$- \sin k L_{sl} \frac{\sin 2k_g L_{sl} - 2k_g L_{sl}}{2(k_g/k)},$$

Z_V

$$= \frac{4\pi}{ab} \sum_{m=1}^{\infty} \sum_{n=0}^{\infty} \frac{\varepsilon_n (k^2 - k_y^2) \tilde{k}_V^2}{k k_z (\tilde{k}_V^2 - k_y^2)^2} e^{-k_z r_V} \sin^2 k_x x_{0V} [\sin \tilde{k}_V L_V \cos k_y L_V - (\tilde{k}_V/k_y) \cos \tilde{k} L_V \sin k_y L_V]^2,$$

$$Z_{Wg} = \frac{4\pi}{ab} \sum_{m=1}^{\infty} \sum_{n=0}^{\infty} \frac{\varepsilon_n k_x \tilde{k}_V}{k k_z (\tilde{k}_V^2 - k_y^2)} e^{k_z L_{sl}} \sin k_x x_{0V} \cos k_x x_{0sl}$$

$$\times F_e^s [\sin \tilde{k}_V L_V \cos k_y L_V - (\tilde{k}_V/k_y) \cos \tilde{k} L_V \sin k_y L_V],,$$

$$H_{sl}^{s(a)} = \frac{1(-i)}{k} H_0 \cos \frac{\pi x_{0sl}}{a} F_k^{s(a)},$$

$E_V = H_0 \frac{k}{k_g \tilde{k}_V} \sin \frac{\pi x_{0V}}{a} f(\tilde{k}_V L_V), \ F_V = \sin \tilde{k}_V L_V - \tilde{k}_V L_V \cos \tilde{k}_V L_V.$

Then, the final expressions for the currents based on (6.3) can be written as:

$$J_v(s_v) = -\frac{i\omega}{2k^2} H_0 J_v f_v(s_v), \ J_V(s_V) = -\frac{i\omega}{2k^2} H_0 J_V f_V(s_V), \ J_{sl}(s_{sl}),$$

$$= -\frac{i\omega}{2k^2} H_0 \left[J_{sl}^s f_{sl}^s(s_{sl}) + i J_{sl}^a f_{sl}^a(s_{sl}) \right] \tag{6.8}$$

where

$$J_v = -\frac{Z_c F_{sl,V}}{\tilde{Z}_{sl} \tilde{Z}_v + Z_c^2}, \ J_V = \frac{k^2 \sin(\pi x_{0V}/a) F_V}{k_g \tilde{k}_V Z_V^\Sigma} + \frac{\tilde{Z}_v Z_{Wg} F_{sl,V}}{(\tilde{Z}_{sl} \tilde{Z}_v + Z_c^2) Z_V^\Sigma},$$

$$J_{sl}^s = \frac{\tilde{Z}_v F_{sl,V}}{\tilde{Z}_{sl} \tilde{Z}_v + Z_c^2}, \ J_{sl}^a = -\cos \frac{\pi x_{0sl}}{a} \frac{F_k^a}{Z_{sl}^{a\Sigma}},$$

$$F_{sl,V} = \cos\frac{\pi x_{0sl}}{a}F_k^s - \frac{k^2}{k_g\tilde{k}_V}\sin\frac{\pi x_{0V}}{a}F_V.$$

These expressions make it possible to define the electrodynamic characteristics of the Clavin element. The field reflection and transmission coefficients, S_{11} and S_{12}, in the main waveguide and the power radiation coefficient $|S_\Sigma|^2$ can be obtained as:

$$S_{11} = -\frac{2\pi}{iabk_gk}\left[\frac{k_c^2}{k^2}\cos\frac{\pi x_{0sl}}{a}\left(J_{sl}^sF_k^s + J_{sl}^aF_k^a\right) - \frac{2k_g}{\tilde{k}_V}\sin\frac{\pi x_{0V}}{a}J_VF_V\right]e^{2ik_gz},$$

$$(6.9)$$

$$S_{12} = 1 - \frac{2\pi}{iabk_gk}\left[\frac{k_c^2}{k^2}\cos\frac{\pi x_{0sl}}{a}\left(J_{sl}^sF_k^s - J_{sl}^aF_k^a\right) + \frac{2k_g}{\tilde{k}_V}\sin\frac{\pi x_{0V}}{a}J_VF_V\right], \quad (6.10)$$

$$|S_\Sigma|^2 = 1 - |S_{11}|^2 - |S_{12}|^2. \quad (6.11)$$

The formulas $(6.8 - 6.11)$ were obtained assuming that $H_0 = 1$ and the vibrators are losses. In the spherical coordinate system shown in Fig. 6.2, the far-zone electric field of the Clavin element is defined by the expression.

$$\vec{E}(R,\theta,\varphi) = \frac{ik^2}{\omega}\frac{e^{-ikR}}{R}\left[\vec{\theta}^{\,0}\sin\theta\left(\tilde{E}_1e^{-ikx_{d1}\sin\theta\sin\varphi} + \tilde{E}_2e^{ikx_{d2}\sin\theta\sin\varphi}\right)\right.$$
$$\left. +\left(\vec{\varphi}^{\,0}\cos\theta\cos\varphi + \vec{\theta}^{\,0}\sin\varphi\right)2\tilde{E}_3\right], \quad (6.12)$$

where $\vec{\theta}^{\,0}$ and $\vec{\varphi}^{\,0}$ are unit vectors, $\tilde{E}_1 = J_{01}f_{C1}$, $\tilde{E}_2 = J_{02}f_{C2}$, $\tilde{E}_3 = J_{03}^sf_{C3}^s + J_{03}^af_{C3}^a$;

$$f_{C1} = \int_{-L_1}^{L_1}f_1(z)e^{ikz\cos\theta}dz, \quad f_{C2} = \int_{-L_2}^{L_2}f_2(z)e^{ikz\cos\theta}dz, \quad f_{C3}^{s(a)} = \int_{-L_3}^{L_3}f_3^{s(a)}(x)e^{ikx\sin\theta\cos\varphi}dx.$$

Fig. 6.2 Coordinate system for determining the electric field vibrator-slot structure

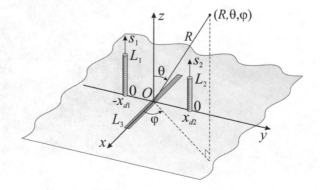

Then, in accordance with formulas (6.3), we obtain:

$$f_{Cm} = \frac{2}{\tilde{k}_m^2 - (k\cos\theta)^2}\left[\tilde{k}_m \cos(kL_m \cos\theta)\sin\left(\tilde{k}_m L_m\right) - k\sin(kL_m \cos\theta)\cos\left(\tilde{k}_m L_m\right)\cos\theta\right]$$
$$- 2L_m \cos\left(\tilde{k}_m L_m\right)\frac{sin(kL_m \cos\theta)}{kL_m \cos\theta}, \quad m = 1, 2;$$

$$f_{C3}^s$$
$$= \frac{2\cos(k_g L_3)}{k - k(\sin\theta\cos\varphi)^2}[\cos(kL_3 \sin\theta\cos\varphi)\sin(kL_3) - \sin(kL_3 \sin\theta\cos\varphi)\cos(kL_3)\sin\theta\cos\varphi]$$
$$- \frac{2\cos(kL_3)}{k_g^2 - (k\sin\theta\cos\varphi)^2}\left[k_g \cos(kL_3 \sin\theta\cos\varphi)\sin(k_g L_3)\right.$$
$$\left. - k\sin(kL_3 \sin\theta\cos\varphi)\cos(k_g L_3)\sin\theta\cos\varphi\right],$$

$$f_{C3}^a$$
$$= \frac{2i\sin(k_g L_3)}{k - k(\sin\theta\cos\varphi)^2}[-\sin(kL_3 \sin\theta\cos\varphi)\cos(kL_3) + \cos(kL_3 \sin\theta\cos\varphi)\sin(kL_3)\sin\theta\cos\varphi]$$
$$- \frac{2i\sin(kL_3)}{k_g^2 - (k\sin\theta\cos\varphi)^2}\left[-k_g \sin(kL_3 \sin\theta\cos\varphi)\cos(k_g L_3)\right.$$
$$\left. + k\cos(kL_3 \sin\theta\cos\varphi)\sin(k_g L_3)\sin\theta\cos\varphi\right].$$

6.3 Vibrator-Slot Radiator Based on a Hollow Rectangular Waveguide

If the combine radiator do not include vibrator in the hollow waveguide (Fig. 6.3), the SLAE (6.4) can be simplified to

Fig. 6.3 The geometry of the problem and accepted notations

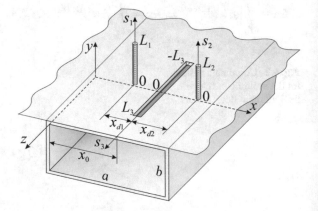

$$\begin{cases} J_{01} Z_{11}^{\Sigma} + J_{02} Z_{12} + J_{03}^{s} Z_{13} = 0, \\ J_{02} Z_{11}^{\Sigma} + J_{01} Z_{21} + J_{03}^{s} Z_{23} = 0, \\ J_{03}^{s} Z_{33}^{s\Sigma} + J_{01} Z_{31} + J_{02} Z_{32} = -\frac{i\omega}{2k} H_3^s, \\ J_{03}^{a} Z_{33}^{a\Sigma} = -\frac{i\omega}{2k} H_3^a, \end{cases} \quad (6.13)$$

where $Z_{11}^{\Sigma} = Z_{11} + F_1^{\overline{Z}}$, $Z_{22}^{\Sigma} = Z_{22} + F_2^{\overline{Z}}$, $Z_{33}^{s(a)\Sigma} = Z_{33}^{s(a)Wg} + Z_{33}^{s(a)Hs}$.
Solution of equations system (6.13) can be written as

$$J_{01} = \frac{i\omega}{2k} H_3^s \frac{Z_{22}^{\Sigma} Z_{13} - Z_{23} Z_{12}}{Z_{33}^{s\Sigma} \tilde{Z}_{12} + Z_{31} \tilde{Z}_{22} + Z_{32} \tilde{Z}_{11}}, \quad J_{02} = \frac{i\omega}{2k} H_3^s \frac{Z_{11}^{\Sigma} Z_{23} - Z_{13} Z_{21}}{Z_{33}^{s\Sigma} \tilde{Z}_{12} + Z_{31} \tilde{Z}_{22} + Z_{32} \tilde{Z}_{11}}, \quad (6.14a)$$

$$J_{03}^{s} = -\frac{i\omega}{2k} H_3^s \frac{\tilde{Z}_{12}}{Z_{33}^{s\Sigma} \tilde{Z}_{12} + Z_{31} \tilde{Z}_{22} + Z_{32} \tilde{Z}_{11}}, \quad J_{03}^{a} = -\frac{i\omega}{2k} H_3^s \frac{1}{Z_{33}^{a\Sigma}}. \quad (6.14b)$$

Since $H_3^{s(a)} = \frac{1(-i)}{k} H_0 \cos \frac{\pi x_0}{a} F_k^{s(a)}$, the final expressions for the currents can be presented as:

$$J_1(s_1) = \frac{i\omega}{2k^2} H_0 J_1 f_1(s_1), \ J_2(s_2) = \frac{i\omega}{2k^2} H_0 J_2 f_2(s_2), \quad (6.15a)$$

$$J_3(s_3) = -\frac{i\omega}{2k^2} H_0 \big[J_3^s f_3^s(s_3) + i J_3^a f_3^a(s_3) \big], \quad (6.15b)$$

where

$$J_1 = \cos \frac{\pi x_0}{a} F_k^s \frac{Z_{22}^{\Sigma} Z_{13} - Z_{23} Z_{12}}{Z_{33}^{s\Sigma} \tilde{Z}_{12} + Z_{31} \tilde{Z}_{22} + Z_{32} \tilde{Z}_{11}},$$

$$J_2 = \cos \frac{\pi x_0}{a} F_k^s \frac{Z_{11}^{\Sigma} Z_{23} - Z_{13} Z_{21}}{Z_{33}^{s\Sigma} \tilde{Z}_{12} + Z_{31} \tilde{Z}_{22} + Z_{32} \tilde{Z}_{11}}, \quad (6.16a)$$

$$J_3^s = \cos \frac{\pi x_0}{a} F_k^s \frac{\tilde{Z}_{12}}{Z_{33}^{s\Sigma} \tilde{Z}_{12} + Z_{31} \tilde{Z}_{22} + Z_{32} \tilde{Z}_{11}}, \ J_3^a = -\cos \frac{\pi x_0}{a} F_k^a \frac{1}{Z_{33}^{a\Sigma}}, \quad (6.16b)$$

$\tilde{Z}_{12} = Z_{11}^{\Sigma} Z_{22}^{\Sigma} - Z_{12} Z_{21}$, $\tilde{Z}_{11} = Z_{13} Z_{21} - Z_{11}^{\Sigma} Z_{23}$, $\tilde{Z}_{22} = Z_{23} Z_{12} - Z_{22}^{\Sigma} Z_{13}$.
After substitution the amplitude $H_0 = 1$ in formulas (6.15a) and (6.15b), the radiation field is determined by the formula (6.12) and the energy characteristics of this structure can be obtained in the form:

$$S_{11} = -\frac{2\pi k_c^2}{iabk_g k^3} \cos \frac{\pi x_0}{a} \big(J_3^s F_k^s + J_3^a F_k^a \big) e^{2ik_g z}, \quad (6.17)$$

$$S_{12} = 1 - \frac{2\pi k_c^2}{iabk_g k^3} \cos \frac{\pi x_0}{a} \big(J_3^s F_k^s - J_3^a F_k^a \big), \quad (6.18)$$

$$|S_\Sigma|^2 = 1 - |S_{11}|^2 - |S_{12}|^2, \tag{6.19}$$

The components of the Green's functions for this case can be written in the form:

$$G_{s_{1(2)}}^{HsE}(s_{1(2)}, s'_{1(2)}) = \frac{e^{-ik\sqrt{(s_{1(2)}-s'_{1(2)})^2+r_{1(2)}^2}}}{\sqrt{(s_{1(2)} - s'_{1(2)})^2 + r_{1(2)}^2}},$$

$$G_{s_{1(2)}}^{HsE}(s_{1(2)}, s'_{2(1)}) = \frac{e^{-ik\sqrt{(s_{1(2)}-s'_{2(1)})^2+(x_{d1}+x_{d2})^2}}}{\sqrt{(s_{1(2)} - s'_{2(1)})^2 + (x_{d1} + x_{d2})^2}},$$

$$G_{s_3}^{HsM}(s_3, s'_3) = 2\frac{e^{-ik\sqrt{(s_3-s'_3)^2+(d_e/4)^2}}}{\sqrt{(s_3 - s'_3)^2 + (d_e/4)^2}},$$

$$\tilde{G}_{s_{1(2)}}^{HsE}(s_3, s'_{1(2)}) = \frac{\partial}{\partial x_{Hs}} G_{s_{1(2)}}^{HsE}[x_{Hs}, 0, z_{Hs}(s_3); x_{d1(2)}, y'_{Hs}(s'_{1(2)}), 0] \text{ at } x_{Hs} = 0,$$

$$\tilde{G}_{s_3}^{HsM}(s_{1(2)}, s'_3) = \frac{\partial}{\partial x_{Hs}} G_{s_3}^{HsM}[x_{Hs}, y_{Hs}(s_{1(2)}), 0; 0, 0, z'_{Hs}(s'_3)] \text{ at } x_{Hs} = x_{d1(2)}.$$

Then, the corresponding coefficients in formulas (6.14) and (6.16) are:

$Z_{11(22)}$

$$= \frac{1}{2k} \int_{-L_{1(2)}}^{L_{1(2)}} f_{1(2)}(s_{1(2)}) \left[\int_{-L_{1(2)}}^{L_{1(2)}} f_{1(2)}(s'_{1(2)}) \left(\frac{d^2}{ds_{1(2)}^2} + k^2 \right) G_{s_{1(2)}}^{HsE}(s_{1(2)}, s'_{1(2)}) ds'_{1(2)} \right] ds_{1(2)}$$

$$= \left(\frac{\tilde{k}_{1(2)}}{k} \right) \sin \tilde{k}_{1(2)} L_{1(2)} F_{1(2)}(L_{1(2)}) - \frac{k}{2} \cos \tilde{k}_{1(2)} L_{1(2)} \int_{-L_{1(2)}}^{L_{1(2)}} F_{1(2)}(s_{1(2)}) ds_{1(2)}$$

$$+ \frac{k^2 - \tilde{k}_{1(2)}^2}{2k} \int_{-L_{1(2)}}^{L_{1(2)}} \cos \tilde{k}_{1(2)} s_{1(2)} F_{1(2)}(s_{1(2)}) ds_{1(2)},$$

$$F_{1(2)}(s_{1(2)}) = \int_{-L_{1(2)}}^{L_{1(2)}} f_{1(2)}(s'_{1(2)}) \frac{e^{-ik\sqrt{(s_{1(2)}-s'_{1(2)})^2+r_{1(2)}^2}}}{\sqrt{(s_{1(2)} - s'_{1(2)})^2 + r_{1(2)}^2}} ds'_{1(2)},$$

$$Z_{12} = Z_{21} = \frac{1}{2k} \int_{-L_1}^{L_1} f_1(s_1) \left[\int_{-L_2}^{L_2} f_2(s'_2) \left(\frac{d^2}{ds_1^2} + k^2 \right) G_{s_2}^{HsE}(s_1, s'_2) ds'_2 \right] ds_1$$

$$= \left(\frac{\tilde{k}_1}{k}\right) \sin \tilde{k}_1 L_1 F_{12}(L_1) - \frac{k}{2} \cos \tilde{k}_1 L_1 \int_{-L_1}^{L_1} F_{12}(s_1) ds_1$$

$$+ \frac{k^2 - \tilde{k}_1^2}{2k} \int_{-L_1}^{L_1} \cos \tilde{k}_1 s_1 F_{12}(s_1) ds_1,$$

$$F_{12}(s_{12}) = \int_{-L_2}^{L_2} f_2(s_2') \frac{e^{-ik\sqrt{(s_1 - s_2')^2 + (x_{d1} + x_{d2})^2}}}{\sqrt{(s_1 - s_2')^2 + (x_{d1} + x_{d2})^2}} ds_2',$$

$$F_{1(2)}^{\overline{Z}} = -\frac{i}{r_{1(2)}} \int_0^{L_{1(2)}} f_{1(2)}^2(s_{1(2)}) \overline{Z}_{S1(2)}(s_{1(2)}) ds_{1(2)}, \tag{6.20}$$

where $\overline{Z}_{S1(2)}(s_{1(2)}) = \overline{R}_{S1(2)} + i\overline{X}_{S1(2)}\phi_{1(2)}(s_{1(2)})$, $\phi_{1(2)}(s_{1(2)})$ are predefined functions,

$$Z_{1(2)3}$$

$$= 2ix_{d1(2)} \int_{-L_{1(2)}}^{L_{1(2)}} f_{1(2)}(s_{1(2)}) \left\{ \int_{-L_3}^{L_3} f_3^s(s_3') \left[\frac{e^{-ik\sqrt{s_{1(2)}^2 + s_3'^2 + x_{d1(2)}^2}}}{(s_{1(2)}^2 + s_3'^2 + x_{d1(2)}^2)^{3/2}} \times \left(ik\sqrt{s_{1(2)}^2 + s_3'^2 + x_{d1(2)}^2} + 1\right) \right] ds_3' \right\} ds_{1(2)},$$

$$Z_{31(2)}$$

$$= ix_{d1(2)} \int_{-L_3}^{L_3} f_3(s_3) \left\{ \int_{-L_{1(2)}}^{L_{1(2)}} f_{1(2)}(s_{1(2)}') \left[\frac{e^{-ik\sqrt{s_3^2 + s_{1(2)}'^2 + x_{d1(2)}^2}}}{(s_3^2 + s_{1(2)}'^2 + x_{d1(2)}^2)^{3/2}} \times \left(ik\sqrt{s_3^2 + s_{1(2)}'^2 + x_{d1(2)}^2} + 1\right) \right] ds_{1(2)}' \right\} ds_3,$$

The expressions $Z_{33}^{s(a)\Sigma} = Z_{33}^{s(a)Wg} + Z_{33}^{s(a)Hs}$ and $F_k^{s(a)}$ are analogues to that defined in Sect. 6.2 can be used after replacing the index sl by 33.

Thus, the equation system for the Clavin element can be written as

$$\begin{cases} J_{0v} Z_v^\Sigma - J_{0v} Z_{vv} + J_{0sl}^s Z_c = 0, \\ J_{0sl}^s Z_{sl}^{s\Sigma} - J_{0v} Z_c = -\frac{i\omega}{2k} H_{sl}^s, \\ J_{0sl}^a Z_{sl}^{a\Sigma} = -\frac{i\omega}{2k} H_{sl}^a. \end{cases} \tag{6.21}$$

The complex current amplitudes obtained by solving the equation system (6.21) are presented as:

$$J_{0v1} = -\frac{i\omega}{2k} H_{sl}^s \frac{Z_c}{Z_{sl}^{s\Sigma}(Z_v^\Sigma - Z_{vv}) - Z_c^2}, \quad J_{0v2} = -J_{0v1},$$

$$J_{0sl}^s = -\frac{i\omega}{2k} H_{sl}^s \frac{Z_v^\Sigma - Z_{vv}}{Z_{sl}^{s\Sigma}(Z_v^\Sigma - Z_{vv}) - Z_c^2}, \quad J_{0sl}^a = -\frac{i\omega}{2k} H_{sl}^a \frac{1}{Z_{sl}^{a\Sigma}}. \quad (6.22)$$

The current distribution along the vibrators and slot are determined by the following expressions:

$$J_v(s) = \frac{i\omega}{2k^2} H_0 \cos\frac{\pi x_0}{a} J_v f_v(s), \quad J_{sl}(s) = -\frac{i\omega}{2k^2} H_0 \cos\frac{\pi x_0}{a}[J_{sl}^s f_{sl}^s(s) + iJ_{sl}^a f_{sl}^a(s)], \quad (6.23)$$

where $J_v = \frac{Z_c F_k^s}{Z_{sl}^{s\Sigma}(Z_v^\Sigma + Z_{vv}) + Z_c^2}$, $J_{sl}^s = \frac{(Z_v^\Sigma + Z_{vv})F_k^s}{Z_{sl}^{s\Sigma}(Z_v^\Sigma + Z_{vv}) + Z_c^2}$, $J_{sl}^a = -\frac{F_k^a}{Z_{sl}^{a\Sigma}}$.

All coefficients defined in Sect. 6.2 and the energy characteristics for this structure, are defined by the expressions (6.17)–(6.19) if the index 33 is replaced by the index sl.

6.3.1 Numerical and Experimental Results

The combined Clavin radiator consists of two identical perfectly conducting monopoles located symmetrically to the slot axis [23]. Clavin has shown experimentally that RPs in the H-plane ($\varphi = 0°$) and E-planes ($\varphi = 90°$) are approximately identical if the monopoles with length $L_v = 0.375\lambda$ are placed symmetrical relative to the slot axis at distances $x_d = 0.086\lambda$. It is quite clear that, according to the expression (6.12), the RP in the H-plane has only the E_φ component, which coincides with the RP of stand along slot since the monopole currents with equal amplitudes are in anti-phase relative to each other. The RP in the plane $\varphi = 0°$ obtained under the assumption of radiator with infinite screen are shown in Fig. 6.4a marked by a solid

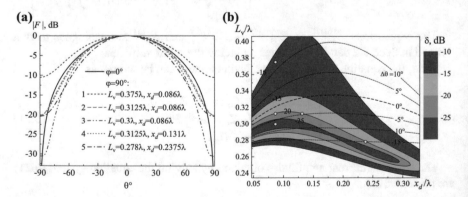

Fig. 6.4 Clavin radiator RPs **a**, and dependences of the relative level of lateral radiation δ and the difference in width RP in orthogonal planes on the electric length of monopoles and their removal from the slot **b**

curve. According to (6.12), the RP in the E-plane has only the E_θ component, and it is variation of vibrator currents that makes it possible to bring its shape closer to the H-plane RP. The RP for this case is shown in Fig. 6.4a by dashed curve 1. Better coincidence of the RP in the two planes as compared to that shown in Fig. 6.4a was achieved through the use of L-shaped vibrators [23].

In [25], based on the solution of an external problem using trigonometric approximations for slot and vibrator currents, the radiator parameters $L_v = 0.365\lambda$, $x_d = 0.065\lambda$, where attained that a little bit different from that obtained by Clavin [23, 24]. This was achieved due to the lower radiation level along the plane as compared to that in [23]. An attempt to determine the relation between the parameters L_v/λ and x_d/λ, which allows to obtain single-lobe RPs with a side lobe level less than -16 dB was undertaken in [28]. The simulation results have also showed that other pairs of parameters L_v/λ and x_d/λ can be found which RPs differ in shapes but have the same level of lateral radiation. This can be seen from curves 2 and 5 in Fig. 6.4a. In this regard, these results cannot claim to be complete and need further study.

The influence of the relative vibrator lengths L_v/λ and distance $2x_d/\lambda$ on the directivity characteristics of the Clavin type radiators where studied by measuring both the relative level δ of lateral radiation, i.e., the maximum of E-plane RP, and the difference $\Delta\theta$ between the widths of RP in E- and H-plane at -3 dB level.

The plots of L_v/λ as functions of x_d/λ are shown Fig. 6.4b, where the δ and $\Delta\theta$ are represented by the gray-level scale and the series of level curves. The numerical results presented in Fig. 6.4b allows us to state there exist two variants of choosing the monopole lengths and the distance between the monopoles: (1) to obtain RPs of the same width in the E- and H-plane with the lowest level of lateral radiation; (2) to provide the E- and H-plane RPs closest in width for predefined levels of lateral radiation. The RPs corresponding to these variants are shown in Fig. 6.4a: curve 2 is plotted for the first variant with the side lobe level -20 dB ($L_v = 0.3125\lambda$, $x_d = 0.086\lambda$); curve 3 is plotted for the lowest level of lateral radiation $\delta = -31$ dB. The parameter pairs for curves shown in Fig. 6.4a are marked by circles.

The radiator energy characteristics were obtained with following parameters: $\lambda = 32.0$ mm, $\{a \times b\} = 23.0 \times 10.0$ mm^2, $h = 1.0$ mm, $2L_{sl} = 2L_s = 16.0$ mm ($2L_s = 0.5\lambda$), $d = 1.5$ mm, $x_0 = 2.5$ mm, $r = 0.17$ mm. Since the relations $2r_v/L_{vv}/L_{vv}/L_{vv}/L_{vv}/L_v$ and $[d/(2L_s)]$ do not exceed 0.1, the thin wire and narrow slot approximations were used during simulation. The energy characteristics of the combined Clavin radiators, whose RPs are presented in Fig. 6.4a are summarized in Table 6.1.

Table 6.1 The energy characteristics of combine Clavin radiators

Geometric parameters	$|S_{11}|$	$|S_{12}|$	$|S_\Sigma|^2$	D
$L_v = 0.375\lambda$, $x_d = 0.086\lambda$	0.251	0.854	0.157	6.366
$L_v = 0.3125\lambda$, $x_d = 0.086\lambda$	0.207		0.074	7.485
$L_v = 0.3\lambda$, $x_d = 0.086\lambda$	0.184		0.057	7.854

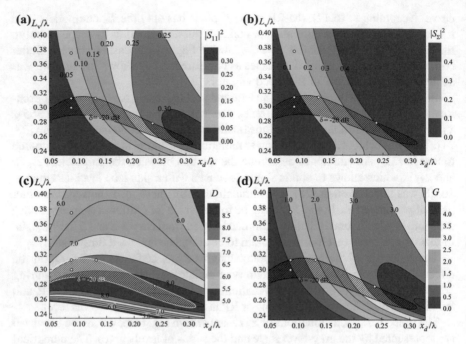

Fig. 6.5 Dependences of the coefficients $|S_{11}|$ **a**, $|S_\Sigma|^2$ **b**, D **c** and G **d** of the radiator on electric length of monopoles and their removal from the slot

The numerical results are summarized in Fig. 6.5, where the radiation, reflection coefficients, directivity factor (D) and the gain (G) of the Clavin radiator are shown as function of monopole electric length and the distance between the slot and monopole. These plots allow to obtain the required energy characteristics and directivity factor by selecting the geometric parameters of the radiator. The characteristic pairs of parameters are marked by circles as in Fig. 6.4b.

It turned out, that the lowest level of lateral radiation ($\delta = -31$ dB) for the radiator with the geometric parameters ($L_v = 0.3\lambda$, $x_d = 0.086\lambda$) can be obtained if radiation coefficient is rather low. This can be explained by the phasing conditions for the radiation fields of the slot and vibrators. Really, in far zone, the slot radiation field along the plane can be completely compensated if the both vibrators induce in the geometric slot center an equivalent electric field whose amplitude is equal to the slot field amplitude and these fields are anti-phased. Consequently, the compensation of intrinsic slot field will take place at the same time significantly reducing the radiating capacity of the slot. As follows from Fig. 6.5b, the $|S_\Sigma|^2$ level can be increased if the distance between the vibrators is also increased. This inevitable violates the phase relations and reduces requirements to the level of lateral radiation. As an example, consider the RP of the radiator with geometric parameters $L_v = 0.3\lambda$, $x_d = 0.131\lambda$, presented in Fig. 6.4a (curve 4). Comparing curves 1 and 4 shows that the radiation coefficient increases up to $|S_\Sigma|^2 = 0.184$, while the E-plane RP grow narrower $\left(\Delta\theta = -5^0\right)$. If the distance between the vibrators is further increased under conditions that $\delta = -20\,\mathrm{dB}$, the radiation and reflection coefficients become relatively large, $|S_\Sigma|^2 = 0.403$, $|S_{11}| = 0.525$ the directivity factor also increases ($D = 8.273$), since the E- plane RP becomes narrower, $\Delta\theta = -15°$ (curve 5 in Fig. 6.4b). In Figs. 6.4b and 6.5, the points corresponding to L_v and x_d, which RPs are shown in Fig. 6.4a are marked by the circles.

The analysis of the plots in Fig. 6.5 shows that the energy characteristics of the combined radiator can be controlled by varying the electric length of vibrators. As is known [29], the vibrator electric length can be varied by coating its surface with imaginary constant impedance. In this case, it can be assumed that the RP shape will not be substantially change, since the RP of monopoles with electric lengths in the range $0 < L_v/\lambda < 0.3$ is similar to that of dipole radiator. This assumption for the vibrator with constant impedance distribution can be verified for one a priori chosen impedance value.

The numerical results have shown that monopoles with inductive surface impedance makes it possible to realize specified radiator characteristics with shorter monopoles. For example, the electric length of the monopole with constant impedance $\overline{Z}_S = 0.1i$ can be reduced relative their physical length by about 30%. The characteristics of the Clavin radiators are presented in Figs. 6.6 and 6.7. As can be seen the minimum RP width $\Delta\theta$, with predefined level of lateral radiation can be obtained by changing the length of the monopoles and distance between them. The dependence of energy characteristics and RP width on the lateral radiation level is illustrated in Table 6.2.

In this case, if the distance between the vibrators is increased and the length of the monopoles is decreased, according to Fig. 6.7, it is possible to increase both the radiation coefficient D and, hence, the gain G. But the value of $\Delta\theta$ increases, since the RP in the plane of the vibrators as can be seen from the curve 3 in Fig. 6.6a, where $\Delta\theta = -23.7°$. Thus, the reflection and radiation coefficients can be changed over a wide range by varying the length of the vibrators, the distance between them,

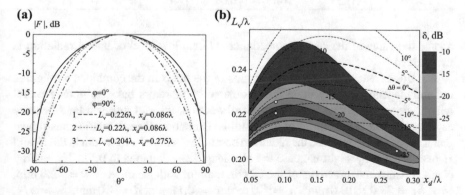

Fig. 6.6 Clavin radiator RP (F) with impedance monopoles **a** and the dependence of the relative level δ of lateral radiation and the difference in the width RP in the orthogonal planes on the electric length of the monopoles and their distance from the slot **b** at $\overline{Z}_S = 0.1i$

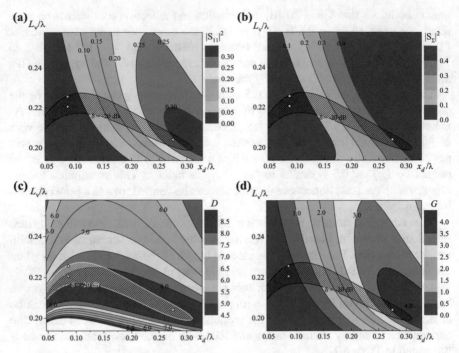

Fig. 6.7 Dependences of reflection coefficients $|S_{11}|$ **a**, radiation $|S_\Sigma|^2$ **b**, D **c** and G **d** from the electric length of the monopoles and their removal from the slot at $\overline{Z}_S = 0.1i$

Table 6.2 The energy characteristics and RP width as function of the lateral radiation level

δ, dB	$\Delta\theta$,	$|S_{11}|$	$|S_\Sigma|^2$	D	Curve in Fig. 6.6a, No
-20	-7	0.17	0.05	7.74	1
-32.5	-11.3	0.15	0.04	8.1	2

and/or the value of the surface impedance while a low level of lateral radiation is maintained.

It is quite clear, that phase relationships of the fields in the combined waveguide radiator depend not only upon the geometry of vibrators but also upon the slot length, since if it differs from resonant dimension, the slot intrinsic field become asymmetric due to conditions of longitudinal excitation. The simulation has shown that it is possible to increase the radiation coefficient when the RP width in the E- and H-planes equal by slight increase the slot length as compared to 0.5λ. The energy characteristics and directivity of the Clavin radiator with parameters: $\lambda = 32.0$ mm, $\{a \times b\} = 23.0 \times 10.0$ mm^2, $h = 1.0$ mm, $r = 0.17$ mm, $d = 1.0$ mm, $x_0 = a/4$, $x_d = 0.086\lambda$ and perfectly conducting and impedance $(\overline{Z}_S = 0.1i)$ vibrators as function of the electric length of the slot and vibrators are presented in Figs. 6.8, 6.9, 6.10 and 6.11. Thus, the RPs of equal width in the E- and H-planes with the

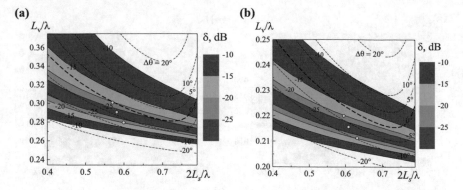

Fig. 6.8 Dependences of the relative level of lateral radiation δ and the difference in width RP Δθ (dashed curves) in orthogonal planes on the electric length of the slot and monopoles: **a**—$\overline{Z}_S = 0$, **b**—$\overline{Z}_S = 0.1i$

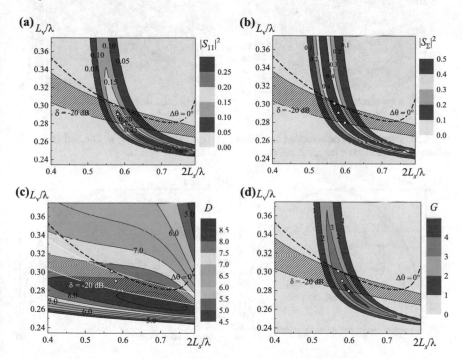

Fig. 6.9 Dependences of reflection coefficients $|S_{11}|$ **a**, radiation $|S_\Sigma|^2$ **b**, D **c** and G **d** of the radiator from the electric length of the slot and monopoles at $\overline{Z}_S = 0$

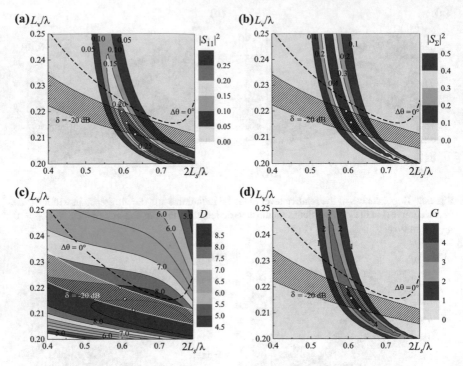

Fig. 6.10 Dependences of reflection coefficients $|S_{11}|$ **a**, radiation $|S_\Sigma|^2$ **b**, D **c** and G **d** of the radiator from the electric length of the slot and monopoles at $\overline{Z}_S = 0.1i$

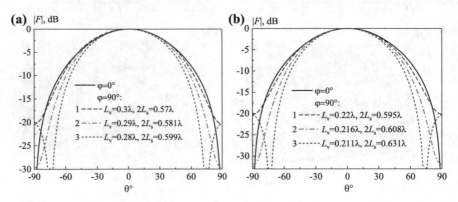

Fig. 6.11 RPs ($|F|$) of the Clavin radiator in the planes $\varphi = 0°$ (solid lines) and $\varphi = 90°$: **a**—$\overline{Z}_S = 0$, **b**—$\overline{Z}_S = 0.1i$.

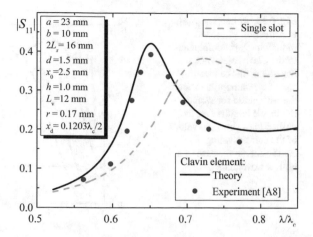

Fig. 6.12 Dependence of the reflection coefficient of the Clavin radiator on the wavelength (λ_c—critical wavelength H_{10}—wave)

lateral radiation level $\delta = -20$dB value can be obtained if the perfectly conducting monopoles are used with parameters $L_v = 0.3\lambda$, $2L_s = 0.57\lambda$ (curve 1 in Fig. 6.11a). The RP closest in width ($\Delta\theta = -3.4°$) can be realized if the impedance monopoles with parameters $L_v = 022\lambda$, $2L_s = 0.595\lambda$ are used (curve 1 in Fig. 6.11b). The radiation coefficients in the first and second cases are equal $|S_\Sigma|^2 = 0.493$ and $|S_\Sigma|^2 = 0.497$. The RPs with the lowest levels of lateral radiation with the maximal radiation coefficient $|S_\Sigma|^2$ and the highest gain are presented in Fig. 6.11, curve 2 and curves 3. The parameters L_v/λ and $2L_s/\lambda$ used for simulation are indicated on Fig. 6.8 by circles.

All the curves presented above were plotted by using the parameters normalized at free space wavelength. This ensures simple evaluating the radiator characteristics in the operating wavelength range. The reliability of the results and the correctness of the proposed mathematical model of the Clavin radiator are confirmed by comparison with experimental data and the results found known in the literature for special cases. For example, Fig. 6.12 shows the simulated and experimental reflection coefficients of the waveguide radiator, which are characterized by satisfactory agreement between themselves. The plots of the reflection coefficient as functions of relative wavelength for the single slot with corresponding geometrical parameters are presented in Fig. 6.12.

To summarize the simulation results, we note that the optimal RP of the combined vibrator-slot structure can be realized only with small radiation coefficient, since compensation of the slot field (along the plane) in the far zone by the diffraction fields of vibrators is accompanied by suppression of intrinsic slot fields. On the other hand, if the influence of vibrators on the slot is decreased by varying the geometric or electrophysical parameters, the slot radiation coefficients are increasing, and the conditions for obtaining the optimal RP shape are violated. In this case, the increase of the radiation coefficient will also lead to increase of the reflection coefficient since the combined radiator that can be interpreted as a waveguide loaded by a slot inhomogeneity. This contradiction can be overcame and the radiating structure with

Fig. 6.13 Reflection $|S_{11}|$, transmission $|S_{12}|$, and radiation $|S_{\Sigma}|^2$ coefficients of the Clavin element depending on the location x_{0v} of the monopole inside the waveguide for various monopole lengths L_V; the dashed lines show the values of the corresponding coefficients in the case of a hollow waveguide

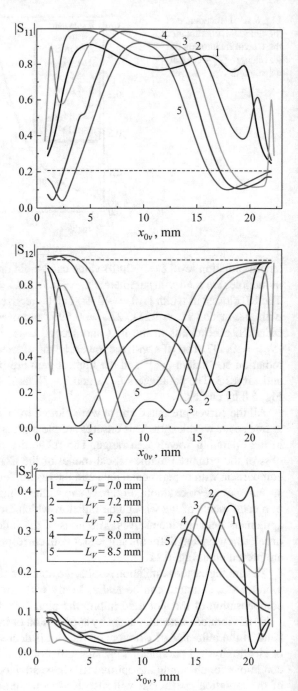

Fig. 6.14 The geometry of
the waveguide vibrator-slot
structure and the accepted
notations

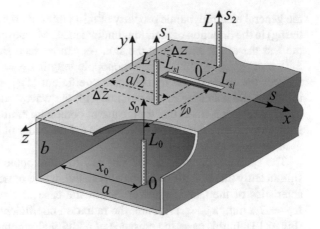

the optimal RP, high radiation coefficient, and low reflection coefficient according to
the results of Sect. 4.1 can be realized by using the additional tuning element—the
vibrator located inside the waveguide [13]. Of course, in this case, the maximum
radiation coefficient can be attained at a frequency close to the frequency at which
the condition $|\arg S_{11}| = 0$ is satisfied with S_{11} defined by the expression (6.17).

6.4   Vibrator-Slot Radiator Based on a Rectangular Waveguide with the Tuning Vibrator (Numerical Results)

Figure 6.13 shows the energy coefficients of the Clavin radiator based on a longitudinal slot in the broad wall of a rectangular waveguide with a vibrator inside the
waveguide, depending on the parameters of the monopole x_{0v}, L_V, located parallel to
its narrow walls in the plane $\{x0y\}$ inside the waveguide (Fig. 6.1). The parameters of
the radiator are as follows: $\{a \times b\} = 23.0 \times 10.0\,\text{mm}^2$, $h = 1.0\,\text{mm}$, $\lambda = 32.0\,\text{mm}$,
$2L_s = 16\,\text{mm}$, $d = 1.5\,\text{mm}$, $x_0 = 2.5\,\text{mm}$, $r = 0.17\,\text{mm}$, $L_v = 0.3125\lambda\,\text{mm}$,
$x_d = 0.086\lambda\,\text{mm}$, the radius and surface impedance of the monopole located in the
waveguide, $r_V = 0.25\,\text{mm}$, $\overline{Z}_{SV} = 0$.

In Sect. 6.3.1, it was shown that the radiator of the indicated geometry on $\lambda =$
32.0 mm forms optimal RPs with the same width in the E- and H-planes, but in the
case of a hollow waveguide it has a low radiation coefficient $|S_{\Sigma}|^2 = 0.074$ at a
sufficiently high level of reflection coefficient. As one would expect, the location
of a monopole inside the waveguide (for a certain length L_V and displacement x_{0v}
relative to the waveguide wall) allows a significant change in this ratio of energy
characteristics.

The calculation results showed that the passive monopole has a weak effect on
the slot, being directly in the region of the waveguide under its aperture. This affects

the general electrodynamic property of thin vibrators: the absence of radiation (scattering) in the direction of their longitudinal axes. Moreover, the longer the monopole, the less the effect exerted on the slot. As can be seen from Fig. 6.13, in the range $1.75\ mm \leq x_{0v} \leq 3.25\ mm$ at monopole length $L_V = 8.5\ mm$ non-significantly reduces the value of the transmission coefficient $|S_{12}|$: within $\approx 5\%$. However, it improves the matching of the radiator with the waveguide by almost 10 times. The observed effect is important for the use of combined emitters in multi-element linear antenna arrays of significant dimensions (of the order of hundreds λ), for example, used on spacecraft.

The spatial separation of the slot and the monopole inside the waveguide can significantly increase the radiation coefficient, while preserving the directional characteristics of the Clavin radiator. So, in the case of a monopole with parameters $L_V = 7.2$ mm, $x_{0v} = 17.5$ mm, the radiation coefficient increases to $|S_\Sigma|^2 = 0.4$ (Fig. 6.13). In this case, the presence of additional inhomogeneity in the waveguide (in the form of a monopole) leads to an increase in the reflection coefficient (relative to the isolated resonant slot) to a level $|S_{11}| = 0.55$.

In general, the calculations confirm the possibility of effective control of the energy coefficients of the Clavin radiator by introducing a tuning monopole into the waveguide. However, the multi-parameter tuning of the radiator to resonance, which is difficult to achieve experimentally, remains a separate problem. The next section discusses this issue in more detail for the case of the transverse position of the slot relative to the longitudinal axis of the waveguide, when the radiator coefficient $|S_\Sigma|^2$ can be significantly increased compared with the analyzed case of the longitudinal location of the slot, as shown in Chap. 4.

6.5 Problem Formulation and Initial Equations in the General Case for Transverse Slot Element

Let us consider the radiating structure with the Clavin element, transverse slot cut in the wide wall of rectangular waveguide symmetrically relative to its longitudinal axis, and impedance vibrator inside the waveguide (Fig. 6.14). The external vibrators with lengths $2L_1 = 2L_2 = 2L_v = 2L$ are located at equal distances $x_{d1} = x_{d2} = x_d = \Delta z$ from the longitudinal slot axis. The internal vibrator with length $2L_4 = 2L_0$ is offset along the axes $\{0x\}$ and $\{0z\}$ at distances x_0 and z_0. The impedance of the internal vibrator is $\overline{Z}_{S4}(s_4) = \overline{Z}_{SV}(s_0) = 2\pi r_V z_{iV}(s_0)/Z_0$. The slot length and width are $2L_3 = 2L_{sl}$ and d, and its center is located at a point $z = 0$ at the distance $x_{0sl} = a/2$ from the waveguide narrow wall. In this case, redefinitions $s_3 = s$, $s_4 = s_0$ have been introduced. The waveguide and half-space above the plane are characterized as air-filled spatial regions with $\varepsilon_{1,2} = \mu_{1,2} = 1$.

The analysis of this structure should be made by taking into account the anti-symmetric slot current component $\left(J_{0sl}^a(s) = 0\right)$, since the waveguide radiator is symmetric relative to the excitation field vector. Therefore, the system of Eq. (6.6) can be simplified to

$$
\begin{cases}
J_{0v} Z_v^{\Sigma} + J_{0v} Z_{vv} + J_{0sl}^s Z_c = 0, \\
J_{0sl}^s Z_{sl}^{s\Sigma} - J_{0v} Z_c + J_{0v} Z_{Wg} = -\dfrac{i\omega}{2k} H_{sl}^s, \\
J_{0V} Z_V^{\Sigma} - J_{0sl}^s Z_{Wg} = -\dfrac{i\omega}{2k} E_V,
\end{cases}
\tag{6.24}
$$

where the coefficients transformed due to the change of the slot orientation are:

$$
Z_{sl}^{s\Sigma} = \frac{8\pi}{ab} \sum_{m=1,3\ldots}^{\infty} \sum_{n=0,1\ldots}^{\infty} \frac{\varepsilon_n k e^{-k_z \frac{d_e}{4}}}{k_z (k^2 - k_x^2)} \left[\sin k L_{sl} \cos k_x L_{sl} - \frac{k}{k_x} \cos k L_{sl} \sin k_x L_{sl} \right]^2
$$

$$
+ (\mathrm{Si}4k L_{sl} - i\,\mathrm{Cin}4k L_{sl})
$$

$$
- 2\cos k L_{sl} \left[
\begin{array}{l}
2(\sin k L_{sl} - k L_{sl} \cos k L_{sl}) \\
\times \left(\ln \frac{16 L_{sl}}{d_e} - \mathrm{Cin}2k L_{sl} - i\,\mathrm{Si}2k L_{sl} \right) + \sin 2k L_{sl} e^{-ik L_{sl}}
\end{array}
\right],
$$

$$
Z_{Wg} = \frac{4\pi}{ab} \sum_{m=1,3\ldots}^{\infty} \sum_{n=0,1\ldots}^{\infty} \frac{\varepsilon_n k \tilde{k}_V e^{-k_z z_0}}{i(k^2 - k_x^2)(\tilde{k}_V^2 - k_y^2)} \sin k_x x_0
$$

$$
\times \left[\sin \tilde{k}_V L_0 \cos k_y L_0 - (\tilde{k}_V/k_y) \cos \tilde{k}_V L_0 \sin k_y L_0 \right]
$$

$$
\left[\sin k L_{sl} \cos k_x L_{sl} - \frac{k}{k_x} \cos k L_{sl} \sin k_x L_{sl} \right],
$$

$$
Z_V^{\Sigma}
= \frac{4\pi}{ab} \sum_{m=1}^{\infty} \sum_{n=0}^{\infty} \frac{\varepsilon_n (k^2 - k_y^2)\tilde{k}_V^2}{kk_z (\tilde{k}_V^2 - k_y^2)^2} e^{-k_z r_V} \left[\sin \tilde{k}_V L_0 \cos k_y L_0 - \frac{\tilde{k}_V}{k_y} \cos \tilde{k}_V L_0 \sin k_y L_0 \right]^2 \sin^2 k_x x_0
$$

$$
- \frac{i}{r_V} \int_0^{L_0} \left(\cos \tilde{k}_V s_0 - \cos \tilde{k}_V L_0 \right) \overline{Z}_{SV}(s_0) \mathrm{d}s_0,
$$

$$
H_{sl}^s = \frac{2H_0}{k} (\sin k L_{sl} - k L_{sl} \cos k L_{sl}),
$$

$$
E_V = \frac{2k H_0}{k_g \tilde{k}_V} e^{-ik_g z_0} \sin \frac{\pi}{a} x_0 \left(\sin \tilde{k}_V L_0 - \tilde{k}_V L_0 \cos \tilde{k}_V L_0 \right).
$$

The unknown current amplitudes in the vibrators and slot can be found from the solution of the equation system (6.24). Hence, the reflection and transmission coefficients of the vibrator-slot radiator, S_{11} and S_{12} are defined as:

$$
S_{11} = \frac{4\pi i}{abk k_g} \left\{ \frac{2k_g^2}{k} \tilde{J}_{0sl} f(k L_{sl}) - \frac{kk_g}{\tilde{k}_V} \tilde{J}_{0v} \sin(\pi x_0/a) e^{-ik_g z_0} f(\tilde{k}_V L_0) \right\} e^{2ik_g z},
$$

$$S_{12} = 1 + \frac{4\pi i}{abkk_g}\left\{ \frac{2k_g^2}{k}\tilde{J}_{0sl}f(kL_{sl}) + \frac{kk_g}{\tilde{k}_V}\tilde{J}_{0V}\sin(\pi x_0/a)e^{-ik_g z_0}f(\tilde{k}_V L_0) \right\},$$

$$(6.25)$$

where $\tilde{J}_{0V} = \frac{2ik_1}{\omega}J_{0V}$, $\tilde{J}_{0sl} = \frac{2ik}{\omega}J_{0sl}^s$, $f(k_1 L_{sl}) = \sin k_1 L_{sl} - k_1 L_{sl}\cos k_1 L_{sl}$ and $f(\tilde{k}_V L_0) = \sin \tilde{k}_V L_0 - \tilde{k}_V L_0 \cos \tilde{k}_V L_0$.

6.5.1 Resonant Radiation Conditions for Structure with Clavin Element and Transverse Slot

Of course, the energy coefficients of the combined radiator can be correctly modelled based on the expressions (6.25). Conditions of the resonant radiation by the structure can be defined by analyzing the frequency curves of the energy characteristics. However, these conditions cannot be easily determined due to sufficiently large number of parameters in the corresponding electrodynamic problem. An alternative approach to the problem solution consists in obtaining these conditions in the form of analytical relations. Assuming that the resonant radiation control can be realized by changing the impedance of the internal vibrator, we can obtain such relations based on the solution of the equations system (6.24).

The system of Eq. (6.24) can be easily reduced to one equation relative to the current J_{0sl}^s by the successive exclusion method to the form

$$J_{0sl}^s\left(Z_{sl}^{s\Sigma} - \frac{Z_c^2}{Z_v^\Sigma + Z_{vv}} + \frac{Z_{Wg}^2}{Z_V^\Sigma} \right) = -\frac{i\omega}{2k}\left(H_{sl}^s + \frac{E_V Z_{Wg}}{Z_V^\Sigma} \right). \qquad (6.26)$$

If the multiplier in the brackets on the left-hand side of the Eq. (6.26) is interpreted as the total slot conductivity, the resonance condition known from the slot theory [30] can be written as

$$\mathrm{Im}\left(Z_{sl}^{s\Sigma} - \frac{Z_c^2}{Z_v^\Sigma + Z_{vv}} + \frac{Z_{Wg}^2}{Z_V^\Sigma} \right) = 0. \qquad (6.27)$$

Since the terms $Z_{sl}^{s\Sigma}$ and $\frac{Z_c^2}{Z_v^\Sigma + Z_{vv}}$ do not depend upon the impedance $\overline{Z}_{SV}(s_0)$ the expression (6.27) can be presented as

$$\mathrm{Im}\left(\frac{Z_{Wg}^2}{Z_V^\Sigma} \right) = -P_1 + P_2, \qquad (6.28)$$

where $P_2 = \mathrm{Im}\left(\frac{Z_c^2}{Z_v^\Sigma + Z_{vv}} \right)$,

$$P_1 = \text{Im}\left(Z_{sl}^{s\Sigma}\right)$$

$$= -\frac{8\pi}{ab}\frac{k\cos\left(\frac{d_e}{4}\sqrt{k^2-(\pi/a)^2}\right)}{\left(k^2-(\pi/a)^2\right)^{\frac{3}{2}}}\left[\sin kL_{sl}\cos\left(\frac{\pi L_{sl}}{a}\right)-\frac{2a}{\lambda}\cos kL_{sl}\sin k_x L_{sl}\right]$$

$$-\left[\text{Cin}4kL_{sl}-2\left(\sin(2kL_{sl})-2kL_{sl}\cos^2 kL_{sl}\right)\text{Si}2kL_{sl}+\sin^2 2kL_{sl}\right].$$

Expressions Z_{Wg} and Z_V^{Σ} can be presented in the form $Z_{Wg} = RD + i \cdot JD$ and $Z_V^{\Sigma} = RDz + i \cdot JDz$ where the real and imaginary parts are presented explicitly. Then, we assume that the constant vibrator impedance $\overline{Z}_{SV}(s_0) = i\overline{X}_{SV}$ is purely reactive. This assumption does not limit the common problem objectives, since the control vibrator element should not contribute losses to the radiating structure. Hence, the real part of impedance should be vanishingly small. Taking in account that $\tilde{k}_V = k - \alpha 2\pi \overline{X}_{SV}$, we can write

$$RD = \frac{4\pi}{ab\tilde{k}_V}\frac{k\sin\left(z_0\sqrt{k^2-(\pi/a)^2}\right)}{k^2-(\pi/a)^2}$$
$$\times f(\tilde{k}_V L_0)\left[\sin kL_{sl}\cos\left(\frac{\pi L_{sl}}{a}\right)-\frac{2a}{\lambda}\cos kL_{sl}\sin\frac{\pi L_{sl}}{a}\right], \qquad (6.29a)$$

$$JD = -\frac{4\pi}{ab\tilde{k}_V}\frac{k\cos\left(z_0\sqrt{k^2-(\pi/a)^2}\right)}{k^2-(\pi/a)^2}$$
$$\times f(\tilde{k}_V L_0)\left[\sin kL_{sl}\cos\left(\frac{\pi L_{sl}}{a}\right)-\frac{2a}{\lambda}\cos kL_{sl}\sin\frac{\pi L_{sl}}{a}\right]$$
$$-\frac{8\pi}{ab}\sum_{m=1,3\dots}^{\infty}\sum_{n=1,2\dots}^{\infty}\frac{k\tilde{k}_V e^{-k_z z_0}}{(k^2-k_x^2)(\tilde{k}_V^2-k_y^2)}$$
$$\times\left[\sin\tilde{k}_V L_0\cos k_y L_0-(\tilde{k}_V/k_y)\cos\tilde{k}_V L_0\sin k_y L_0\right]\left[\begin{array}{c}\sin kL_{sl}\cos k_x L_{sl}\\-\frac{k}{k_x}\cos kL_{sl}\sin k_x L_{sl}\end{array}\right], \qquad (6.29b)$$

$$RDz = \frac{8\pi}{ab}\sum_{m=1,3\dots}^{\infty}\sum_{n=1,2\dots}^{\infty}\frac{(k^2-k_y^2)\tilde{k}_V^2}{kk_z(\tilde{k}_V^2-k_y^2)^2}e^{-k_z rv}\left(\begin{array}{c}\sin\tilde{k}_V L_0\cos k_y L_0\\-\frac{\tilde{k}_V}{k_y}\cos\tilde{k}_V L_0\sin k_y L_0\end{array}\right)^2$$
$$-\frac{4\pi}{ab\tilde{k}_V^2}\frac{k\sin\left(rv\sqrt{k^2-(\pi/a)^2}\right)}{\sqrt{k^2-(\pi/a)^2}}f^2(\tilde{k}_V L_0)+\frac{\overline{X}_{SV}}{\tilde{k}_V^3 rv}f(\tilde{k}_V L_0), \qquad (6.29c)$$

$$JDz = \frac{4\pi}{ab\tilde{k}_V^2}\frac{k\cos\left(rv\sqrt{k^2-(\pi/a)^2}\right)}{\sqrt{k^2-(\pi/a)^2}}f^2(\tilde{k}_V L_0). \qquad (6.29d)$$

The summation in (6.29b) and (6.29c) is carried out over odd indices m. It is not difficult to see that the following equality holds

$$\mathrm{Im}\left(\frac{Z_{Wg}^2}{Z_V^\Sigma}\right) = \frac{2RD \cdot JD \cdot RDz - JDz \cdot \left[(RD)^2 - (JD)^2\right]}{(RDz)^2 - (JDz)^2}.$$

Then

$$\frac{2RD \cdot JD \cdot RDz - JDz \cdot \left[(RD)^2 - (JD)^2\right]}{(RDz)^2 - (JDz)^2} = P_2 - P_1. \qquad (6.30)$$

This equality is the parametric equation with respect to the real unknown $\overline{X}_{SV}$. Since coefficients in (6.30) are real, and the unknown $\overline{X}_{SV}$ belongs to the interval $-0.35 < \overline{X}_{SV} < 0.35$ defined by the impedance concept (see Sect. 1.3.1), the equation can be effectively solved by numerical method.

Let us obtain the approximate analytical solution of the problem for the vibrator-slot structure with various parameters. Without losing physical correctness, the multi-mode interaction between the slot and vibrators can be modeled taking into account the restrictions $z_0 \gg r_V$, $z_0 \gg d$, $z_0 > 0.1\lambda$ concerning the distance between the slot and controlling monopole. In this case, the expression (6.29b) takes the following form

$$JD \approx -\frac{4\pi}{ab\tilde{k}_V} \frac{k\cos\left(z_0\sqrt{k^2 - (\pi/a)^2}\right)}{k^2 - (\pi/a)^2}$$
$$\times f(\tilde{k}_V L_0)\left[\sin kL_{sl}\cos\left(\frac{\pi L_{sl}}{a}\right) - \frac{2a}{\lambda}\cos kL_{sl}\sin\frac{\pi L_{sl}}{a}\right]. \qquad (6.31)$$

It is not difficult to see considering the expressions (6.29a) and (6.31), that when the condition $z_0\sqrt{k^2 - (\pi/a)^2} = \pi/2$ is fulfilled, the imaginary part $JD = 0$, and the real part RD is determined as

$$RD = \frac{4\pi}{ab\tilde{k}_V}\frac{kf(\tilde{k}_V L_0)}{k^2 - (\pi/a)^2}\left[\sin kL_{sl}\cos\left(\frac{\pi L_{sl}}{a}\right) - \frac{2a}{\lambda}\cos kL_{sl}\sin\frac{\pi L_{sl}}{a}\right]. \qquad (6.32)$$

Under these assumptions, we obtain

$$Im\left(\frac{Z_{Wg}^2}{Z_V^\Sigma}\right) = Im\left(\frac{RDz \cdot (RD)^2 - iJDz \cdot (RD)^2}{(RDz)^2 - (JDz)^2}\right) = -\frac{JDz \cdot (RD)^2}{(RDz)^2 - (JDz)^2}.$$
$$(6.33)$$

Let us analyze the numerator of this expression

$$JDz \cdot (RD)^2 = \left(\frac{4\pi}{ab}\right)^3 \frac{k^3 \cos\left(r_V \sqrt{k^2 - (\pi/a)^2}\right)}{\tilde{k}_V^4 \sqrt{k^2 - (\pi/a)^2}} f^4(\tilde{k}_V L_0)$$

$$\times \left[\sin k L_{sl} \cos\left(\frac{\pi L_{sl}}{a}\right) - \frac{2a}{\lambda} \cos k L_{sl} \sin \frac{\pi L_{sl}}{a}\right]^2 \qquad (6.34)$$

taking into account that the monopole length $L_0 \approx \lambda/4$ and the parameter $\alpha \ll 1$. Then, omitting terms proportional to α^2 we obtain the following expressions for the multipliers in (6.34):

$$\tilde{k}_V^4 \approx k^3(k - \alpha 8\pi \overline{X}_{SV}), \; \sin \tilde{k}_V L_0 \approx \sin k L_0 + \alpha 2\pi L_0 \overline{X}_{SV} \cos k L_0 \big|_{L_0=\lambda/4} = 1,$$

$$\cos \tilde{k}_V L_0 \approx \cos k L_0 - \alpha 2\pi L_0 \overline{X}_{SV} \sin k L_0 \big|_{L_0=\lambda/4} = -\alpha 2\pi L_0 \overline{X}_{SV}, \qquad (6.35)$$

$$f(\tilde{k}_V L_0) \approx 1 + \alpha \frac{\pi^3}{2} \overline{X}_{SV}, \; f^4(\tilde{k}_V L_0) \approx 1 + \alpha 2\pi^3 \overline{X}_{SV}.$$

With notations $RDz = RDz(\overline{X}_{SV})$ and $JDz = JDz(\overline{X}_{SV})$, based on the expressions (6.28) and (6.34), we obtain

$$\overline{X}_{SV} = -\frac{1}{2\alpha\pi^3} + \left(\frac{ab}{4\pi}\right)^3 \frac{(k - \alpha 8\pi \overline{X}_{SV})\sqrt{k^2 - (\pi/a)^2}}{2\alpha\pi^3 \cos\left(r_V \sqrt{k^2 - (\pi/a)^2}\right)}$$

$$\times \frac{(P_1 - P_2)\left[(RDz(\overline{X}_{SV}))^2 - (JDz(\overline{X}_{SV}))^2\right]}{\left[\sin k L_{sl} \cos\left(\frac{\pi L_{sl}}{a}\right) - \frac{2a}{\lambda} \cos k L_{sl} \sin \frac{\pi L_{sl}}{a}\right]^2}. \qquad (6.36)$$

Since $\tilde{k}_V = k - \alpha 2\pi \overline{X}_{SV}$, Eq. (6.36) can be solved by the method of successive iterations (see Sect. 2.2.2.2). The zero approximation can be written as

$$\overline{X}_{SV} = -\frac{1}{2\alpha\pi^3} + \left(\frac{ab}{4\pi}\right)^3 \frac{k\sqrt{k^2 - (\pi/a)^2}}{2\alpha\pi^3 \cos\left(r_V \sqrt{k^2 - (\pi/a)^2}\right)}$$

$$\times \frac{(P_1 - P_2)\left[(RDz(0))^2 - (JDz(0))^2\right]}{\left[\sin k L_{sl} \cos\left(\frac{\pi L_{sl}}{a}\right) - \frac{2a}{\lambda} \cos k L_{sl} \sin \frac{\pi L_{sl}}{a}\right]^2}. \qquad (6.37)$$

Thus, in this Section, the problem of determining the currents in the elements of combined waveguide vibrator-slot structures based on the Clavin radiator is solved by the generalized method of induced EMMF. These structures include the impedance vibrator elements, two of which are located outside and one inside the waveguide. The possibility of controlling the resonant radiation of combined vibrator-slot structures by changing the impedance of vibrator located in the rectangular waveguide under the condition of single-mode excitation is also shown. In the case of the transverse slot, the parametric equation for determining the reactive impedance of the monopole is obtained providing the resonant regime radiation of the structure. The approximate analytical solution of this equation is proposed for the quarter-wave monopoles and slot.

References

1. Naiheng Y, Harrington R (1983) Electromagnetic coupling to an infinite wire through a slot in a conducting plane. IEEE Trans Antennas Propag 31:310–316
2. Levin ML (1951) Slot antenna with guide device. J Tech Phys **21**(7):795−801 (1951) (in Russian)
3. King RWP, Owyang GH (1960) The slot antenna with coupled dipoles. IRE Trans Antennas Propag 8:136–143
4. Butler CM, Umashankar KR (1976) Electromagnetic excitation of a wire through an aperture-perforated conducting screen. IEEE Trans Antennas Propag 24:456–462
5. Harrington RF (1982) Resonant behavior of a small aperture backed by a conducting body. IEEE Trans Antennas Propag 30:205–212
6. Hsi SW, Harrington RF, Mautz JR (1985) Electromagnetic coupling to a conducting wire behind an aperture of arbitrary size and shape. IEEE Trans Antennas Propag 33:581–587
7. Min K-S, Hirokawa J, Sakurai K, Ando M (1998) Phase control of circular polarization from a slot with a parasitic dipole. IEICE Trans Commun **E81-B**(3):668–673
8. Morioka T, Komiyama K, Hirasawa K (2001) Effects of a parasitic wire on coupling between two slot antennas. IEICE Trans Commun **E84-B**(9):2597–2603
9. Kim K-C, Lim SM, Kim MS (2005) Reduction of electromagnetic penetration through narrow slots in conducting screen by two parallel wires. IEICE Trans Commun **E88-B**(4):1743–1745
10. Levin ML (1953) Passive radiating systems in waveguides. Rep USSR Acad Sci **XCI**(4):807–810 (in Russian)
11. Seidel DB (1978) Aperture excitation of a wire in a rectangular cavity. IEEE Trans Microw Theor Tech 26:908–914
12. Lee Y-H, Hong D-H, Ra J-W (1983) Waveguide slot antenna with a coupled dipole above the slot. Electron Lett 19(8):280–282
13. Yatsuk LP, Zhironkina AV (1988) Scattering of a type wave by a slot-hole inhomogeneity in a rectangular waveguide. Radioeng Electron **33**(10):2185–2189 (in Russian)
14. Hashemi-Yeganeh S, Elliott RS (1990) Analysis of untilted edge slots excited by tilted wires. IEEE Trans Antennas Propag 38:1737–1745
15. Yao H-W, Zaki KA, Atia AE, Hershtig R (1995) Full wave modeling of conducting posts in rectangular waveguides and its applications to slot coupled combline filters. IEEE Trans Microw Theor Tech 43:2824–2830
16. Hirokawa J, Kildal P-S (1997) Excitation of an untilted narrow-wall slot in a rectangular waveguide by using etched strips on a dielectric plate. IEEE Trans Antennas Propag 45:1032–1037
17. Hirokawa J, Manholm L, Kildal P-S (1997) Analysis of an untilted wire-excited slot in the narrow wall of a rectangular waveguide by including the actual external structure. IEEE Trans Antennas Propag 45:1038–1044
18. Wongsan R, Phongcharoenpanich C, Krairiksh M, Takada J-I (2003) Impedance characteristic analysis of an axial slot antenna on a sectoral cylindrical cavity excited by a probe using method of moments. IEICE Trans Fundament **E86-A**(6):1364–1373
19. Park S-H, Hirokawa J, Ando M (2003) Simple analysis of a slot and a reflection-canceling post in a rectangular waveguide using only the axial uniform currents on the post surface. IEICE Trans Commun **E86-B**(8):2482–2487
20. Lim K-S, Koo V-C, Lim T-S (2007) Design, simulation and measurement of a post slot waveguide antenna. J Electromagn Waves Appl 21(12):1589–1603

21. Hashimoto K, Hirokawa J, Ando M (2010) A post-wall waveguide center-feed parallel plate slot array antenna in the millimeter-wave band. IEEE Trans Antennas Propag 58:3532–3538
22. Yamaguchi S, Aramaki Y, Takahashi T, Otsuka M, Konishi Y (2012) A slotted waveguide array antenna covered by a dielectric slab with a post-wall cavity. In: Proceedings of IEEE APS URSI international symposium. Chicago, USA, pp 559–560
23. Clavin A, Huebner DA, Kilburg FJ (1974) An improved element for use in array antennas. IEEE Trans Antennas Propag 22:521–526
24. Clavin A (1975) A multimode antenna having equal E- and H-planes. IEEE Trans Antennas Propag 23:735–737
25. Papierz AB, Sanzgiri SM, Laxpati SR (1977) Analysis of antenna structure with equal E- and H-plane patterns. Proc IEE 124(1):25–30
26. Elliott RS (1980) On the mutual admittance between Clavin elements. IEEE Trans Antennas Propag 28:864–870
27. Kominami M, Rokushima K (1984) Analysis of an antenna composed of arbitrarily located slots and wires. IEEE Trans Antennas Propag 32:154–158
28. Penkin YuM, Semenikhin VA, Yatsuk LP (1987) Investigation of the internal and external characteristics of radiators such as a Clavin radiator. Radio Eng 83:3–10 ((in Russian))
29. Nesterenko MV, Katrich VA, Penkin YuM, Dakhov VM, Berdnik SL (2011) Thin impedance vibrators. Theory and applications. Springer Science+Business Media, New York
30. Nesterenko MV, Katrich VA, Penkin YuM, Berdnik SL (2008) Analytical and hybrid methods in theory of slot-hole coupling of electrodynamic volumes. Springer Science+Business Media, New York

Chapter 7
Combined Vibrator-Slot Structures Located on a Perfectly Conducting Sphere

In this chapter, the problem of electromagnetic wave radiation by combined vibrator-slot structures with impedance vibrator elements located on a perfectly conducting sphere is considered. The relevance of such studies is determined by urgent need of new radiating structures for mobile vehicles with spherical or close to it shape. The solution is based on the results obtained earlier for spherical antennas with resonant slots and radial impedance vibrators placed on a sphere of arbitrary radius.

7.1   Resonant Slot Radiator on a Sphere

As is known [1], non-protruding slotted radiators are preferred when antennas are placed on mobile vehicles, since these antennas do not significantly influence upon mass-dimensional parameters and aerodynamic properties of vehicles. The application of such antennas ranges from spacecrafts [1] to autonomous microdevices [2]. Typically, surfaces of mobile vehicles or their structural part can be approximated by spherical surfaces, which radii are comparable with operating wavelengths There-fore, interest in spherical antennas with resonant slots has not disappeared during several decades.

Annular and sectorial narrow slots on spherical scatterers have been detailly studied [3–13]. As it is usually assumed, the slot antennas with symmetric excitation are used. The plane wave diffraction by a hollow conducting shell with an annular slot or hole with a circular aperture was studied in [14, 15]. The characteristics of spherical antennas with rectangular slot radiators were investigated in [5, 16–18]. The external electrodynamic characteristics of slot antennas were considered in [16, 18] by using predefined cosine distributions of magnetic current along the half-wave narrow slot radiator. The far zone RPs of spherical antennas in the equatorial plane were obtained in [16]. External intrinsic and mutual conductivities of half-wave slot radiators located along parallels on conducting spheres were studied in [18]. The

© The Author(s), under exclusive license to Springer Nature Switzerland AG 2020
M. V. Nesterenko et al., *Combined Vibrator-Slot Structures: Theory and Applications*, Lecture Notes in Electrical Engineering 689,
https://doi.org/10.1007/978-3-030-60177-5_7

characteristics of narrow rectangular slots cut in a conducting infinitely thin spherical shell were analyzed by using the method of moments in [16, 17]. In the first case, a free space outside a spherical scatterer and inner antenna region presented by a hollow spherical resonator coupling through a slot is considered [16]. In the second case, a free space and a spherical resonator with a concentric conducting sphere inside it is studied. In both cases, the slot is excited by a hypothetical point voltage generator [17].

Currently, microwave devices controlled by concentrated semiconductor elements, such as *p-i-n* diodes, are widely used in practice. These low-power control devices are thoroughly studied theoretically and mastered in production. However, the further development of microwave technology is characterized by a transition from the traditional elements to structures using film hybrid microcircuits. Thus, the technical and operational performance of microwave equipment can be essentially improved. It determines miniaturization and efficiency of its operation in automatic control mode. But these film elements have not yet been fully studied. Therefore, the development of controlled microwave devices based on the film elements is difficult due to the lack of adequate mathematical models. This is especially true for combined radiating waveguide devices with control film elements which are a surface part affecting formation of the radiator fields. We emphasize that the paramount role of mathematical modeling in this case is determined by the multi-parameter nature of problems concerning device development and optimization become long and costly and sometimes even completely impossible.

The integration of film elements into microwave devices requires new mathematical models for the design and analysis of these devices. As a rule, each element used in combined devices, require some electrodynamically rigorous numeric-analytic method of analysis that takes into account the geometric parameters and physical features of the local electrodynamic problem. These methods can usually be selected from previously developed ones. Moreover, the direct combination of such methods defining electromagnetic fields in most cases becomes impossible, that necessitate direct numerical simulations. It should be bear in mind that the use of direct numerical simulation for open surface antennas is also limited by their electrical dimensions. In some cases, approximate one-sided boundary conditions can be used during the general formulation of the problem. It may be, for example, impedance-type boundary conditions allowing to reduce in the solution a number of coupling electrodynamic volumes. The main advantage of applying the impedance boundary conditions consists in eliminating the need to determine the fields inside the metal-dielectric structural elements during the problem formulation [Chap. 1, 19].

The integrating of resonators with pronounced frequency-selective properties and slot radiators makes it possible to form the required energy and spatial characteristics of such antennas [20–23]. However, it should be noted that the results presented in these publications are limited only to the cases of slot excitation by a voltage generator.

7.1.1 Problem Formulation and Solution

Consider a hollow semi-infinite rectangular waveguide with perfectly conducting walls (region index V^i). A narrow transverse slot is cut in the end wall of the waveguide section symmetrically with respect to the longitudinal waveguide axis. Let us introduce the Cartesian coordinate system associated with the waveguide (Fig. 7.1a) and the spherical coordinate system related to the spherical scatterer adjoint to the waveguide end wall, as shown in Fig. 7.1c. Cross-section of waveguide is $\{a \times b\}$. The slot width and length, d and $2L_i$, satisfy the inequalities $[d/(2L_i)] \ll 1$, $[d/\lambda] \ll 1$, (λ is the wavelength in free space). The coordinates of the geometric center of the internal slot aperture S_i in the Cartesian system and the center of the external aperture in the spherical system are $(a/2, y_0, 0)$ and $(R, \pi/2, 0)$. The length of the slot external aperture along the arc in the spherical coordinate system is $2L_e$. The inner side of the waveguide end wall is characterized by a constant distributed surface impedance Z_S. The H_{10}-wave propagates in the waveguide from the direction $z = \infty$. The slot radiates into the space outside of the perfectly conducting sphere (region index V^e). The sphere radius is R and the material parameters of the medium outside the sphere are ε_1 and μ_1.

The slot tunnel cavity bounded by the apertures S_i and S_e, has the complex spatial shape (index region V^v). The cavity volume is defined by the sphere radius and dimensions of the waveguide cross section, since these geometric parameters determine the mutual spatial positions of the slot apertures S_i and S_e (Fig. 7.1b).

The initial equation system for the spherical antenna can be formulated based on the continuity of the tangential magnetic field components on the internal and external slot apertures as

$$
\begin{cases}
\text{at} \quad S_i: \vec{H}_\tau^i(\vec{e}_{si}) + \vec{H}_{0\tau}^i = \vec{H}_\tau^v(\vec{e}_{si}) + \vec{H}_\tau^v(\vec{e}_{se}), \\
\text{at} \quad S_e: \vec{H}_\tau^v(\vec{e}_{si}) + \vec{H}_\tau^v(\vec{e}_{se}) = \vec{H}_\tau^e(\vec{e}_{se}),
\end{cases}
\tag{7.1}
$$

where $\vec{e}_{si}, \vec{e}_{se}$ are the electric fields on the surfaces S_i and S_e; $\vec{H}_\tau^i(\vec{e}_{si})$, $\vec{H}_\tau^e(\vec{e}_{se})$, $\vec{H}_\tau^v(\vec{e}_{si})$, and $\vec{H}_\tau^v(\vec{e}_{se})$ are tangential magnetic field components relative to the slot apertures in the corresponding electrodynamic volumes; $\vec{H}_{0\tau}^i$ is the component of

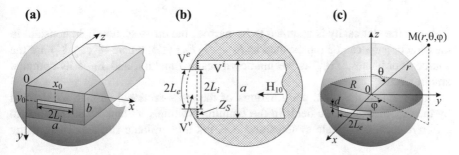

Fig. 7.1 The local coordinate systems used for analysis of the slotted spherical antenna

intrinsic magnetic field in the waveguide exciting the slot. The Eqs (7.1) allow using magnetic field representations in arbitrarily local coordinate systems for each coupling volumes.

A mathematically rigorous justification allowing to reduce the equation system (7.1) to a single equation

$$\vec{H}^i_\tau(\vec{e}_{si}) + \vec{H}^i_{0\tau} = \vec{H}^e_\tau(\vec{e}_{se}),\tag{7.2}$$

was proposed in [24]. As can be seen, fields in the slot cavity are not included into this equation. This approach was confirmed for the coupling problem between two arbitrary electrodynamic volumes through a narrow rectangular slot under condition that the region V^v is a rectangular parallelepiped $2L_i \times d \times h$. It has also been proved that if the inequality $(hd/\lambda^2) \ll 1$ holds, the problem solution for a slot cut in a wall of finite thickness $h((h/\lambda) \ll 1)$ and infinitely thin wall approximately coincide if the actual slot width d is replaced by the equivalent width d_e. This concept for narrow rectangular slots cut in flat sections of screen surfaces was introduced in [24]. It was also concluded based on the physical principles that this approach is valid for curved narrow slot cut on an arbitrary smooth surface if the main curvature radii of the surface are much larger than d and h. The same arguments are also relevant for the case considered here, when the slot cavity is a part of rectangular parallelepiped with one face subjected to small perturbations which becomes infinitesimal for large sphere radius R.

Thus, the problem for the slotted spherical antenna can be rigorously solved based on the Eq. (7.2). To ensure the mathematical correctness, the problem solution of this equation can be represented in a generalized coordinates system (ξ_1, ξ_2, ξ_3), and then the solution should be found on some cross-section S located in the vicinity of the cavity aperture S_e belonging to the slot cavity V^v. Without loss of generality, let us assume that unit vectors $\vec{e}_{\xi1}$ and $\vec{e}_{\xi2}$ are directed along the longitudinal and transverse axes of this cross-section. First, let us multiply the left and right-hand sides of Eq. (7.2) by the predefined scalar function $\psi(\xi_1)$, and then integrate the result over the cross-sectional surface S

$$\int_S \vec{H}^i_\tau[\vec{e}_s]\psi(\xi_1)\mathrm{d}s + \int_S \vec{H}^i_{0\tau}(s)\psi(\xi_1)\mathrm{d}s = \int_S \vec{H}^e_\tau[\vec{e}_s]\psi(\xi_1)\mathrm{d}s.\tag{7.3}$$

Since the slot cavity is assumed to be narrow, the cross-section field constant in the direction $\vec{e}_{\xi2}$ can be represented as $\vec{e}_s = \vec{e}_{\xi2}E_0 f(\xi'_1)$, where (ξ'_1, ξ'_2, ξ'_3) are the source coordinates, $f(\xi'_1)$ is the unknown scalar function, and E_0 is its complex amplitude.

Suppose that the small perturbations of the cross-sectional surface are defined by the fact that magnetic fields in the coupling volumes V^i and V^e are defined in different local coordinate systems and S is not a coordinate surface. Then, using the summation operation in the formal sense, we can introduce representations of perturbed surfaces as $S_e = S + \Delta s_e$ and $S_i = S + \Delta s_i$, where $\Delta S_{e(i)}$ are the small perturbations. The Eq. (7.3) can be rewritten in the following form

$$E_0 \int\limits_{S+\Delta s_e} \vec{H}_\tau^e [\vec{e}_{\xi 2} f(\xi_1')] \psi(\xi_1) ds$$

$$- E_0 \int\limits_{S+\Delta s_i} \vec{H}_\tau^i [\vec{e}_{\xi 2} f(\xi_1')] \psi(\xi_1) ds$$

$$= \int\limits_{S+\Delta s_i} \vec{H}_{0\tau}^i(s) \psi(\xi_1) ds. \tag{7.4}$$

This equation is approximate, and it becomes exact under condition $\Delta s_{e(i)} \to 0$. The small perturbations can be interpreted by similarity with the classical problem of a thin vibrator excitation: the electric current in the vibrator is supposed to be concentrated on its longitudinal axis, and the boundary conditions for electromagnetic fields should be satisfied only on its generatrix [25]. Similarly, in our case, the surface S can be considered as the cross section where the secondary magnetic current equivalent to the field $\vec{e}_s$ is concentrated, and the boundary conditions for the fields should be satisfied on the surfaces S_i and S_e.

It should be noted that application of the local coordinate systems (7.4) force us to represent magnetic fields $\vec{H}_\tau^e[\vec{e}_{\xi 2} f(\xi_1)]$, $\vec{H}_\tau^i[\vec{e}_{\xi 2} f(\xi_1)]$ also in these coordinate system, and, hence, to disturb the surface S where secondary excitation sources are assumed to be concentrated. Then, the magnetic fields can be determined by using integral-differential operators, where integration must be carried out over the surface cross-sections $S_e = S + \Delta s_e$ and $S_i = S + \Delta s_i$ instead of integration over the cross-section S, as in Eq. (7.3). Moreover, if the function $f(\xi_1)$ presents the exact solution of the Eq. (7.2) on the cross-section S, then it will also define an approximate solution of the Eq. (7.4) with a fairly small error. As is known from the general antenna theory, the small errors of the current distributions do not lead to significant errors in the integral quantities, such as excited electromagnetic fields. That is, if the basis and weight functions $f(\xi_1)$ and $\psi(\xi_1)$ are selected to be equal, while the inequality $|\Delta s_{e(i)}| \ll \lambda^2$ holds, using the Eq. (7.4) instead (7.3) will not violate the correctness of the electrodynamic problem solution.

The above consideration can be used to justify application of the generalized method of induced MMF for analyzing the characteristics of the spherical antennas. The initial equation based on the Eq. (7.4) can be written in the selected local coordinate systems with unit vectors $\vec{e}_\theta$ and $\vec{e}_y$ (Fig. 7.1) as

$$E_0 \int\limits_{-L_e/R}^{L_e/R} \vec{H}_\tau^e [\vec{e}_\theta f(\varphi)] \psi(\varphi) d\varphi$$

$$- E_0 \int\limits_{x_0-L_i}^{x_0+L_i} \vec{H}_\tau^i \big[\vec{e}_y f(x) \big] \psi(x) dx$$

$$= \int\limits_{x_0-L_i}^{x_0+L_i} \vec{H}_{0\tau}^i(x) \psi(x) dx, \tag{7.5}$$

where it was taken into account that the field $\vec{e}_s$ in the spherical and rectangular coordinate systems are $\vec{e}_s = \vec{e}_\theta J_0 \delta(r' - R) f(\varphi')$ and $\vec{e}_s = \vec{e}_y J_0 \delta(z') f(x')$ under condition that the field $\vec{e}_s$ is constant in the transvers direction relative the slot and the parameters $\theta_0 = \pi/2$ and $x_0 = a/2$ are fixed during the problem formulation.

According to the generalized method of induces MMF, the analytical solution of the Eq. (7.3) obtained by the asymptotic averaging method of the key problem concerning electromagnetic waves radiation through the slot in the impedance end wall of the semi-infinite rectangular waveguide into half-space above the perfectly conducting plane can be used as the basis functions f. Then this function can be written as follows:

$$f(\varphi) = \cos(kR\varphi)\cos(\pi L_e/a) - \cos kL_e \cos(\pi R\varphi/a)$$

and

$$f(x) = \cos k\left(x - \frac{a}{2}\right)\cos\left(\frac{\pi L_i}{a}\right) - \cos kL_i \cos\left[\frac{\pi(x - a/2)}{a}\right]$$

in the spherical and rectangular coordinate systems, respectively. Here $k = 2\pi/\lambda$ is the wave number, ω is the circular frequency. The electromagnetic fields depend on time t as $e^{i\omega t}$.

The complex field amplitude is the solution of the Eq. (7.5)

$$E_0 = F_0^m/(Y^e + Y^i), \tag{7.6}$$

where the magnetomotive force is determined by the expression

$$F_0^m = \int\limits_{a/2-L_i}^{a/2+L_i} H_{0x}^i(x) f(x) dx, \tag{7.7}$$

and the slot conductivity in the corresponding electrodynamic volumes are:

$$Y^e = \int\limits_{-L_e/R}^{L_e/R} H_\varphi^e \big[\vec{e}_\theta f(\varphi) \big] f(\varphi) d\varphi,$$

$$Y^i = - \int\limits_{a/2-L_i}^{a/2+L_i} H_x^e [\vec{e}_y f(x)] \psi(x) dx. \qquad (7.8)$$

Next, let us take into account the relationships between the magnetic Hertz vectors $\vec{\Pi}_{e(i)}^m(\vec{r})$ and the magnetic field $\vec{H}^{e(i)}(\vec{r}) = (\text{graddiv} + k_1^2)\vec{\Pi}_{e(i)}^m(\vec{r})$, valid under the conditions $\Pi_{e(i)r}^m(\vec{r}) = 0$, and magnetic currents $\vec{J}_{se(si)}^m(\vec{r}')$ which can be written as $\vec{\Pi}_{e(i)}^m(\vec{r}) = \frac{1}{i\omega\mu_1} \int_{S_{e(i)}} \hat{G}^{e(i)}(\vec{r}, \vec{r}') \vec{J}_{se(si)}^m(\vec{r}') dr'$, where $\hat{G}^{e(i)}(\vec{r}, \vec{r}')$ are the magnetic tensor Green's functions for the corresponding volumes, $\vec{r}$ and $\vec{r}'$ are the radius vectors of observation and source points, and $k_1 = k\sqrt{\varepsilon_1\mu_1}$.

The external slot conductivity Y^e can be found by using the Green's tensor components, which was constructed for space outside the perfectly conducting sphere in the monograph [11, 12] (see Appendix A). First let us determine components of the Hertz magnetic vector $\vec{\Pi}_e^m$:

$$\Pi_{e\theta}^m(r, \theta, \varphi) = \frac{1}{i\omega\mu_1} \sum_{n=0}^{\infty} \sum_{m=1}^{n} \left\{ \frac{m Q_n(r) FS_m(\varphi)}{n(n+1)C_{nm}} \left[\frac{d P_n^m(\cos\theta)}{d\theta} F_n^m + \frac{P_n^m(\cos\theta)}{\sin\theta} \Phi_n^m \right] \right\},$$

$$\Pi_{e\varphi}^m(r, \theta, \varphi) = \frac{1}{i\omega\mu_1} \sum_{n=0}^{\infty} \sum_{m=0}^{n} \left\{ \frac{\varepsilon_m Q_n(r) FC_m(\varphi)}{2n(n+1)C_{nm}} \left[m^2 \frac{P_n^m(\cos\theta)}{\sin\theta} F_n^m + \frac{d P_n^m(\cos\theta)}{d\theta} \Phi_n^m \right] \right\}, \qquad (7.9)$$

where

$$Q_n(r) = \frac{h_n^{(2)}(k_1 r)}{(n+1)h_n^{(2)}(k_1 R) - k_1 R h_{n+1}^{(2)}(k_1 R)},$$

$$F_n^m = P_n^m(\cos\theta')\big|_{\theta'=\pi/2} \quad \Phi_n^m = \frac{d P_n^m(\cos\theta')}{d\theta'}\bigg|_{\theta'=\pi/2},$$

$$FC_m(\varphi) = \int\limits_{-L_e/R}^{L_e/R} f(\varphi') \cos(m(\varphi - \varphi')) d\varphi',$$

$$FS_m(\varphi) = \int\limits_{-L_e/R}^{L_e/R} f(\varphi') \sin(m(\varphi - \varphi')) d\varphi',$$

$C_{nm} = \frac{2\pi (n+m)!}{(2n+1)(n-m)!}$, $\varepsilon_m = \begin{cases} 1, & m = 0, \\ 2, & m \neq 0, \end{cases}$, m and n are integers, $P_n^m(\cos\theta)$ are associated functions of Legendre of the first kind, $h_n^{(2)}(k_1 r) = \sqrt{\frac{\pi}{2k_1 r}} J_{n+1/2}(k_1 r) - i\sqrt{\frac{\pi}{2k_1 r}} N_{n+1/2}(k_1 r) = \sqrt{\frac{\pi}{2k_1 r}} H_{n+1/2}^{(2)}(k_1 r)$ are spherical Hankel function of the second kind, $J_{n+1/2}(k_1 r)$, $N_{n+1/2}(k_1 r)$ are Bessel and Neumann functions of half integer indices.

The external slot conductivity can be obtained using the expression

$$H_\varphi(r, \theta, \varphi) = \frac{1}{i\omega\mu_1} \sum_{n=0}^{\infty} \sum_{m=0}^{n} \frac{Q_n(r) F C_m(\varphi)}{2n(n+1)C_{nm}} \left\{ \varepsilon_m k_1^2 \frac{d P_n^m(\cos\theta)}{d\theta} \Phi_n^m \right.$$

$$\left. - 2m^2 \left[\frac{1}{r^2} n(n+1) - k_1^2 \right] \frac{P_n^m(\cos\theta)}{\sin\theta} F_n^m \right\}$$

in the form

$$Y^e = Y^e(k_1 L_e, k_1 R) = -\frac{4}{k_1 R} \sum_{n=1}^{\infty} \frac{1}{n(n+1)} \cdot \frac{1}{(n+1) - k_1 R h_{n+1}^{(2)}(k_1 R)/h_n^{(2)}(k_1 R)}$$

$$\times \left\{ (k_1 R)^2 C_0^2 \left(A_n^0\right)^2 - 2 \sum_{m=1}^{n} C_m^2 \left[m^2 \left(n(n+1) - (k_1 R)^2\right) (B_n^m)^2 - (k_1 R)^2 (A_n^m)^2 \right] \right\}, \tag{7.10}$$

$$C_m = \frac{\cos(\pi L_e/a)}{m^2 - (k_1 R)^2} \left[m \sin\frac{m L_e}{R} \cos k_1 L_e - k_1 R \cos\frac{m L_e}{R} \sin k_1 L_e \right]$$

$$- \frac{\cos k_1 L_e}{m^2 - (\pi R/a)^2} \left[m \sin\frac{m L_e}{R} \cos\frac{\pi L_e}{a} - \frac{\pi R}{a} \cos\frac{m L_e}{R} \sin\frac{\pi L_e}{a} \right]$$

$$= C_m^I - C_m^{II},$$

$$C_m^I \big|_{m \to k_1 R} = \left(\frac{L_e}{2R} + \frac{\sin(2k_1 L_e)}{4k_1 R} \right) \cos\frac{\pi L_e}{a}, \quad C_m^{II} \big|_{m \to \frac{\pi R}{a}}$$

$$= \left(\frac{L_e}{2R} + \frac{\sin(2\pi L_e/a)}{4\pi R/a} \right) \cos k_1 L,$$

$\bar{P}_n^m(\cos\theta) = \frac{(2n+1)(n-m)!}{2\pi(n+m)!} P_n^m(\cos\theta)$ are normalized associated Legendre functions of the first kind.

The internal slot conductivity of the waveguide section is

$$Y^i = Y^i(k L_i, \bar{Z}_S)$$

$$= \frac{4\pi}{ab} \sum_{m=1,3...}^{\infty} \sum_{n=0}^{\infty} \frac{\varepsilon_n(k^2 - k_x^2)}{k k_z}$$

$$\cos k_y y_0 \cos k_y(y_0 + d_e/4) F_Z(k_z, \bar{Z}_S) g^2(k L_i), \tag{7.11}$$

where

$$k_x = \frac{m\pi}{a}, k_y = \frac{n\pi}{b}, k_z$$

$$= \sqrt{k_x^2 + k_y^2 - k^2},$$

$$\varepsilon_n = \begin{cases} 1, n = 0 \\ 2, n \neq 0 \end{cases},$$

$$F_Z(k_z, \bar{Z}_S) = \frac{k k_z (1 + \bar{Z}_S^2)}{(ik + k_z \bar{Z}_S)(k \bar{Z}_S - ik_z)} \left(1 - i\frac{k k_z \bar{Z}_S}{k^2 - k_x^2} \right),$$

$$g(kL_i) = 2\left\{ \frac{k \sin kL_i \cos k_x L_i - k_x \cos kL_i \sin k_x L_i}{k^2 - k_x^2} \cos \frac{\pi L_i}{a} \right.$$

$$\left. - \frac{\left(\frac{\pi}{a}\right) \sin \frac{\pi L_i}{a} \cos k_x L_i - k_x \cos \frac{\pi L_i}{a} \sin k_x L_i}{(\pi/a)^2 - k_x^2} \cos kL_i \right\}.$$

When the waveguide section is excited by the main wave mode $H_{10}(x, z) = H_0 \sin \frac{\pi x}{a} e^{-ik_g z}$, where H_0 is the amplitude, $k_g = \sqrt{k^2 - (\pi/a)^2}$ is the propagation constant, the formula for determining the magnetic current in the slot aperture after integration in (7.4) can be written as

$$J(s) = -\frac{i\omega}{k^2} H_0 F(kL_i) \frac{[\cos ks \cos(\pi L_{i(e)}/a) - \cos kL_{i(e)} \cos(\pi s/a)]}{Y^i(kL_i, \bar{Z}_S) + Y^e(kL_e, kR)}, \quad (7.12)$$

where

$$F(kL_i) = 2 \cos \frac{\pi L_i}{a} \frac{\sin kL_i \cos \frac{\pi L_i}{a} - \left(\frac{\pi}{ka}\right) \cos kL_i \sin \frac{\pi L_i}{a}}{1 - (\pi/ka)^2}$$

$$- \cos kL_i \frac{\sin \frac{2\pi L_i}{a} + \frac{2\pi L_i}{a}}{(2\pi/ka)},$$

and the local coordinate $s = R\varphi'$ or $s = x' - a/2$ are used in the spherical or rectangular coordinate systems to determine the fields in space outside the sphere or inside the waveguide section.

Thus, the asymptotic solution (7.12) of the integral Eq. (7.4) allows us to define the energy and directional characteristics of the waveguide-slot radiator. The field reflection coefficient in the waveguide can be defined by the following formula

$$S_{11} = \left\{ \frac{1 - (k_g/k)\bar{Z}_S}{1 + (k_g/k)\bar{Z}_S} - \frac{8\pi k_g F^2(kL_i)}{iabk\left[Y^i(kL_i, \bar{Z}_S) + Y^e(kL_e, kR)\right]} \cdot \frac{1 + \bar{Z}_S^2}{1 + \frac{k_g}{k}\bar{Z}_S} \right\} e^{-2ik_g z}.$$

$$(7.13)$$

The power radiation coefficient of the spherical antenna is defined by the formula

$$|S_\Sigma|^2 = \frac{P_\Sigma}{P_{10}} = \frac{|E_0|^2}{2} \text{Im} Y^e(kL_e, kR), \quad (7.14)$$

where P_Σ is the average power radiated through the slot aperture, P_{10} is the input power of the H_{10}-wave, and $\text{Im} Y^e(kd_e, kL_e, kR)$ is the imaginary part of the external slot conductivity (7.10).

The equivalent slot width d_e was calculated by the formula $d_e \approx d \exp\left(-\frac{\pi h_e}{2d}\right)$, where $h_e = V^v/S_i$, V^v is the volume of the slot cavity, S_i is the area of the internal slot aperture under condition that inequality $[(h + h_{e\mu})d/\lambda^2] \ll 1$ holds. In this inequality $(h + h_{e\mu})$ is the maximum size of the tunnel slot cavity in the radial direction, which takes into account the actual thickness $h_{\varepsilon\mu}$ of the waveguide end wall impedance coating.

The power losses in the impedance coating can be defined by using the energy balance equation $|S_{11}|^2 + |S_\Sigma|^2 + P_\sigma = 1$. This equation will also be used to verify algorithms for calculating the energy parameters of spherical slot antennas with imaginary surface impedance $\bar{Z}_S$, when the losses in the impedance element are absent, i.e., $P_\sigma = 0$.

If a resonant slot diaphragm is placed in the waveguide section at a distance H from the waveguide end wall, a reentrant resonator is formed in the waveguide path. The slot length, width, thickness, and position relative broad waveguide wall are $2L_d$, d_d, h_d, y_{0d} and the resonator dimensions are $\{a_r \times b_r \times H\}$. The impedance coating may be applied to the diaphragm surface facing the incident wave source. The reflection coefficient in the waveguide is determined by the expression

$$
S_{11} = \left\{ 1 - \frac{8\pi k_g F^2(kL_d)}{iabk^3} \right.
$$
$$
\left. \times \frac{[Y^e(kL_e, kR) + Y^r(kL_i, kH)]}{[(Y^i(kL_d, \bar{Z}_S) + Y^r(kL_d, kH))(Y^e(kL_e, kR) + Y^r(kL_i, kH)) - Y^r(kL_{di})Y^r(kL_{id})]} \right\} e^{-i2k_g z} \quad (7.15)
$$

where $Y^i(kL_d, \bar{Z}_S)$ is calculated by using the formula (7.11) with $k_x = \frac{m\pi}{a\{a_r\}}$, $k_y = \frac{n\pi}{b\{b_r\}}$, and $k_z = \sqrt{k_x^2 + k_y^2 - k^2}$,

$$
Y^r(kL_{d(i)}, kH) = \frac{4\pi}{a_r b_r} \sum_{m=1,3\dots}^{\infty} \sum_{n=0}^{\infty} \frac{\varepsilon_n(k^2 - k_x^2)}{kk_z} \coth k_z H
$$
$$
\times \cos k_y y_{d(i)} \cos k_y (y_{d(i)} + d_{d(i)}/4) g^2(kL_{d(i)}), \quad (7.16)
$$

$$
Y^r(kL_{di(id)}) = \frac{4\pi}{a_r b_r} \sum_{m=1,3\dots}^{\infty} \sum_{n=0}^{\infty} \frac{\varepsilon_n(k^2 - k_x^2)}{kk_z \mathrm{sh} k_z H}
$$
$$
\times \cos k_y y_{0d(i)} \cos k_y (y_{0i(d)} + d_{i(d)}/4) g(kL_d) g(kL_i). \quad (7.17)
$$

7.1.2 Radiation Fields of the Slotted Spherical Antenna

The magnetic current distribution (7.12) obtained as solution of the boundary value problem allows us to calculate the electrodynamic characteristics of the antenna in space outside the sphere. According to the model of spherical antenna (Fig. 7.1c), the total radiation field can be determined by the components of magnetic Hertz vectors $\Pi^m_{e\theta}(r, \theta, \varphi)$ and $\Pi^m_{e\varphi}(r, \theta, \varphi)$ (7.9). Then, the expressions for the components of the total radiation field of the spherical surface antenna after substitution of the current distribution $J(s) = J\left(R\varphi'\right)$ (7.12) in the formula (7.9) can be written as:

$$
E_{er}(r, \theta, \varphi) = -\frac{1}{r} \sum_{n=0}^{\infty} Q_n(r) \left(\frac{F C_0(\varphi)}{2 C_{n0}} P_n(\cos\theta) \frac{d P_n(\cos\theta')}{d\theta'} \bigg|_{\theta'=\pi/2} \right.
$$

$$
\left. + \sum_{m=1}^{\infty} \frac{F C_m(\varphi)}{C_{nm}} P_n^m(\cos\theta) \Phi_n^m \right),
$$

$$
E_{e\theta}(r, \theta, \varphi) = \frac{1}{r} \sum_{n=0}^{\infty} \sum_{m=0}^{n} \frac{(2 - \delta_{0m}) Q_n^*(r) F C_m(\varphi)}{2n(n+1) C_{nm}}
$$

$$
\left[m^2 \frac{P_n^m(\cos\theta)}{\sin\theta} F_n^m + \frac{d P_n^m(\cos\theta)}{d\theta} \Phi_n^m \right],
$$

$$
E_{e\varphi}(r, \theta, \varphi) = -\frac{1}{r} \sum_{n=0}^{\infty} \sum_{m=1}^{n} \frac{m Q_n^*(r) F S_m(\varphi)}{n(n+1) C_{nm}}
$$

$$
\left[\frac{d P_n^m(\cos\theta)}{d\theta} F_n^m + \frac{P_n^m(\cos\theta)}{\sin\theta} \Phi_n^m \right],
$$

$$
H_{er}(r, \theta, \varphi) = -\frac{1}{i\omega\mu_1 r^2} \sum_{n=0}^{\infty} \sum_{m=1}^{n} \left\{ \frac{m F S_m(\varphi)}{C_{nm}} \left(Q_n^*(r) - 2 Q_n(r) \right) P_n^m(\cos\theta) F_n^m \right\},
$$

$$
H_{e\theta}(r, \theta, \varphi) = \frac{k_1^2}{i\omega\mu_1} \sum_{n=0}^{\infty} \sum_{m=1}^{n} \left\{ \frac{Q_n(r) m F S_m(\varphi)}{n(n+1) C_{nm}} \right.
$$

$$
\left. \left[\frac{P_n^m(\cos\theta)}{\sin\theta} \Phi_n^m + \left(1 - \frac{n(n+1)}{(k_1 r)^2} \right) \frac{d P_n^m(\cos\theta)}{d\theta} F_n^m \right] \right\},
$$

$$
H_{e\varphi}(r, \theta, \varphi) = \frac{k_1^2}{i\omega\mu_1} \sum_{n=0}^{\infty} \sum_{m=0}^{n} \left\{ \frac{Q_n(r) F C_m(\varphi)}{2n(n+1) C_{nm}} \left[(2 - \delta_{0m}) \frac{d P_n^m(\cos\theta)}{d\theta} \Phi_n^m \right. \right.
$$

$$
\left. \left. + 2m^2 \left(1 - \frac{n(n+1)}{r^2} \right) \frac{P_n^m(\cos\theta)}{\sin\theta} F_n^m \right] \right\}, \tag{7.18}
$$

where $Q_n^*(r) = \frac{\partial}{\partial r}(r Q_n(r)) = \frac{(n+1) h_n^{(2)}(k_1 r) - k_1 r h_{n+1}^{(2)}(k_1 r)}{(n+1) h_n^{(2)}(k_1 R) - k_1 R h_{n+1}^{(2)}(k_1 R)}$.

The formulas (7.18) allows us to find the radiated electromagnetic fields at any distance from the antenna, that is, for arbitrary $r \geq R$. If the external homogeneous medium is lossless and ε_1 is purely real, the formulas (7.18) defining the antenna fields in far zone ($r \gg \lambda$) can be simplified, since the terms of orders $1/r^2$ and $1/r^3$ can be omitted.

As an example, let us represent in an explicit form the components of the magnetic field radiated by spherical antenna:

$$
H_{er}(r, \theta, \varphi) = 0,
$$

$$H_{e\theta}(r, \theta, \varphi) = \frac{k_1^2}{i\omega\mu_1} \sum_{n=0}^{\infty} \sum_{m=1}^{n} \frac{Q_n(r) m F S_m(\varphi)}{n(n+1) C_{nm}}$$

$$\left[\frac{P_n^m(\cos\theta)}{\sin\theta} \Phi_n^m + \frac{dP_n^m(\cos\theta)}{d\theta} F_n^m \right],$$

$$H_{e\varphi}(r, \theta, \varphi) = \frac{k_1^2}{i\omega\mu_1} \sum_{n=0}^{\infty} \sum_{m=0}^{n} \frac{Q_n(r) F C_m(\varphi)}{2n(n+1) C_{nm}}$$

$$\left[(2 - \delta_{0m}) \frac{dP_n^m(\cos\theta)}{d\theta} \Phi_n^m + 2m^2 \frac{P_n^m(\cos\theta)}{\sin\theta} F_n^m \right]. \tag{7.19}$$

Since the relations $k_1 r \to \infty$ and $|k_1 r| >> n$ are valid in the antenna far zone, the Hankel spherical functions of the second kind can be replaced by the known asymptotic representation $h_n^{(2)}(kr_1) \approx (i)^{n+1} \frac{e^{-ikr_1}}{kr_1}$, the functions $Q_n(r)$ and $Q_n^*(r)$ in (7.18) and (7.19) can be written as

$$Q_n(r) \approx \frac{e^{-k_1 r}}{k_1 r} \cdot \frac{(i)^{n+1}}{(n+1) h_n^{(2)}(k_1 R) - k_1 R h_{n+1}^{(2)}(k_1 R)},$$

$$Q_n^*(r) \approx \frac{e^{-k_1 r}}{k_1 r} \cdot \frac{(i)^{n+1}(n+1-ik_1 r)}{(n+1) h_n^{(2)}(k_1 R) - k_1 R h_{n+1}^{(2)}(k_1 R)}.$$

Of course, the expressions (7.19) allow us to analyze the antenna directional characteristics in the main polarization planes, which in the coordinate system presented in Fig. 7.1c are defined as follows: H-plane ($r \in [0, \infty]$, $\theta = \pi/2$, $\varphi \in [0, 2\pi]$) and E-plane, consisting of two half-planes ($r \in [0, \infty]$, $\varphi = 0$, $\theta \in [0, \pi]$) and ($r \in [0, \infty]$, $\varphi = \pi$, $\theta \in [0, \pi]$). It is easy to verify that the E-plane of the magnetic field components $H_{e\theta}(r, \theta, \varphi) = 0$, therefore, the functional coefficients $F S_m(\varphi)$ are equal to zero if $\varphi = 0$ and $\varphi = \pi$, since the coefficients are determined by the integral of odd functions (7.9) in symmetric limits. The H-plane consists of the main field component $H_{e\theta}(r, \theta, \varphi)$ and cross component $H_{e\varphi}(r, \theta, \varphi)$ arising due to the influence of the sphere on the antenna directional properties which becomes a vanishingly small if the diffraction sphere radii is sufficiently large.

7.1.3 Numerical and Experimental Results

7.1.3.1 Spherical Antenna with Perfectly Conductive Surfaces

The results of numerical modeling and corresponding software were verified by comparing the calculated and experimental data. The prototype model, which was used to obtain the experimental research is presented in Fig. 7.2. The geometric parameters of the prototype model were as follows: $R = 50.0$ mm, $2L_i = 18.0$ mm,

Fig. 7.2 The prototype model of the spherical antenna

$d = 1.5$ mm, $a = 23.0$ mm, $b = 10.0$ mm. The calculated effective thickness and width of the internal slot cavity were $h_e = 0.1$ mm and $d_e = 1.35$ mm, respectively.

The experimental and calculated plots of the reflection coefficient modulus in the waveguide $|S_{11}|$ and power radiation coefficient $|S_\Sigma|^2$ are shown by circles and solid curves in Fig. 7.3. As can be seen from Fig. 7.3, the calculated and experimental data are in satisfactory agreement with each other. The maximum normalized RP, close to unity, is observed if $\lambda \approx 37.5$ mm, $2L_i \approx 0.48\lambda$ (that is, $2L_e \approx 0.484\lambda$) and the sphere diffraction radius $kR = 8.38$. The RP maximum is achieved by the optimal matching mode of the spherical antenna and waveguide transmission line. As can be seen, the wavelength maxima of the calculated and experimental curves are slightly displaced in the wavelength relative one another. This can be explained by alignment error of the longitudinal slot axis and waveguide cross section during the prototype manufacturing. It should be noted that the prototype model is sufficiently broadband, since the ratio of the antenna bandwidth at the half radiation power $\Delta\lambda \in [29.0;\ 42.5]$ mm to the operating wavelength $\lambda \approx 37.5$ mm is about 35%.

Fig. 7.3 Energy characteristics of the spherical antenna: $R = 50.0$ mm, $2L_i = 18.0$ mm, $d = 1.5$ mm, $a = 23.0$ mm, $b = 10.0$ mm

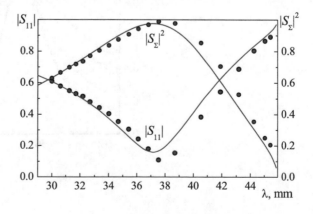

The decrease of the resonant slot length, determined by the achieved maximum level of radiated power, as compared with the half-wavelength can be explained by the influence of the spherical scatterer and dimensions of the internal slot cavity. If the slot is cut in an infinite plane screen, the effect of the actual slot thickness on the slot radiation can be directly taken into account, but this cannot be done for the spherical antenna with fixed dimensions of the waveguide cross-section. Indeed, in this case, the dimensions of the internal slot cavity vary when the sphere radius is changed. Therefore, it is interesting to study the energy characteristics of the spherical antenna as function of the sphere radius for determining the antenna electrical parameters which ensure the maximum radiation at the operating wavelength in the single-mode regime of the waveguide excitation.

The plots of energy characteristics of the spherical antenna as a function of the wavelength are shown in Fig. 7.4 for the spheres of various radii and standard waveguide cross-section $\{23.0 \times 10.0\}$ mm^2. Numerical simulation has revealed the following listed below relationships between the antenna parameters. The maximum

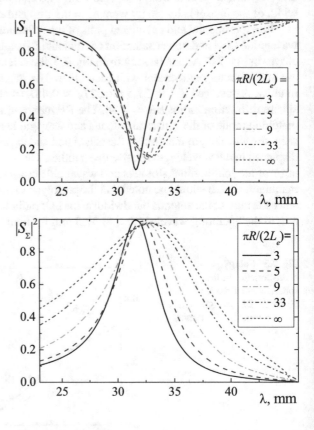

Fig. 7.4 The energy characteristics of the spherical antenna with parameters: $2L_e = 16.0$ mm, $d/(2L_e) = 0.05$, and the waveguide cross section $\{23.0 \times 10.0\}$ mm^2

level of antenna radiation at any wavelength in the range of the single-mode waveguide operation, excluding the region close to the critical wavelength, can be obtained by selecting the slot length. If the sphere radius is small, for example, $\pi R/(2L_e) = 3$, the slot resonant length is close to half-wave $2L_e \approx 0.5\lambda$, and if it is increased, the resonant slot length decreases. The maximum of the resonant slot length is observed for the infinite screen and it corresponds to the slot length $2L_e \approx 0.48\lambda$. The bandwidth of the antenna radiation coefficient measured at the half power level is also maximal for the infinite screen, and it is significantly reduced when the sphere radius and the operating wavelength are reduced. In near the resonance, the perfectly conducting screen approximation becomes correct for the diffraction radius $kR \geq 10$. However, far from the romance, the energy characteristics cannot be calculated using this approximation even for sufficiently large sphere radii, $kR \approx 33$. Additional calculations have showed that satisfactory modeling accuracy of the slotted spherical antennas by using the formula for the slot conductivity of infinite screen can be obtained only for $kR \geq 50$. It should be noted that for other ratios $d/(2L_e)$ within the framework of the accepted restrictions on the slot width, all the described trends are preserved, but only the numerical estimates of the resonance length shortening are varying.

An important question concerning the practical application of the spherical antenna design, is a possibility of using waveguides with an underestimated height of the waveguide cross-section b. The calculated radiation coefficient as function of the wavelength for the spherical antennas with parameters: $b = 5$ mm, $b = 7.5$ mm, $b = 10$ mm, and $\pi R/(2L_e) = 3$ are presented in Fig. 7.5. As can be seen from the plots, if the waveguide height is reduced, the resonant wavelength of the spherical antenna can be also reduced, and the maximum radiation can be achieved by adjusting the slot length. The energy characteristics of the spherical antenna with other sphere radii behave in the similar form. However, all relations inherent to the spherical antenna with the waveguide of standard cross section are also valid for the underestimated waveguide height.

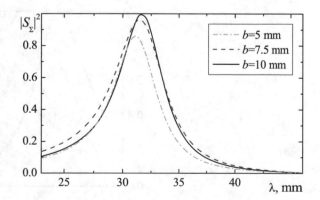

Fig. 7.5 The energy characteristics of the spherical antenna with parameters: $2L_e = 16.0$ mm, $\pi R/(2L_e) = 3$, for the waveguide with different waveguide cross-sections $\{23.0 \times b\}$ mm^2

7.1.3.2 Spherical Antenna with an Impedance End-Wall of a Waveguide Section

As noted above, the maximum radiation level of the spherical antenna can be achieved at any operating wavelength in the waveguide single-mode range. For example, if the sphere radius is $\pi R/(2L_e) = 3$, the resonant slot length is close to half-wave, $2L_e \approx 0.5\lambda$, and it decreases when the sphere radius is increased. Therefore, a question important for practical applications arises to what limit the slot resonant length can be tuned by varying the impedance distributed over the end-wall of the waveguide section. The calculated plots of the radiation coefficient of the antenna with the sphere radii $R = 5L_e/\pi$, and the slot length $2L_e = 14.0$ mm, are shown in Fig. 7.6. The calculations were performed for the waveguide with cross-section $\{a \times b\} = 23.0 \times 10.0$ mm^2 operating in the single-mode excitation range.

Fig. 7.6 The slot radiation coefficient as function of the wavelength for varying imaginary part of the surface impedance of the spherical antenna with parameters $2L_e = 14.0$ mm, $d/(2L_e) = 0.05$, and the waveguide cross-section $\{23.0 \times 10.0\}$ mm^2: **a** $\pi R/(2L_e) = 5$, **b** $\pi R/(2L_e) = 10$, 1 $\bar{Z}_S = 0$, 2 $\bar{Z}_S = i0.01$, 3 $\bar{Z}_S = i0.05$, 4 $\bar{Z}_S = i0.2$

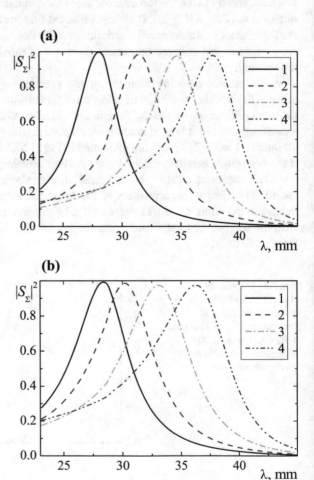

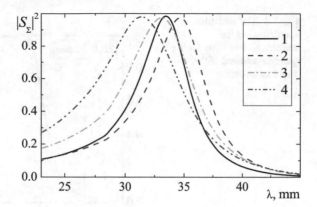

Fig. 7.7 The radiation coefficient as function of the operating wavelength for the spherical antenna with parameters $2L_e = 14.0$ mm, $d/(2L_e) = 0.05, \bar{Z}_S = i0.05$. The sphere radii are: 1 $\pi R/2L_e = 3, 2$ $\pi R/2L_e = 5, 3$ $\pi R/2L_e = 10, 4$ $\pi R/2L_e = 20$

As can be seen from Fig. 7.6, the imaginary part of the impedance element shifts the maximum of the radiated power to the long-wave part of the operating band. The impedance variation in the range $\bar{Z}_S \in [0; \; i0.2]$ allows tuning resonant wavelength of the spherical antenna within $(30 \div 35)\%$ as compared with the case $\bar{Z}_S = 0$. Despite the fact that the range of impedance tuning is almost half of the single-mode operating band of the waveguide section, the antenna reflection coefficient does not significantly increase, and the radiation coefficient maximum does not decrease. Thus, it can be stated that the antenna and waveguide transmission line are satisfactory matched in the entire waveguide operating range. As can be seen, if the sphere radius is increased, a tendency to slight narrowing of the antenna tuning range is observed (Fig. 7.6b). It can be also stated that the width of the antenna operating band at the half power level for any value $\bar{Z}_S$ of the impedance element is maximal for large sphere radius, and it significantly decreases when the sphere radius is decreased. On the other hand, if the sphere radius is increased, the resonant slot length is shifted first to the long-wavelength band when ($\pi R/2L_e = 5$), and then it is shifted to the short-wavelength band when ($\pi R/2L_e = 10; \; 20$). For the fixed slot length and surface impedance of the coating, $2L_e = 14.0$ mm and $\bar{Z}_S = i0.05$, these trends can also be observed in Fig. 7.7.

7.1.3.3 Spherical Antenna with the Reentrant Resonator in the Waveguide

The radiation coefficients $|S_\Sigma|^2$ for the system consisting of a resonant diaphragm, reentrant resonator, and radiating slot as function of the wavelength in the single-mode range of the rectangular waveguide are plotted in Fig. 7.8. The parameters of the antenna are as follows: $\{a \times b\} = 23.0 \times 10.0$ mm^2, $R = 80/\pi$ mm; slot diaphragm length, width, thickness, and position are $2L_d = 2L_i$, $d_d = d = 0.8$ mm, $h_d = 1.0$ mm, and $y_{0d} = b/8$.

As can be seen from Fig. 7.8, the reentrant resonator placed in the waveguide significantly increases the quality factor of the system. In this case, the resonant

Fig. 7.8 The radiation coefficients as the wavelength functions for the spherical antenna with the reentrant resonator

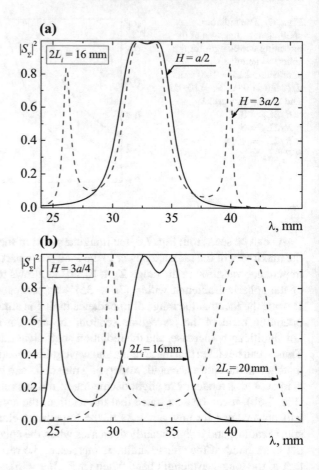

curve of the radiation coefficient has large slopes, and its shape becomes closed to a rectangular one (Fig. 7.8a). If several reentrant resonators are included in the radiating system, even greater variation of the resonant curve slopes can be observed. If the waveguide conductivity is not taken into account, the maximum antenna radiation at two or three frequencies in the range of the single-mode regime of the H_{10}-wave propagation can be obtained by varying the length of the reentrant cavity ($H = 3a/4$, $H = 3a/2$), while the shape of the main resonance curve practically coincide with the case $H = a/2$ (Fig. 7.8a). The extrema of the frequency characteristics are displaced practically without changing the curve shape to one or another part of the operating range by varying the slot length and, hence, the resonator length (Fig. 7.8b).

7.2 Radial Impedance Vibrator Located on a Sphere

Radially oriented radiators located near the conducting spheres were considered in [11, 12, 26–31]. Studies of spherical antenna characteristics have always been of practical interest, since the vibrator radiators are widely used in mobile vehicle, including aircraft [1]. It should be noted that the impedance vibrators were investigated only in [11, 12, 30, 31], the radiators based on the Hertz dipole and perfectly conductive vibrators were studied in [26–29].

The publications [11, 12, 30, 31] are devoted to the approximate analytical solution of the equation for the current on a radial impedance vibrator located near or on a conducting sphere using a method of successive iterations. To achieve this goal, a traditional approach was used, based on the application of the Green function constructed in [11, 12] for space outside the perfectly conducting sphere. However, as further studies have showed [32], the solution obtained in [30, 31] turns out to be physically correct only for spheres of sufficiently large radii as compared to the wavelength. Of course, the spherical scatterers of resonant dimensions, which are most interesting for practical applications, cannot be analyzed by using this approach. In this regard, the problem of analytical solving the integral equation for the current on the radial impedance monopole, physically correct for arbitrary sizes of a spherical antenna, is relevant for further researches. Also, no less relevant is studying the influence of the monopole surface impedance on the antenna radiation characteristics. These issues are the subject of the next Subsection.

7.2.1 Problem Formulation and Solution

Consider a radiation system consisting of a perfectly conducting sphere and thin cylindrical impedance vibrator whose axis is oriented in the direction $\theta' = \theta_0$, $\varphi' = \varphi_0$ (Fig. 7.9). The sphere radius r and length L are selected so that inequalities $(r/L) \ll 1$, $(r/\lambda) \ll 1$ hold.

Fig. 7.9 Radial impedance monopole located on a sphere

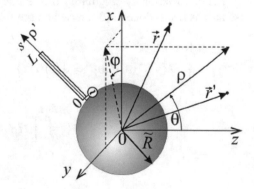

In this case, the electric Hertz vector potential $\vec{\Pi}(\vec{r})$ has only the radial component

$$\Pi_\rho(\vec{r}) = \frac{1}{i\omega\varepsilon_1} \int\limits_{\tilde{R}}^{\tilde{R}+L} J(\rho')G_{\rho\rho'}(\rho, \theta, \varphi; \rho', \theta_0, \varphi_0)d\rho', \quad (7.20)$$

where $\vec{r}$ is the radius vector of the observation point. If the intrinsic linear impedance of the vibrator is constant, the initial integral equation presented on the vibrator generatrix approximated by a segment of the radial beam $\rho \in [\tilde{R}; \tilde{R} + L]$ in the direction $\theta = \theta_0, \varphi = \varphi_0 + r/(\tilde{R} + L/2)$, can be written as [33]

$$\frac{d^2[k_1\rho\Pi_\rho(\rho)]}{d\rho^2} + k_1^2[k_1\rho\Pi_\rho(\rho)] = -E_{0\rho}(\rho) + z_i J(\rho), \quad (7.21)$$

where $E_{0_\rho}(\rho)$ is the radial component of the external excitation field. The Eq. (7.21) is written in terms of the differential operator $\mathrm{rot} -\left[\vec{\rho}^0, \vec{\tilde{\Pi}}(\rho, \theta, \varphi)\right]$ under condition that the Hertz pseudovector is defined as $\vec{\Pi}(\rho, \theta, \varphi) = k_1\rho\Pi_\rho\vec{\rho}^0 + \vec{\theta}^0\Pi_\theta + \vec{\varphi}^0\Pi_\varphi$ ($\vec{\rho}^0, \vec{\theta}^0, \vec{\varphi}^0$ are unit vectors) through the true electrical Hertz vector, whose components are expressed similarly to (7.20). It is easy to verify using the relations known from electrodynamics theory that self-consistency of electromagnetic fields can be satisfied for the E-type spherical waves which are considered here. If the expression (7.20) and notation $G_{\rho\rho'}(\rho, \rho') = G\left(\rho, \theta_0, \varphi_0 + r/(\tilde{R} + L/2); \rho', \theta_0, \varphi_0\right)$ are taken into account, the Eq. (7.21) can be written as

$$\left[\frac{d^2}{d\rho^2} + k_1^2\right]\left(k_1\rho \int\limits_{\tilde{R}}^{\tilde{R}+L} J(\rho')G(\rho, \rho')d\rho'\right) = -i\omega\varepsilon_1 E_{0\rho}(\rho) + i\omega\varepsilon_1 z_i J(\rho), \quad (7.22)$$

which coincides with the Poclington equation, widely used in the theory of thin wire antennas [34].

The quasistationary singularity of the integral equation kernel in (7.22) can be isolated as it was done in [34] based on the following identical transformations

$$\int\limits_{\tilde{R}}^{\tilde{R}+L} J(\rho')G(\rho, \rho')d\rho' = J(\rho)\Omega(\rho)$$

$$+ \int\limits_{\tilde{R}}^{\tilde{R}+L}\left[J(\rho')G(\rho, \rho') - \frac{J(\rho)}{R(\rho,\rho')}\right]d\rho', \quad (7.23)$$

where $R(\rho, \rho') = \sqrt{(\rho - \rho')^2 + r^2}$, $\Omega(\rho) = \int_{\tilde{R}}^{\tilde{R}+L} \frac{d\rho'}{\sqrt{(\rho-\rho')^2+r^2}} = \ln$

$\left[\frac{\sqrt{(L-\rho)^2+r^2}+(L-\rho)}{\sqrt{\rho^2+r^2}-\rho} \right]$, and the average value of the integral is $\bar{\Omega}(\rho) = 2\ln(L/r) - 0.614$. Let us introduce the functional

$$F[k_1\rho, J(\rho)] = \left[\frac{d^2}{d\rho^2} + k_1^2 \right] \left(k_1\rho \int\limits_{\tilde{R}}^{\tilde{R}+L} \left[J(\rho')G(\rho, \rho') - \frac{J(\rho)}{R(\rho,\rho')} \right] d\rho' \right), \quad (7.24)$$

and a small parameter $\alpha = -1/\bar{\Omega}(\rho) \approx \frac{1}{2\ln(r/L)}$. Then the Eq. (7.22) can be represented as follows

$$\left[\frac{d^2}{d\rho^2} + k_1^2 \right] [k_1\rho\, J(\rho)] = i\omega\varepsilon_1\alpha\, E_{0\rho}(\rho)$$

$$- i\omega\varepsilon_1 z_i J(\rho) - \alpha F[k_1\rho, J(\rho)]. \quad (7.25)$$

This equation can be rewritten as

$$\left[\frac{d^2}{d\rho^2} + \tilde{\tilde{k}}_1^2 \right] [k_1\rho\, J(\rho)] = \alpha\left\{ i\omega\varepsilon_1 E_{0\rho}(\rho) + \tilde{F}[k_1\rho, J(\rho)] \right\}, \quad (7.26)$$

where $\tilde{\tilde{k}}_1 = k_1\sqrt{1 + i\alpha\omega\varepsilon_1 z_i/k_1} = k_1\sqrt{1 + i2\alpha\bar{Z}_S/(kr)}$, $\tilde{F}[k_1\rho, J(\rho)] = i\omega\varepsilon_1 z_i(\rho-1)J(\rho) - F[k_1\rho, J(\rho)]$, and $\bar{Z}_S = \bar{R}_S + i\bar{X}_S$ is the normalized distributed surface impedance. The general solution of the Eq. (7.26) relative to the current $J(\rho)$ valid for field $E_{0\rho}(\rho)$ of arbitrary external sources can be obtained by inverting the operator in the left-hand side of the equation

$$k_1\rho\, J(\rho) = C_1 \sin(\tilde{\tilde{k}}_1\rho) + C_2 \cos(\tilde{\tilde{k}}_1\rho)$$

$$- \frac{i\omega\varepsilon_1\alpha}{2\tilde{\tilde{k}}_1}\left\{ \int\limits_{\tilde{R}}^{\rho} E_{0\rho}(\rho') \sin[\tilde{\tilde{k}}_1(\rho-\rho')]d\rho' - \int\limits_{\rho}^{\tilde{R}+L} E_{0\rho}(\rho') \sin[\tilde{\tilde{k}}_1(\rho-\rho')]d\rho' \right\}$$

$$+ \frac{\alpha}{2\tilde{\tilde{k}}_1}\left\{ \int\limits_{\tilde{R}}^{\rho} \tilde{F}[k_1\rho, J(\rho)] \sin[\tilde{\tilde{k}}_1(\rho-\rho')]d\rho' - \int\limits_{\rho}^{\tilde{R}+L} \tilde{F}[k_1\rho, J(\rho)] \sin[\tilde{\tilde{k}}_1(\rho-\rho')]d\rho' \right\}, \quad (7.27)$$

where C_1 and C_2 are arbitrary constants, determined from the boundary conditions at the monopole ends. The exciting field $E_{0\rho}(\rho)$ is specified for the problem under consideration. In accordance with the method of successive iterations [34], the zeroth approximation for the vibrator current can be obtained from the solution (7.27) in the form

$$k_1\rho\, J_0(\rho) = C_1 \sin(\tilde{\tilde{k}}_1\rho) + C_2 \cos(\tilde{\tilde{k}}_1\rho)$$

$$-\frac{i\omega\varepsilon_1\alpha}{2\tilde{\tilde{k}}_1}\left\{\int\limits_{\tilde{R}}^{\rho} E_{0\rho}(\rho')\sin[\tilde{\tilde{k}}_1(\rho-\rho')]d\rho' - \int\limits_{\rho}^{\tilde{R}+L} E_{0\rho}(\rho')\sin[\tilde{\tilde{k}}_1(\rho-\rho')]d\rho'\right\}. \quad (7.28)$$

Consider a case of vibrator excitation by a point voltage δ-generator with amplitude V_0 located at the monopole base, so that the field $E_{0\rho}(\rho) = V_0\delta(\rho-\tilde{R})$. Then the zero approximation of the monopole current can be represented according to (7.28) as

$$k_1\rho\, J_0(\rho) = C_1 \sin(\tilde{\tilde{k}}_1\rho) + C_2 \cos(\tilde{\tilde{k}}_1\rho) - \frac{i\omega\varepsilon_1\alpha\, V_0}{\tilde{\tilde{k}}_1}\sin\tilde{\tilde{k}}_1(\rho-\tilde{R}) \quad (7.29)$$

or

$$J_0(\rho) = \tilde{C}_1 j_0(\tilde{\tilde{k}}_1\rho) + \tilde{C}_2 y_0(\tilde{\tilde{k}}_1\rho) - \frac{i\omega\varepsilon_1\alpha\, V_0}{k_1}\frac{\sin\tilde{\tilde{k}}_1(\rho-\tilde{R})}{\tilde{\tilde{k}}_1\rho}, \quad (7.30)$$

where $j_0(\tilde{\tilde{k}}_1\rho)$ и $y_0(\tilde{\tilde{k}}_1\rho)$ are spherical functions of the zero order:

$$j_0(\tilde{\tilde{k}}_1\rho) = \frac{\sin(\tilde{\tilde{k}}_1\rho)}{\tilde{\tilde{k}}_1\rho}, \ y_0(\tilde{\tilde{k}}_1\rho) = -\frac{\cos(\tilde{\tilde{k}}_1\rho)}{\tilde{\tilde{k}}_1\rho}. \quad (7.31)$$

Since the relation between the constants C_1 and C_2 can be found from the boundary condition $J_0(\tilde{R} + L) = 0$ as

$$C_2 = C_1 tg[\tilde{\tilde{k}}_1(\tilde{R} + L)] + \frac{\alpha\, i\omega\varepsilon_1\, V_0}{k_1}\cdot\frac{\sin(\tilde{\tilde{k}}_1 L)}{\cos[\tilde{\tilde{k}}_1(\tilde{R}+L)]}, \quad (7.32)$$

the formula (7.30) are converted to

$$J_0(\rho) = C_1\left[j_0(\tilde{\tilde{k}}_1\rho) + tg[\tilde{\tilde{k}}_1(\tilde{R} + L)]y_0(\tilde{\tilde{k}}_1\rho)\right]$$

$$+ \frac{\alpha\, i\omega\varepsilon_1\, V_0}{k_1}\cdot\left\{\frac{\sin(\tilde{\tilde{k}}_1 L)}{\cos[\tilde{\tilde{k}}_1(\tilde{R}+L)]}y_0(\tilde{\tilde{k}}_1\rho) - \frac{\sin\tilde{\tilde{k}}_1(\rho-\tilde{R})}{\tilde{\tilde{k}}_1\rho}\right\}. \quad (7.33)$$

Since the boundary condition $J_0(\tilde{R} + L) = 0$ is satisfied for expression (7.33) with the arbitrary constant C_1, which can be found using the excitation condition at the point of monopole contact with the perfectly conducting sphere. In this point, the equality $\mathrm{div}[k_1\rho\, J_0(\rho)]|_{\rho=\tilde{R}} = 0$ should be satisfied due to the current continuity. The product $k_1\rho\, J_0(\rho) = J_{act}(\rho)$ is considered as the magnitude of the effective current. Then, after identical transformations, the following expression for the constant C_1 is obtained

$$C_1 = -\frac{\alpha i \omega \varepsilon_1 V_0}{k_1} \frac{\tilde{\tilde{k}}_1 \tilde{R} \cos[\tilde{\tilde{k}}_1(\tilde{R}+L)] + \sin(\tilde{\tilde{k}}_1 L)[2\cos(\tilde{\tilde{k}}_1 \tilde{R}) - \tilde{\tilde{k}}_1 \tilde{R} \sin(\tilde{\tilde{k}}_1 \tilde{R})]}{2\sin(\tilde{\tilde{k}}_1 L) - \tilde{\tilde{k}}_1 \tilde{R} \cos(\tilde{\tilde{k}}_1 L)}.$$

(7.34)

Thus, the final expression for the monopole current can be represented in a form convenient for calculations

$$J_0(\rho) = -\frac{\alpha i \omega \varepsilon_1 V_0}{k_1}\left\{ C_j j_0(\tilde{\tilde{k}}_1 \rho) + C_y y_0(\tilde{\tilde{k}}_1 \rho) + \frac{\sin[\tilde{\tilde{k}}_1(\tilde{R}-\rho)]}{\tilde{\tilde{k}}_1 \rho} \right\},$$

(7.35)

where

$$C_j = \frac{\tilde{\tilde{k}}_1 \tilde{R} \cos[\tilde{\tilde{k}}_1(\tilde{R}+L)] + \sin(\tilde{\tilde{k}}_1 L)[2\cos(\tilde{\tilde{k}}_1 \tilde{R}) - \tilde{\tilde{k}}_1 \tilde{R} \sin(\tilde{\tilde{k}}_1 \tilde{R})]}{2\sin(\tilde{\tilde{k}}_1 L) - \tilde{\tilde{k}}_1 \tilde{R} \cos(\tilde{\tilde{k}}_1 L)},$$

$$C_y = \frac{\tilde{\tilde{k}}_1 \tilde{R} \sin[\tilde{\tilde{k}}_1(\tilde{R}+L)] + \sin(\tilde{\tilde{k}}_1 L)\left[2\sin(\tilde{\tilde{k}}_1 \tilde{R}) + \tilde{\tilde{k}}_1 \tilde{R} \cos(\tilde{\tilde{k}}_1 \tilde{R})\right]}{2\sin(\tilde{\tilde{k}}_1 L) - \tilde{\tilde{k}}_1 \tilde{R} \cos(\tilde{\tilde{k}}_1 L)}.$$

As follows from (7.35), the solution for the current on the impedance monopole is valid for the tuned ($|\tilde{\tilde{k}}_1 L| = n\frac{\pi}{2}$, $n = 1, 2, \ldots$) and non-tuned ($|\tilde{\tilde{k}}_1 L| \neq n\frac{\pi}{2}$) vibrators, i.e. it is applicable for the vibrators of arbitrary electrical length.

After identical transformations of the expression (7.35), based on the formulas (7.31), the following expression for the current can be obtained

$$J_0(\rho) = C_1 \frac{\sin[\tilde{\tilde{k}}_1(\rho - (\tilde{R}+L))]}{\tilde{\tilde{k}}_1 \rho \cos[\tilde{\tilde{k}}_1(\tilde{R}+L)]}$$

$$- \frac{\alpha l \omega \varepsilon_1 V_0\left[\cos(\tilde{\tilde{k}}_1 \rho)\sin(\tilde{\tilde{k}}_1 L) + \sin[\tilde{\tilde{k}}_1(\tilde{R}-\rho)]\cos[\tilde{\tilde{k}}_1(\tilde{R}+L)]\right]}{k_1 \tilde{\tilde{k}}_1 \rho \cos[\tilde{\tilde{k}}_1(\tilde{R}+L)]}.$$

(7.36)

The expression (7.36) is called by the improved zero approximation. But only the pure zero approximation, presented by the first term in (7.36), is used as the basis function. that is, before determining the constant C_1, the term proportional to the small parameter α should be excluded from the current distribution, and the term $\cos[\tilde{\tilde{k}}_1(\tilde{R}+L)]$ in the denominator should be combined with the current J_0. Therefore, the expression for the current can be presented as

$$J(\rho) = J_0 f(\rho) = J_0 \frac{\sin[\tilde{\tilde{k}}_1(\rho - (\tilde{R}+L))]}{\tilde{\tilde{k}}_1 \rho}.$$

(7.37)

7.2.2   Numerical Results

As known from literature, the approximation of currents on cylindrical vibrators by the function $\sin[k_1(L - |s|)]$ is widely and successfully used in practice for calculating vibrators characteristics, such as the radiation field in all wave zones. Since the characteristics are integral quantities of the current distribution function, small errors of the current approximation do not make a significant contribution to the final result. It must be borne in mind that the vector potential of the spherical scatterer at any point on the monopole is determined not only by the local current at this point, but by the total effect of the currents induced both in all other monopole and spherical scatterer points. This effect allows us to consider the monopole located on the perfectly conducting sphere as a single antenna. Therefore, it can be expected that the current distribution along the monopole can differ from the sinusoidal distribution the greater, the smaller is the diffraction sphere radius $k\tilde{R}$, i.e., the greater the is difference of mutual influence between the monopole and spherical scatterer from that for the monopole and infinite screen. This conclusion is true both for the perfectly conducting and impedance monopolies.

The curves of the normalized current J_N as function of the local coordinate s/L ($s = \rho - \tilde{R}$) for the perfectly conducting monopole located on the sphere (Fig. 7.9) are presented in Fig. 7.10. The curves were plotted based on the formula (7.37) with following parameters: $\bar{R}_S = 0.0001, r/\lambda = 0.0033$, and $\lambda = 10.0$ cm. The analysis has confirmed that the spheres of small and resonant diffraction radii have greater effect on the current distribution as compared for the spheres of large electrical dimensions. Therefore, the shortening effect of the electric monopole length due to the spheres of indicated radii.

The normalized current distributions on the quarter-wave monopole ($L = \lambda/4$), located on the sphere with the diffraction radius $k\tilde{R} = \pi$ characterized by the surface impedance $(\bar{Z}_S = i\bar{X}_S)$ of the inductive $(\bar{X}_S > 0)$ or capacitive $(\bar{X}_S < 0)$ types are shown in Fig. 7.11. As can be seen from the plots, the resonant characteristics of the perfectly conducting and impedance monopoles are quite different. The resonant length of the monopoles depends on the type and magnitude of the distributed surface impedance, it can be smaller or greater than a quarter of operating wavelength, hence, the monopole shortening or lengthening occur if $\bar{X}_S < 0$ or $\bar{X}_S > 0$. It should be noted that the monopole lengthening due to the inductive type impedance can compensate the monopole shortening due to the sphere influence. The similar trend is also observed for other dimensions of the spherical scatterers.

7.3   Combined Vibrator-Slot Structure Located on a Sphere

Consider a radiating system consisting of perfectly conducting sphere with a narrow arc slot cut along the parallel and of two radial cylindric impedance thin vibrators placed on the sphere near the slot (Fig. 7.12). The sphere radius is $\tilde{R}$, slot length and

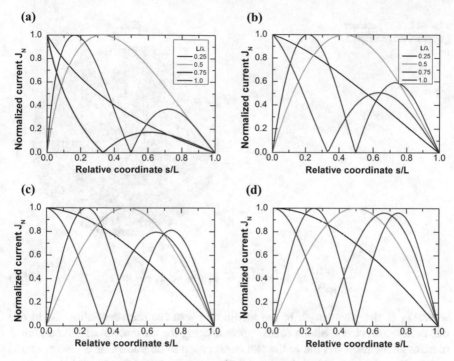

Fig. 7.10 Current distributions on the monopoles of various lengths: **a** $\tilde{R} = 0.1\lambda$ ($k\tilde{R} = 0.2\pi$), **b** $\tilde{R} = 0.5\lambda$ ($k\tilde{R} = \pi$), **c** $\tilde{R} = 2\lambda$ ($k\tilde{R} = 4\pi$), **d** $\tilde{R} = 12\lambda$ ($k\tilde{R} = 24\pi$)

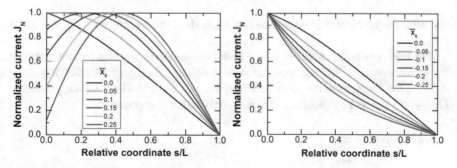

Fig. 7.11 Current distributions on the impedance monopoles located on the sphere ($L = \lambda/4$, $k\tilde{R} = \pi$)

width are $2L$ and d, slot geometric center is in the point with spherical coordinates $\left(\rho' = \tilde{R}; \theta' = \theta_0; \varphi' = 0\right)$. The vibrator radii and lengths are $r_{V1}, r_{V2}, L_{V1}, L_{V2}$, and their axes are oriented along the directions $\theta_{V1}, \varphi_{V1}$ and $\theta_{V2}, \varphi_{V2}$. The homogeneous medium outside the sphere is characterized by material parameters (ε_1, μ_1). The following inequalities are supposed to be satisfied

Fig. 7.12 The problem
geometry and accepted
notations

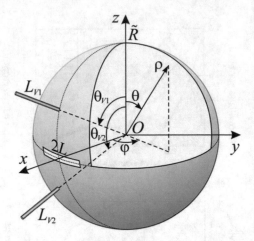

$$\frac{r_{V1(V2)}}{L_{V1(V2)}} \ll 1, \frac{r_{V1(V2)}}{\lambda_1} \ll 1, \frac{d}{2L} \ll 1, \frac{d}{\lambda_1} \ll 1, \tag{7.38}$$

where λ_1 is the wavelength in the medium. These requirements allow us to use physically correct one-dimensional approximating the currents in of the combined radiator elements. The field in the slot is proposed to be known. It is also assumed that the boundary conditions for the electric currents at the vibrator ends

$$J_{V1(V2)}(\rho)\big|_{\rho=\tilde{R}+L_{V1(V2)}} = 0, \ \text{div}[k_1\rho\, J_{V1(V2)}(\rho)]\big|_{\rho=\tilde{R}}=0 \tag{7.39}$$

and for the equivalent magnetic current at the ends of the slot $J_S(\varphi)\big|_{\varphi=\pm L/\tilde{R}} = 0$ are fulfilled.

The thin conductor model accepted in the antenna theory states that the fields of the vibrator surface currents are equivalent to fields of linear currents flowing along the vibrator longitudinal axes [34]. Then, the true electric Hertz vector potentials $\vec{\Pi}_{V1(V2)}(\vec{r})$ excited by these currents have only radial components

$$\Pi_{\rho\,V1(V2)}(\vec{r}) = \frac{1}{i\omega\varepsilon_1}$$

$$\int\limits_{\tilde{R}}^{\tilde{R}+L_{V1(V2)}} J_{V1(V2)}(\rho')G_{\rho\rho'}(\rho, \theta, \varphi;\, \rho', \theta_{V1(V2)}, \varphi_{V1(V2)})d\rho', \tag{7.40}$$

where $\vec{r}$ is the radius vector of the observation point, ω is the angular frequency, and $G_{\rho\rho'}(\rho, \theta, \varphi;\, \rho', \theta', \varphi')$ is the component of the electric tensor Green's function for space outside the perfectly conducting sphere [11, 12] (Appendix A),

$$G^e_{\rho\rho'}(\rho, \theta, \varphi; \rho', \theta', \varphi') = -\sum_{n=0}^{\infty} \frac{n + 1/2}{2\pi}$$

$$h_n(\rho, \rho') P_n(\cos\theta\cos\theta' + \sin\theta\sin\theta'\cos(\varphi - \varphi')),$$

where

$$h_n(\rho, \rho') = \begin{cases} k_1 h_n^{(2)}(k_1\rho')\left[h_n^{(2)}(k_1\rho)\overline{Q}_n\left(y_n\left(k_1\tilde{R}\right)\right) - y_n(k_1\rho)\right], \tilde{R} \le \rho < \rho', \\ k_1 h_n^{(2)}(k_1\rho)\left[h_n^{(2)}(k_1\rho')\overline{Q}_n\left(y_n\left(k_1\tilde{R}\right)\right) - y_n(k_1\rho')\right], \rho > \rho', \end{cases}$$

$$\overline{Q}_n\left(y\left(k_1\tilde{R}\right)\right) = \frac{k_1\tilde{R}y_{n-1}\left(k_1\tilde{R}\right) - ny_n\left(k_1\tilde{R}\right)}{k_1\tilde{R}h_{n-1}^{(2)}\left(k_1\tilde{R}\right) - nh_n^{(2)}\left(k_1\tilde{R}\right)},$$

and $P_n(\cos\theta)$ are Legendre polynomials. Noted that the boundary conditions for electric type functions in the form $\left.\frac{d(\rho h_n(\rho,\rho'))}{d\rho}\right|_{\rho=\tilde{R}} = 0$ should be fulfilled on the surface of the perfectly conducting sphere.

7.3.1 Solution of the External Electrodynamic Problem

The initial system of two integral equations on the on generatrixes of the vibrators approximated by segments of radial rays $\rho \in [\tilde{R}; \tilde{R} + L_{V1(V2)}]$ in the directions $\theta = \theta_{V1(V2)}$ and $\varphi_{1(2)} = \varphi_{V1(V2)} + r/(\tilde{R} + L_{V1(V2)}/2)$ can be written in the following form [32]:

$$\begin{cases} \left(\frac{d^2}{d\rho^2} + k_1^2\right)\left(\begin{array}{l} k_1\rho \int\limits_{\tilde{R}}^{\tilde{R}+L_{V1}} J_{V1}(\rho')G_{\rho\rho'}(\rho, \theta_{V1}, \varphi_1; \rho', \theta_{V1}, \varphi_{V1})d\rho' \\ + k_1\rho \int\limits_{\tilde{R}}^{\tilde{R}+L_{V2}} J_{V2}(\rho')G_{\rho\rho'}(\rho, \theta_{V1}, \varphi_1; \rho', \theta_{V2}, \varphi_{V2})d\rho' \end{array}\right) = -E_{0S}^{V1}(\rho) + i\omega\varepsilon_1 z_{i1}J_{V1}(\rho), \\[20pt] \left(\frac{d^2}{d\rho^2} + k_1^2\right)\left(\begin{array}{l} k_1\rho \int\limits_{\tilde{R}}^{\tilde{R}+L_{V1}} J_{V1}(\rho')G_{\rho\rho'}(\rho, \theta_{V2}, \varphi_2; \rho', \theta_{V1}, \varphi_{V1})d\rho' \\ + k_1\rho \int\limits_{\tilde{R}}^{\tilde{R}+L_{V2}} J_{V2}(\rho')G_{\rho\rho'}(\rho, \theta_{V2}, \varphi_2; \rho', \theta_{V2}, \varphi_{V2})d\rho' \end{array}\right) = -E_{0S}^{V2}(\rho) + i\omega\varepsilon_1 z_{i2}J_{V2}(\rho). \end{cases}$$

$$(7.41)$$

where $Z_{i1(2)} = const$ [Ω/m] is the constant internal linear impedances of the vibrators. The external fields on the vibrators are determined as [35]

$$E_{0S}^{V1(V2)}(\rho) = -\frac{i\omega\varepsilon_1}{\rho}\sum_{n=0}^{\infty}Q_n(\rho)\left(\begin{array}{l}\dfrac{FC_0(\varphi_{1(2)})}{2C_{n0}}P_n(\cos\theta_{V1(V2)})\left.\dfrac{dP_n(\cos\theta')}{d\theta'}\right|_{\theta'=\theta_0}\\[3mm]+\displaystyle\sum_{m=1}^{\infty}\dfrac{FC_m(\varphi_{1(2)})}{C_{nm}}P_n^m(\cos\theta_{V1(V2)})\left.\dfrac{dP_n^m(\cos\theta')}{d\theta'}\right|_{\theta'=\theta_0}\end{array}\right),\qquad(7.42)$$

where $Q_n(\rho)$ and $FC_m(\varphi)$ are defined in (7.9). Here, it was taken into account that the magnetic current $J_S(\varphi')$ specified in the slot was approximated by the function $f(\varphi') = \cos\left(k_1\tilde{R}\varphi'\right) - \cos k_1 L$ and the unit amplitude [36].

7.3.2 Solution of the Equation System for Currents

The system of Eq. (7.41) can be solved by the generalized method of induced EMF, assuming that the currents on the vibrators have the form (7.37). The system of linear algebraic equations for unknown current amplitudes can be obtained by following the standard procedures of the generalized method of induced EMF and by using the weight functions (7.37) in the form

$$\begin{cases} J_1 Z_{\Sigma}^{V1} + J_2 Z_{V2}^{V1} = F_{V1}, \\ J_1 Z_{V1}^{V2} + J_2 Z_{\Sigma}^{V2} = F_{V2}, \end{cases}\qquad(7.43)$$

where the matrix coefficients are defined as follows:

$$F_{V1(V2)} = \frac{1}{2k_1}\int_{\tilde{R}}^{\tilde{R}+L_{V1(V2)}}f_{V1(V2)}(\rho)E_{0s}^{V1(V2)}(\rho)d\rho,\qquad(7.44)$$

$$Z_{V2}^{V1} = \frac{1}{2k_1}\int_{\tilde{R}}^{\tilde{R}+L_{V1}}f_{V1}(\rho)\left\{\left[\frac{d^2}{d\rho^2}+k_1^2\right]\left(k_1\rho\int_{\tilde{R}}^{\tilde{R}+L_{V2}}f_{V2}(\rho')G_{\rho\rho'}(\rho,\theta_{V1},\varphi_1;\rho',\theta_{V2},\varphi_{V2})d\rho'\right)\right\}d\rho,\qquad(7.45a)$$

$$Z_{V1}^{V2} = \frac{1}{2k_1}\int_{\tilde{R}}^{\tilde{R}+L_{V2}}f_{V2}(\rho)\left\{\left[\frac{d^2}{d\rho^2}+k_1^2\right]\left(k_1\rho\int_{\tilde{R}}^{\tilde{R}+L_{V1}}f_{V1}(\rho')G_{\rho\rho'}(\rho,\theta_{V2},\varphi_2;\rho',\theta_{V1},\varphi_{V1})d\rho'\right)\right\}d\rho,\qquad(7.45b)$$

$$Z_{\Sigma}^{V1(V2)} = -\frac{ik\varepsilon_1\tilde{Z}_{V1(V2)}}{k_1 r_{V1(V2)}}\int_{\tilde{R}}^{\tilde{R}+L_{V1(V2)}}f_{V1(V2)}^2(\rho)d\rho + \frac{1}{2k_1}\int_{\tilde{R}}^{\tilde{R}+L_{V1(V2)}}f_{V1(V2)}(\rho)$$

$$\times \left\{ \left[\left[\frac{d^2}{d\rho^2} + k_1^2 \right] k_1 \rho \left[\int_{\tilde{R}}^{\tilde{R}+L_{V1(V2)}} \left(\begin{array}{c} f_{V1(V2)}(\rho') \\ \times G_{\rho\rho'}(\rho, \theta_{V1(V2)}, \varphi_{1(2)}; \rho', \theta_{V1(V2)}, \varphi_{V1(V2)}) d\rho' \end{array} \right) \right] \right] \right\} d\rho.$$

$$(7.46)$$

The solution of the equations system (7.43) with account of (7.37), can be written as

$$J_{V1(V2)}(\rho) = J_{1(2)} f_{V1(V2)}(\rho)$$

$$= J_{1(2)} \frac{\sin(\tilde{k}_1 [\rho - \tilde{R} - L_{V1(V2)}])}{\tilde{k}_1 \rho}, \qquad (7.47)$$

where $J_1 = \frac{F_{V1} Z_\Sigma^{V2} - F_{V2} Z_{V2}^{V1}}{Z_\Sigma^{V1} Z_\Sigma^{V2} - Z_{V2}^{V1} Z_{V1}^{V2}}$, $J_2 = \frac{F_{V2} Z_\Sigma^{V1} - F_{V1} Z_{V1}^{V2}}{Z_\Sigma^{V1} Z_\Sigma^{V2} - Z_{V2}^{V1} Z_{V1}^{V2}}.$

7.3.3 Fields Radiated by the Vibration-Slot Structure

When the current amplitudes $J_{1(2)}$ are known, the radiation fields of the Clavin type spherical antenna at any observation point outside the sphere can be found by using well-known formulas. If the slot is located on the sphere equator $(\theta' = \theta_0 = \pi/2)$, the expressions for the electric field radiated by the antenna can be presented as:

$$E_\rho(\rho, \theta, \varphi) = \frac{\partial^2 \left(k_1 \rho \prod_{\rho V1}(\vec{r}) + k_1 \rho \prod_{\rho V2}(\vec{r}) \right)}{\partial \rho^2}$$

$$+ k^2 \varepsilon_1 \mu_1 \left(k_1 \rho \prod_{\rho V1}(\vec{r}) + k_1 \rho \prod_{\rho V2}(\vec{r}) \right)$$

$$- \frac{1}{\rho} \sum_{n=0}^\infty Q_n(\rho) \left(\frac{FC_0(\varphi)}{2C_{n0}} P_n(\cos\theta) \frac{dP_n(\cos\theta')}{d\theta'} \Big|_{\theta'=\pi/2} \right.$$

$$\left. + \sum_{m=1}^\infty \frac{FC_m(\varphi)}{C_{nm}} P_n^m(\cos\theta) \Phi_n^m \right),$$

$$E_\theta(\rho, \theta, \varphi) = \frac{1}{\rho} \frac{\partial^2 \left(k_1 \rho \prod_{\rho V1}(\vec{r}) + k_1 \rho \prod_{\rho V2}(\vec{r}) \right)}{\partial \rho \partial \theta}$$

$$+ \frac{1}{\rho} \sum_{n=0}^\infty \sum_{m=0}^n \frac{\varepsilon_m Q_n^*(\rho) FC_m(\varphi)}{2n(n+1)C_{nm}}$$

$$\left[m^2 \frac{P_n^m(\cos\theta)}{\sin\theta} F_n^m + \frac{dP_n^m(\cos\theta)}{d\theta} \Phi_n^m \right],$$

$$E_\varphi(\rho, \theta, \varphi) = \frac{1}{\rho \sin \theta} \frac{\partial^2 \left(k_1 \rho \prod_{\rho V1}(\vec{r}) + k_1 \rho \prod_{\rho V2}(\vec{r}) \right)}{\partial \rho \partial \varphi}$$

$$-\frac{1}{\rho} \sum_{n=0}^{\infty} \sum_{m=1}^{n} \frac{m \, Q_n^*(\rho) \, FS_m(\varphi)}{n(n+1) C_{nm}}$$

$$\left[\frac{d P_n^m(\cos \theta)}{d\theta} F_n^m + \frac{P_n^m(\cos \theta)}{\sin \theta} \Phi_n^m \right], \tag{7.48}$$

where the functional coefficients $FC_m(\varphi)$ and $FS_m(\varphi)$ coincide with that in (7.9), and the coefficients $Q_n^*(\rho)$, F_n^m and Φ_n^m are defined in (7.18).

7.3.4 Numerical Results

Characteristics of the waveguide vibrator-slot structures radiating into a homogeneous space above an infinite screen having the similar RPs in the E- and H-planes were thoroughly studied in Chap. 6. This Subsection is aimed at numerical modeling the Clavin type radiators located on spheres of resonant dimensions.

The modelling of the electric field in the antenna far zone by using the formulas (7.48) have confirmed the possibility of creating such radiator. The antenna RPs in the main polarization planes for the antenna with the following structure parameters: $R = \lambda$, $2L = 0.5\lambda$, $d = 0.05\lambda$, $L_{v1} = L_{v2} = L_v = 0.29\lambda$, $r_{v1} = r_{v2} = r_v = 0.005\lambda$, $x_{d1} = x_{d2} = x_d = 0.0865\lambda$, $\theta_{v1(2)} = x_{d1(2)}/R$ at operating wavelength $\lambda = 32.0$ mm are shown in Fig. 7.13b. The fields in the E- and H-planes were calculated by using the asymptotics and approximations specified in Sect. 7.1.2 for the single slot on the sphere. As expected, the longitudinal field component in the far zone $E_\rho(\rho, \theta, \varphi) = 0$, and the other two constituents are the main and cross-polarization components. The calculated modules of the total electric field are plotted in Figs. 7.13 and 7.14. For comparison, the RP of the spherical antenna whose geometry coincides with that of the combined radiator is shown in Fig. 7.13. As can be seen from Fig. 7.13, the monopoles of the combined radiator located symmetrically relative to the slot do not influence the H-plane RP. However, in the E-plane, the RP may be narrowed, and its shape may become similar to the H-plane RP by selecting the monopole lengths and their location relative to the slot as it was shown for the Clavin radiator located on the infinite plane. If the distances between the slot center and vibrators are close to that of the Clavin radiator located over the infinite screen ($x_{d1} = x_{d2} = x_d = 0.0865\lambda$ as compared to $x_d = 0.086\lambda$), the length of the monopoles due to the sphere influence is somewhat reduced ($L_{v1} = L_{v2} = L_v = 0.29\lambda$ as compared to $L_v = 0.3125\lambda$). Thus, the similar RP in the E- and H-planes can be obtained by decreasing the sphere radius that may require a stronger shortening of the monopoles.

To assess how significant is the difference when selecting the vibrator lengths for antennas with various sphere radius, the RPs of the combined radiator with the parameters: $2L = 0.5\lambda$, $d = 0.05\lambda$, $L_v = 0.29\lambda$, $x_d = 0.0865\lambda$ and $r_v = 0.005\lambda$

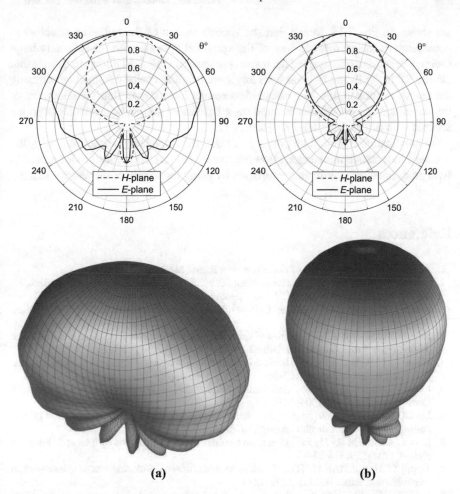

Fig. 7.13 The RPs of the spherical antennas: **a** slot cut in the sphere, **b** clavin radiator on a sphere

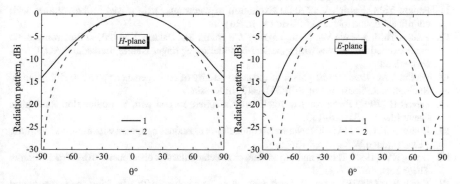

Fig. 7.14 The RPs of the vibrator-slot radiator: 1 on the sphere ($\tilde{R} = \lambda$), 2 on the plane

are shown in Fig. 7.14. Recall that the second variant of the radiating structure is considered as the asymptotic case of the spherical antenna with the infinitely large radius. Comparative analysis of the results presented in Fig. 7.14 allows us to state that similar RP in the E- and H-planes, for any sphere radius, can be obtained by selecting the length of monopoles. The calculation results have shown that the geometry of the combined Clavin radiator located on a flat screen can be successfully used for spheres with diffraction radii $kR \geq 20$.

In conclusion, it should be pointed out that the impedance monopoles with the constant inductance impedance allow us to realize spherical vibrator-slot antennas with predefined directivity characteristics by using monopole of the shorter length.

References

1. Reznikov GB (1967) Aerials of aircraft. Soviet Radio, Moskow (in Russian)
2. Schantz H (2005) Nanoantennas: a concept for efficient electrically small UWB devices. In: IEEE international conference ICU 2005, pp 264–268
3. Ramo S, Whinnery JR, Van Duzer T (1994) Fields and waves in communication electronics. Wiley
4. Karr PR (1951) Radiation properties of spherical antennas as a function of the location of the driving force. J Res Nat Bur Stand 46:422–436
5. Mushiake Y, Webster RF (1957) Radiation characteristics with power gain for slots on a sphere. IRE Trans Antennas Propag 5:47–55
6. Liepa VV, Senior TBA (1965) Modification to the scattering behavior of a sphere by reactive loading. Proc IEEE 112:1004–1011
7. Lin CC, Chen KM (1969) Improved radiation from a spherical antenna by overdense plasma coating. IEEE Trans Antennas Propag 17:675–678
8. Lin CC, Chen KM (1971) Radiation from a spherical antenna covered by a layer of lossy hot plasma. Proc IEEE 118:36–42
9. Towaij SJ, Hamid MAK (1971) Diffraction by a multilayered dielectric-coated sphere with an azimuthal slot. Proc IEEE 119:1209–1214
10. Jang SO, Hyo JE (2009) Radiation of a hertzian dipole in a slotted conducting sphere. IEEE Trans Antennas Propag 57:3847–3851
11. Penkin YuM, Katrich VA (2003) Excitation of electromagnetic waves in the volumes with coordinate boundaries. Fakt, Kharkov (in Russian)
12. Penkin YuM, Katrich VA, Nesterenko MV, Berdnik SL, Dakhov VM (2019) Electromagnetic fields excited in volumes with spherical boundaries. Springer Nature Swizerland AG, Cham, Swizerland
13. Penkin YM (1998) Investigation of the conductivity of an impedance spherical slot antenna. Radiophys Radioastronomy 3(3):341–347 (in Russian)
14. Gavris B (1992) Plane wave diffraction by a sphere loaded with a circular slot. Radiophys Quant Electron 35:126–130
15. Rothwell E, Cloud M (1999) Natural frequencies of a conducting sphere with a circular aperture. J Electromagn Waves Appl 13:729–755
16. Leung KW (1998) Theory and experiment of a rectangular slot on a sphere. IEEE Trans Microw Theor Tech 46:2117–2123
17. Kwok WL (2003) Rectangular and zonal slots on a sphere with a backing shell: theory and experiment. IEEE Trans Antennas Propag 51:1434–1442
18. Penkin YuM, Klimovich RI (2000) Intrinsic and mutual conductivities of slot radiators on a perfectly conducting sphere. Radio Eng 115:75–80 (in Russian)

19. Berdnik SL, Katrich VA, Penkin YM, Nesterenko MV, Pshenichnaya SV (2014) Energy characteristics of a slot cut in an impedance end-wall of a rectangular and radiating into the space over a perfectly conducting sphere. Prog Electromagn Res M 34:89–97
20. Long SA (1975) Experimental study of the impedance of cavity-backed slot antennas. IEEE Trans Antennas Propag 23:1–7
21. Grinev AY, Kotov AY (1978) Machine method of analysis and partial parametric synthesis of resonator-slot structures. Izv Vuz Radioelectron 21(2):30–35 (in Russian)
22. Lee JY, Horng TSh, Alexopoulos NG (1994) Analysis of cavity-backed aperture antennas with a dielectric overlay. IEEE Trans Antennas Propagat 42:1556–1562
23. Kirilenko AA, Rud LA, Senkevich SL, Tkachenko VI (1997) Electrodynamic synthesis and analysis of broadband waveguide filters on resonant diaphragms. Izv Vuz Radioelectron 40(11):54–62 (in Russian)
24. Garb HL, Levinson IB, Fredberg PSh (1968) Effect of wall thickness in slot problems of electrodynamics. Radio Eng Electron Phys 13:1888–1896
25. Mittra R (1973) Computer techniques for electromagnetics. Pergamon Press, Oxford
26. Belkina MG, Weinstein LA (1957) The characteristics of radiation of spherical surface antennas. Diffraction of electromagnetic waves on some bodies of rotation. Soviet Radio, Moscow (in Russian)
27. Bolle DM, Morganstern MD (1969) Monopole and conic antennas on spherical vehicles. IEEE Trans Antennas Propag 17:477–484
28. Tesche FM, Neureuther RE (1976) The analysis of monopole antennas located on a spherical vehicle: part 1. Theory. IEEE Trans. EMC. 18:2–8
29. Tesche FM, Neureuther RE, Stovall RE (1976) The analysis of monopole antennas located on a spherical vehicle: part 2, numerical and experimental results. IEEE Trans EMC 18:8–15
30. Nesterenko MV, Katrich VA, Penkin YuM, Dakhov VM, Berdnik SL (2011) Thin impedance vibrators. Theory and applications. Springer Science + Business Media, New York
31. Nesterenko MV, Penkin DYu, Katrich VA, Dakhov VM (2010) Equation solution for the current in radial impedance monopole on the perfectly conducting sphere. Prog Electromagn Res B 19:95–114
32. Berdnik SL, Katrich VA, Nesterenko MV, Penkin DY, Pshenichnaya SV (2012) Electromagnetic waves excitation by vibrator-slot structure in rectangular waveguide. In: Proceedings of the 6-th international conference ultrawideband and ultrashort impulse signals. Sevastopol, Ukraine, pp 195–197
33. Penkin DY, Katrich VA, Penkin YM, Nesterenko MV, Dakhov VM, Berdnik SL (2015) Electrodynamic characteristics of a radial impedance vibrator on a perfect conduction sphere. Prog Electromagn Res B 62:137–151
34. King RWP (1956) The theory of linear antennas. Harvard University Press, Cambridge
35. Berdnik SL, Katrich VA, Nesterenko MV, Penkin YM (2015) Electrodynamic characteristics of slotted spherical radiators. Appl Electron 14(1):24–35 (in Russian)
36. Penkin DY, Katrich VA, Dakhov VM, Nesterenko MV, Berdnik SL (2013) Radiation fields of radial impedance monopole mounted on a perfectly conducting sphere. In: Proceedings of the IX international conference on antenna theory and techniques. Odessa, Ukraine, pp 123–125

Chapter 8
Combined Vibrator-Slot Radiators in Antenna Arrays

One of important avionics problems is creation of integrated systems combing several functions, such as radio navigation, communication, radar, etc. The possibility of creating scanning antenna arrays with several beams formed at different frequencies is quite essential for the problem [1, 2]. The aperture of combined antenna arrays usually contains two types of radiators, operating at different wavelengths [2]. The integration of multi-frequency radiators into the antenna array aperture may lead to a significant interaction between the radiators of different sub-bands, which, in turn, may cause specific distortions of directivity of multi-frequency antenna arrays. First of all, this concerns antenna gain decrease due to the power loss in radiators of other sub-bands and origination of additional side lobes [2]. The main goal of combined antenna array development consists in avoiding such distortions as much as possible.

The time division switching of antenna array operating modes can be carried out by high-speed semiconductor diode. Electrically controlled microwave switches were proposed more than fifty years ago [3, 4]. The switches have the following properties: a relatively low control power, good compactness, low losses, high operating speed, wide pass band, long lifetime, and high reliability. The switches can be built by using semiconductor diodes, including uncased *p-i-n* diodes which can provide compromises for switching devices operating in the decimeter and centimeter wavelength with pulsed power up to several kilowatts. Of course, in complex combined array elements, in which a structural part of the main radiator isolated by diode switches can be used as an alternative radiator. This is quite natural and most effective for combined antenna arrays.

One of the known combined radiators is the Clavin element consisting of narrow slot cut in an infinite screen and two identical passive vibrators (monopoles) of fixed length placed on both sides of the slot at specified distances from the slot center [5–8]. The slot radiates into upper half-space over the screen. The RPs (radiation patterns) of the Clavin antenna element with perfectly conducting monopoles in *E*- and *H*-planes were shown to be almost identical [5, 7, 8]. The influence of the electric lengths of impedance monopoles and the distance between the Clavin type elements

© The Author(s), under exclusive license to Springer Nature Switzerland AG 2020 257
M. V. Nesterenko et al., *Combined Vibrator-Slot Structures: Theory and Applications*, Lecture Notes in Electrical Engineering 689,
https://doi.org/10.1007/978-3-030-60177-5_8

was analyzed under conditions that the level of lateral radiation in the E-plane and the difference in the widths of the RP in main planes at the level of -3 dB were taken into account [9, 10]. However, application of Clavin radiator in combined antenna array has not been studied.

Radiating arrays known as Yagi-Uda (wave-channel) antennas have been widely used in practice throughout the last century. The successful using of Yagi-Uda structures has started with invention of linear vibrator arrays [11]. In the classic version, the antenna consists of an active element, passive reflector and several vibrator directors. The number of vibrator directors can be quite large, since each of them guides scattered fields to the next vibrator director, providing the necessary conditions for their excitation. The best array directivity can be achieved by selecting its geometrical parameters: the length of vibrator directors and the distances between them along the longitudinal array axis [12]. The electrodynamic model of the Yagi-Uda antennas with perfectly conducting elements was developed in [13] based on the analytical solution of integral equations for currents on thin vibrators. Later, the authors of this monograph generalized the model for analyzing the characteristics of the Yagi-Uda antennas with impedance vibrators [14].

The structural diversity of the Yagi-Uda antennas is not limited by using only electric elements. The principle of the wave channel antenna was applied to magnetic type elements. For example, an antenna consisting of loop elements, i.e., electrodynamic equivalents of magnetic dipoles, was considered in [15]. This loop antenna consists of the reflector, active element, and vibrator directors coaxially placed in parallel on the common rod so that they are dielectrically isolated from one another. All loop elements are open-ended with a minimal clearance, and their planes are. A more complex design of the loop wave channel was proposed in [16], where the symmetric relative to the longitudinal antenna axis antenna elements are made in the form of grouped pairs of loop radiators. The Yagi-Uda antennas with the vibrator or loop elements are usually used in the meter and decimeter wavelength ranges. First of all, this is related to the physical principles of radiation by the active elements and the power supply implemented by using coaxial cables. It should also be noted that to minimize the design dimensions of the Yagi-Uda antennas in the decimeter wavelength range, microstrip device based on low-profile structures have been proposed [17, 18].

The wave-channel radiating structures can be used in the microwave and extremely high frequency (EHF) bands under two conditions: the power should be supplied to the antenna by the waveguides and the magnetic elements must be applied. These conditions were realized in antenna design developed in [19, 20]: apertures of rectangular waveguides were proposed to use as antenna structure elements [19], and slot in microstrip metal patches were excited by a plane dielectric waveguide [20]. However, this type antennas operating in the centimeter wave band are characterized by insufficient isolation of the cross-polarization components, required for the remote sensing and wave polarimetry. To minimize this drawback, electrically thin radiating elements should be used in these antennas.

The microwave antennas with linear elements with good polarization isolation, can be built based on combined vibrator-slot radiating structures [21]. Electrodynamic characteristics of these radiating structures with waveguide excitation can be easily obtained by a generalized method of induced EMMF [21, 22]. Unfortunately, such approach for this problem solution was never discussed earlier. The article is aimed at developing based on the mathematical model a Yagi-Uda combined vibrator-slot radiating structure (YUCVSRS) operating in the centimeter and millimeter wave ranges. The antenna characteristics will be modelled by generalized method of induced EMMF, since modelling conducted by existing commercial programs are resource-intensive and low-efficient.

In this chapter the possibility of application Clavin type radiators in combined two-frequency antenna arrays with diode switching of vibrator and slot elements is validated. The conventional Clavin elements with passive monopoles operating at the main frequency and two active monopoles operating at the alternative frequency are used as the combined array radiator. An electrodynamically rigorous mathematical model of a combined vibrator-slot structure consisting of a narrow radiating slot cut in a rectangular waveguide end wall and several thin impedance vibrators placed over the infinite screen is also presented. Numerical results concerning internal and external electrodynamic characteristics of the antennas with optimized structural parameters have confirmed the possibility of constructing the YUCVSRSs in the microwave and EHF bands.

8.1 Commutation Modes of Clavin Radiators in Antenna Arrays

Consider the structure of Clavin type vibrator-slot radiator shown in Fig. 8.1. The slot cut in an infinite perfectly conducting screen with center located at the origin of rectangular coordinate system (x, y, z) radiates into half-space with material parameters (ε_1, μ_1) over the screen. Two asymmetric vibrators (monopoles) are placed in

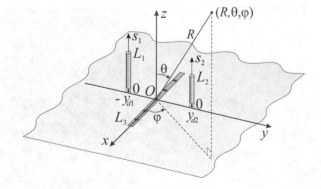

Fig. 8.1 The geometry of the single Clavin vibrator-slot radiator

the plane $\{y, 0, z\}$ at distances y_{d1} and y_{d2} from the slot axis. The vibrator and slot lengths are L_1, L_2, and $2L_3$, the vibrator radii and slot width are r_1, r_2 and d. The structural elements of radiator will be analyzed in approximation thin linear radiators: $\frac{r_{(1,2)}}{L_{(1,2)}} \ll 1$, $\frac{r_{(1,2)}}{\lambda_1} \ll 1$, $\frac{d}{2L_3} \ll 1$, $\frac{d}{\lambda_1} \ll 1$ (λ_1 is the wavelength in the external medium).

Consider the Clavin structure with perfectly conducting vibrators as a radiator of combined antenna array. As in [5, 7, 8], we will assume that following parameters of the radiator are used: vibrator lengths $L_1 = L_2 = L_v$, radii $r_1 = r_2 = r_v$, and displacements $y_{d1} = y_{d2} = y_v$. For the convenience of further analysis, we redefine $2L_3 = 2L_{sl}$. For definiteness, we also assume that the slot cut in the end wall of the rectangular waveguide adjacent to the screen is excited from the lower half-space by the waveguide wave [23]. The waveguide node intended for slot excitation ensures resonant tuning of the combined radiator operating in the two-dimensional flat array shown in Fig. 8.2, where the origin of rectangular coordinate system coincides with the antenna array center. The array consists of several rows of vibrator-slot radiators. The distance between adjacent rows of radiators is d_y, the distance between the phase centers of the radiators located in the row is d_x, the number of rows is N_y, and the number of radiators in a row is N_x. We will introduce the spherical coordinate system which polar axis coincides with the axis $\{0z\}$, and the angle φ is counted out from the axis $\{0x\}$. It is quite clear that the monopoles have electrical contact with the metal screen.

Two group of p-i-n diodes marked as shown in Fig. 8.1 by the red color, are included in the design of the combined array radiator. The diodes of the first group are evenly distributed over the slot aperture so that one diode was obligatory placed in the slot center. If the diodes of this group are in the OFF state, the electromagnetic field radiation passes through the slot and the combined radiator operates in the common mode. If these diodes are in the ON state, the slot aperture is shunted, therefore, it can be considered to be metallized. Synchronization of ON command for the diodes and OFF command controlling the power supply to the waveguide

Fig. 8.2 The antenna array aperture operating at the main frequency

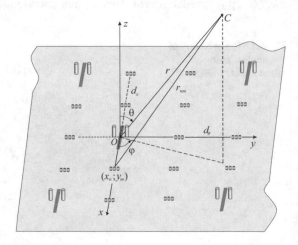

Fig. 8.3 Virtual antenna aperture operating at the alternative frequency

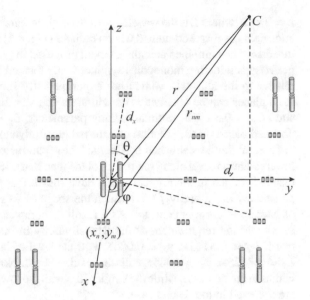

excitation node is required for the correct radiator operation. Thus, if the slot diodes are in the OFF state, the antenna array aperture operates at the main frequency as shown in Fig. 8.2.

The diode switches intended for controlling power applied to the monopoles are also attached to the monopole bases to isolate the excitation nodes and monopoles (Fig. 8.1). If the diodes are in the OFF state, the monopoles operate in the passive mode ensuring the functioning of the combined emitter in the main mode. If the diodes are in the ON state, the monopoles are connected to the excitation nodes allowing operation in the active mode. It is quite clear that the synchronization ON and OFF commands for the vibrator diodes and the monopole excitation generators is also required. In addition, the design of excitation nodes must guarantee resonant monopole tuning taking into account their mutual influence. Thus, if all diode switches in the slot and vibrators are in the ON state, the antenna array aperture shown in Fig. 8.3 operates at the frequency of vibrator excitation. It should be noted that the virtual aperture shown in Fig. 8.3 was built under conditions that the symmetrical vibrators are active, the perfectly conducting screen is absent, and the mirror images of monopoles were not taken into account.

8.1.1 Radiation Fields of the Combined Antenna Array

The radiation fields of the combined antenna array will be analyzed under the assumption that it radiates into free half-space ($\varepsilon_1 = 1$; $\mu_1 = 1$), therefore the wave number

$k = 2\pi/\lambda$, where λ is the wavelength in free space. First, let us characterize the radiation fields of single combined Clavin radiator (Fig. 8.1) based on well-known literature data. The combined antenna element proposed in [7, 8] consists of two identical perfectly conducting monopoles symmetrically located at the distance $y_v = 0.086\lambda$ relative to the slot axis. Clavin has experimentally shown that similar RPs in the main planes can be obtained if the vibrators and slot dimensions are $L_v = 0.375\lambda$ and $2L_{sl} = 0.5\lambda$. The optimal radiator parameters, $L_v = 0.365\lambda$ and $y_v = 0.065\lambda$ obtained based on the solution of external electrodynamics problem in [5] slightly differ from those established by Clavin. These parameters were achieved due to the lower radiation level along the screen plane than that obtained in [7]. The simulation results obtained in the reference [10] have shown that there exists several pairs of parameters L_v/λ and y_v/λ for which the shape of RPs can differ even if the level of lateral radiation is constant. As a result of complex simulation [9], it was also found that the requirement of the RPs similarity in the main planes can be satisfied if $L_v = 0.3125\lambda$, $y_v = 0.086\lambda$ with the level of lateral radiation -20 dB and $L_v = 0.3125\lambda$, $y_v = 0.086\lambda$ with the level -31 dB. Note that the RPs have similar widths in the first case, while the similar RPs with the lowest levels of lateral radiation are observed in the second case.

Analytical formulas applicable for simulation of the vibrator-slot structure (Fig. 8.1) were obtained by the authors based on the generalized method of induced EMMF [9, 10]. Consider the monochromatic excitation of the combined radiator by a plane wave when the fields and currents depend on time t as $e^{i\omega t}$ (ω is the circular frequency). In the spherical coordinate system (R, θ, φ), the electric field of the radiator in the far zone $(R \gg \lambda)$ can be presented as (6.12):

$$\vec{E}(\theta, \varphi) = \frac{ik^2 e^{-ikR}}{\omega R}\left[\vec{\theta}^0 \sin\theta\left(\tilde{E}_1 e^{-iky_v \sin\theta \sin\varphi} + \tilde{E}_2 e^{iky_v \sin\theta \sin\varphi}\right)\right.$$
$$\left. + \left(\vec{\varphi}^0 \cos\theta \cos\varphi + \vec{\theta}^0 \sin\varphi\right)2\tilde{E}_{sl}\right],\tag{8.1}$$

where $\vec{\theta}^0$ and $\vec{\varphi}^0$ are the unit vectors,

$$\tilde{E}_{1(2)} = \frac{2J_{1(2)}}{k(1 - \cos\theta)^2}[\cos(kL_v \cos\theta)\sin(kL_v) - \sin(kL_v \cos\theta)\cos(kL_v)\cos\theta]$$
$$- 2L_v \cos(kL_v)\frac{sin(kL_v \cos\theta)}{kL_v \cos\theta},\tag{8.1a}$$

$$\tilde{E}_{sl} = \frac{2J_{sl}}{k - k(\sin\theta \cos\varphi)^2}[\cos(kL_{sl}\sin\theta\cos\varphi)\sin(kL_{sl})$$
$$- \sin(kL_{sl}\sin\theta\cos\varphi)\cos(kL_{sl})\sin\theta\cos\varphi] - \frac{2\cos(kL_{sl})\sin(kL_{sl}\sin\theta\cos\varphi)}{k\sin\theta\cos\varphi}.\tag{8.1b}$$

The expression (8.1b) was obtained under assumption of symmetrical slot excitation. The complex amplitudes of electric currents $J_{1(2)}$ on monopoles were determined under the condition of predefined amplitude of slot magnetic current [9, 10]

obtain as the rigorous solution of the problem taking into account the mutual influence between the vibrators and slot.

At the main frequency, the H-plane ($\varphi = 0°$) RP of the Clavin radiator according to the formula (8.1) has only the E_φ component which coincides with the single slot RP since $J_1 = -J_2$. The E-plane RP has only the E_φ component, and variation of the monopole currents makes it possible to approximate its shape to the H-plane RP. At the alternative frequency, when the slot is "metallized" and the vibrators are in the active mode, the expression for the electric field of the radiator is simplified to

$$\vec{E}(\theta, \varphi) = \frac{2ik^2 e^{-ikR} \sin \theta}{\omega R} \tilde{E}_{1(2)} \cos(ky_v \sin \theta \cos \varphi)\vec{\theta}^0, \qquad (8.2)$$

where the vibrator currents are assumed to be equal ($J_1 = J_2$).

When RP of the single Clavin radiator defined by the expressions (8.1), (8.2) is known, the total radiation field of the antenna array $\vec{E}(r, \theta, \varphi)$ can be defined as the sum of the radiation fields of each element, taking into account the amplitudes and phases of these fields in the observation point $C(r, \theta, \varphi)$. Assuming that all radiators in the nodes of plane arrays shown in Fig. 8.2 and Fig. 8.3 are similar, we can write

$$\vec{E}(r, \theta, \varphi) = \sum_{n=1}^{N_x} \sum_{m=1}^{N_y} \frac{e^{-ikR_{nm}}}{R_{nm}} \vec{E}_{nm}(\theta, \varphi). \qquad (8.3)$$

It should be recalled that all combined emitters in the arrays are assumed to be resonantly tuned by selecting either their intrinsic parameters or the array parameters, d_x/λ and d_y/λ, which allow us to compensate the mutual influence between the radiators. When the radiators of periodic antenna array are characterized by equal current amplitudes, the double summation in expression (8.3) can be replaced by the normalized array factor [1]

$$\vec{E}(r, \theta, \varphi) = \frac{\vec{E}(\theta, \varphi)}{N_x N_y} \cdot \frac{\sin\left[\frac{N_x}{2}(kd_x \sin \theta \cos \varphi - \psi_x)\right]}{\sin\left[\frac{1}{2}(kd_x \sin \theta \cos \varphi - \psi_x)\right]}$$
$$\cdot \frac{\sin\left[\frac{N_y}{2}(kd_y \sin \theta \sin \varphi - \psi_y)\right]}{\sin\left[\frac{1}{2}(kd_y \sin \theta \sin \varphi - \psi_y)\right]}, \qquad (8.4)$$

where ψ_x and ψ_y are the phase shifts between the currents of neighboring radiators in a row along the $\{0x\}$ axis, and of neighboring radiators in a row along the axis $\{0y\}$.

At the main frequency, the maximum of RP of the single Clavin radiator (Fig. 8.1) defined by the formula (8.1) is directed along the $\{0z\}$ axis. Since monopoles of the Clavin radiator are passive, the amplituds and phases should be related to the magnetic current in the slot. As follows from physical considerations, the normal (transverse) radiation mode of the antenna array consisting of Clavin radiators (Fig. 8.2) can be achieved if $\psi_x = \psi_y = 0$, i.e., with the in-phase operation of the slotted elements.

The form of array factor will be kept invariable in any plane ($\varphi = const$)passing through the $\{0z\}$ axis. Besides, the form of amplitude RPs in the H- and E-planes do not depend upon the number of radiators in a row N_y and the number of rows N_x.

At the alternative frequency, the radiator of the antenna array (Fig. 8.3) can be presented by the two monopoles excited by the current of equal amplitudes. In this case, the phase shifts of the fields radiated by the adjacent radiators in the expression (8.4) should be took into account relative to phase centers monopole pairs. Moreover, from physical considerations it follows that the regime of axial radiation for the array operating at this mode can be realized under the conditions $\psi_x = kd_x$ and $\psi_y = kd_y$.

8.1.2 Numerical Results

The analysis of radiation fields of combined antenna arrays will be carried out based on the simulation results of RP for combined structures presented in Fig. 8.2 and Fig. 8.3 The normalized RPs of single Clavin radiator with parameters $L_v = 0.3125\lambda$, $y_v = 0.086\lambda$ calculated by the formula (8.1) are presented in Fig. 8.4a. As can be seen, the H-plane RPs calculated with parameters $L_v = 0.3125\lambda$, $y_v = 0.086\lambda$ and $L_v = 0.3\lambda$, $y_v = 0.086\lambda$ coincide (curve 1 in Fig. 8.4a), while the E-plane RP calculated with parameters $L_v = 0.3125\lambda$, $y_v = 0.086\lambda$ (curve 2) is closer to H-plane RP as compared with E-plane RP calculated with $L_v = 0.3\lambda$, $y_v = 0.086\lambda$ (curve 3). Therefore, only the parameters $L_v = 0.3125\lambda$, $y_v = 0.086\lambda$ will be used in the further simulations of combine vibrator-slot structure operating at the main frequency.

Simulation of antenna array fields at the main frequency was carried out under conditions that the array periods $d_x = d_y = \lambda/2$, and phases $\psi_x = \psi_y = 0$. This selection minimizes the mutual influence between the antenna array elements [1]. The

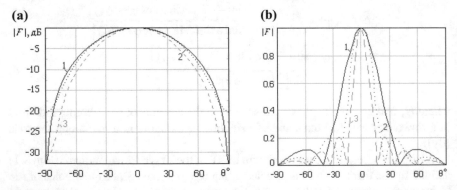

(a)

(b)

Fig. 8.4 The RPs of radiating structures at the main frequency: **a** single Clavin element: 1—in H-plane with parameters $L_v = 0.3125\lambda$, $y_v = 0.086\lambda$; 2—in E-plane with parameters $L_v = 0.3125\lambda$, $y_v = 0.086\lambda$, 3—in E-plane with parameters $L_v = 0.3\lambda$, $y_v = 0.086\lambda$; **b** array of Clavin elements: 1—$N_x = N_y = 3$, 2—$N_x = N_y = 5$, 3—$N_x = N_y = 7$

normalized RPs in the H-plane obtained by using the expression (8.4) are presented in Fig. 8.4b for the following configuration: $N_x = N_y = 3$ (curve 1), $N_x = N_y = 5$ (curve 2), and $N_x = N_y = 7$ (curve 3). Simulation results have also shown that the array RPs in the E-plane and H-plane coincide with good accuracy for arbitrary number of radiators in the array. As expected, we can state based on the well-known electrodynamic principles that the maximum of the RP is oriented normal to the screen plane, the main directional lobes of the RP become narrower, and the number of sides lobes increases if the number of radiators is increased.

At the alternative frequency λ_a, the two identical active monopoles operate as a single radiator (Fig. 8.3), which physical location in the array stays unchanged. Since the electric lengths of the array periods d_x and d_y are changing due to transition to the alternative frequency and differ from $\lambda_a/2$, the resulting phase shifts should be compensated by some or other method. When such compensation is achieved, the directions of the RP maxima for the single radiator and array coincide. The specified compensation can be only achieved by using the phase shifters in the vibrator transmission lines. Since technical possibilities for implementing such phase shifters are quite limited, the main and alternative frequencies should be fairly close to each other. On the other hand, the optimal operating wavelength λ_a of the vibrator radiator is achieved if the monopole length is $L_v = 0.25\,\lambda_a = 0.3125\lambda$. Therefore, the alternative frequency should be selected from the wavelength band $\lambda \leq \lambda_a \leq 1.25\lambda$.

The results of numerical simulation were obtained at the alternative wavelength $\lambda_a = 1.1\lambda$ with the combined radiator parameters, $L_v = 0.284\lambda_a$ and $y_v = 0.078\lambda_a$. The normalized E-plane ($\varphi = \pi/2$) and H-plane ($\varphi = 0$) RPs are presented in Fig. 8.5a by the curves 1 and 2.

As can be seen from Fig. 8.5a, the maximum RP of radiator consisting of two in-phase monopoles is directed in the screen plane ($\theta = \pm 90°$) that it is quite different from that of the Clavin element at the main frequency (Fig. 8.4a). The field amplitude in the plane $\theta = \pm 90°$ as function of the angle φ becomes uneven due to displacements of the monopoles along the axis y at the distance $y_v = 0.078\lambda_a$ (Fig. 8.5b).

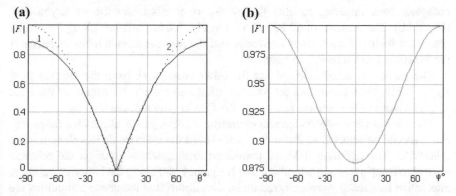

Fig. 8.5 The RPs of two active monopoles with parameters $L_v = 0.284\lambda_a$ and $y_v = 0.078\lambda_a$ at the alternative frequency: **a** 1—in E-plane $\left(\varphi = \pi/2\right)$, 2—in H-plane ($\varphi = 0$); **b** in plane $\theta = \pm 90°$

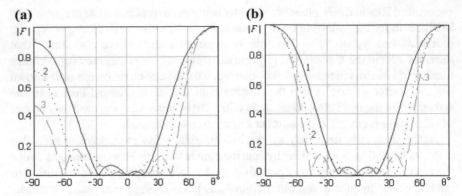

Fig. 8.6 RPs of the antenna arrays with different number of radiators at the alternative frequency: $1—N_x = N_y = 3, 2—N_x = N_y = 5, 3—N_x = N_y = 7$; **a** without phase compensation; **b** with phase compensation

The smallest amplitude is observed in the H-plane ($\varphi = 0°$) and it is lower by 12% as compared with the E-plane amplitude with ($\varphi = 90°$).

If the simulation of antenna array fields based on the formula (8.4) is carried out at the alternative frequency, the array periods are $d_x = d_y = \lambda/2 = \lambda_a/2.2$ and the parameters are $\psi_x = \psi_y = 0.91\pi$. The simulated H-plane RPs of square antenna arrays are shown in Fig. 8.6a for the arrays with $N_x = N_y = 3$ (curve 1), $N_x = N_y = 5$ (curve 2), and $N_x = N_y = 7$ (curve 3). It was also shown that the E-plane RPs are similar to the H-plane RPs represented in Fig. 8.6a for corresponding number of array elements. It should be noted, the E-plane RPs are similar to the H-plane RP shown in Fig. 8.6a for identical number of array radiators. Besides, the field amplitudes decrease in the directions $\theta = \pm 90°$. The RP in the both planes are asymmetric relative to the direction $\theta = 0°$. due to the phase mismatching between the radiators. If the number of radiators is increased, and hence, phase errors in the observation point are accumulated, the skew of the RP increases. The phase compensations, ensuring equality $\psi_x = \psi_y = \pi$ eliminates the RP asymmetry (Fig. 8.6b). The maxima of all RPs shown in Fig. 8.6 are oriented along the screen plane, the width of main lobes decreases and the number of side lobes increases if the number of radiators is increased.

The numerical results carried out for other ratios λ_a/λ from the interval $\lambda \leq \lambda_a \leq 1.25\lambda$ have confirmed the validity of the approach for obtaining the field characteristics of the antenna array with the Clavin type radiators.

Thus, it has been established that operating wavelength of alternative frequency mode should be selected so that it does not exceed the main mode wavelength more than 25%. The radiation fields of combined arrays, whose maxima are oriented normal to its aperture at the fundamental frequency and along the lattice plane at the alternative frequency were analyzed under conditions that the above limitations are taken into account. The described process of the radiation fields formation by the antenna structures allow us to use the combined arrays with diode switching both for horizontal survey of half-space at the alternative frequency and for azimuthal survey at the main frequency. It should be noted that the operating wavelengths of the main and alternative modes of antenna array can coincide.

8.2 Yagi-Uda Combined Radiating Structures

8.2.1 Formulation and Solution of the Diffraction Problem

As the most demanded for practice, we consider the combined radiating vibrator-slot structure of the waveguide type, characterized by the vertical direction of maximum radiation. Let the H_{10}-wave propagates in a semi-infinite rectangular waveguide (marked by index Wg) from the direction $z = -\infty$. A narrow longitudinal slot cut in the waveguide end wall symmetrically with respect to the side walls and oriented along the axis $\{0x\}$ radiates into the half-space above the infinite perfectly conducting plane (marked by index Hs). A system of N thin impedance vibrators each of them is placed at fixed distances z_n from the plane as shown in Fig. 8.7. The cross section of the waveguide is $\{a \times b\}$, the material parameters of the medium filling the waveguide are ε_1, μ_1, the slot length and width are $2L_0$ and d. The length and radius of the vibrator with the number $n \in [1, N]$ are $2L_n$ and r_n. As earlier the monochromatic fields and currents depend on time t as $e^{i\omega t}$.

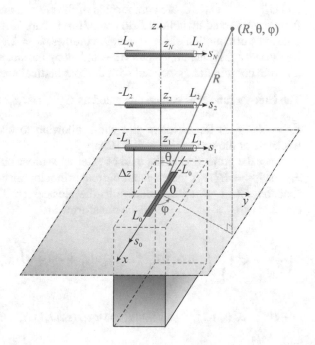

Fig. 8.7 The YUCVSRS geometry and accepted notations

Let us assume that dimensions of the slot and vibrators satisfy the following inequalities $\frac{r_n}{2L_n} \ll 1$, $\frac{d}{2L_0} \ll 1$, $\frac{d}{\lambda_{1(2)}} \ll 1$ (the $\lambda_{1(2)}$ are wavelengths in the media with parameters $\varepsilon_{1(2)}$, $\mu_{1(2)}$). Since the electric currents $J_n(s_n)$ on the vibrators and the equivalent magnetic current in the slot $J_0(s_0)$ should satisfy the boundary conditions $J_n(\pm L_n) = 0$, $J_0(\pm L_0) = 0$, the initial system of integral equations relative to unknown currents [14, 22] can be written as

$$
\left(\frac{d^2}{ds_n^2} + k_2^2 \right) \left\{ \int_{-L_1}^{L_1} J_1(s_1') G_{s_1}^{HsE}(s_n, s_1') ds_1' + \cdots + \int_{-L_N}^{L_N} J_N(s_N') G_{s_N}^{HsE}(s_n, s_N') ds_N' \right\}
$$

$$
- i k \vec{e}_{s_n} \operatorname{rot} \int_{-L_0}^{L_0} J_0(s_0') G_{s_0}^{HsM}(s_n, s_0') ds_0' = i\omega\varepsilon_2 z_{in}(s_n) J_n(s_n), \quad n = 1, 2, \cdots N;
$$

$$
\frac{1}{\mu_1} \left(\frac{d^2}{ds_0^2} + k_1^2 \right) \int_{-L_0}^{L_0} J_0(s_0') G_{s_0}^{WgM}(s_0, s_0') ds_0' + \frac{1}{\mu_2} \left(\frac{d^2}{ds_0^2} + k_2^2 \right) \int_{-L_0}^{L_0} J_0(s_0') G_{s_0}^{HsM}(s_0, s_0') ds_0'
$$

$$
+ i k \vec{e}_{s_0} \operatorname{rot} \sum_{n=1}^{N} \int_{-L_n}^{L_n} J_n(s_n') G_{s_n}^{HsE}(s_0, s_n') ds_n' = -i\omega H_{0s_0}(s_0), \tag{8.5}
$$

where $z_{in}(s_n)$ are the vibrator internal linear impedances, measured in [Ohm/m], $H_{0s_0}(s_0)$ is the projection of the external source field on the slot axis, $G_{s_m}^{HsE}(s_n, s_m')$ and $G_{s_m}^{HsM, WgM}(s_n, s_m')$ are components of the electric and magnetic tensor Green's functions (marked by indices $E(e)$ and $M(m)$ [Chap. 1, Appendix A]) for the vector potentials of coupling electrodynamic volumes, $k = 2\pi/\lambda$, λ is the wavelength in free space, $\vec{e}_{s_0}, \vec{e}_{s_n}$ are unit vectors directed along the slot and vibrator axes, and s_0, s_n are local coordinates associated with the longitudinal axes of the slot and vibrators.

The Green's functions can be presented as $G_{s_0}^{HsM}(s_0, s_0') = \dfrac{2e^{-ik\sqrt{(s_0 - s_0')^2 + (d_e/4)^2}}}{\sqrt{(s_0 - s_0')^2 + (d_e/4)^2}}$ where

$d_e = de^{-\frac{\pi h}{2d}}$ is the effective slot width allowing to take into account the actual thickness of the waveguide wall h [Chap. 3].

Since the projection of the field of external sources on the slot axis is $H_{0s_0}(s_0) = H_0 \cos \frac{\pi s_0}{a} = H_0^s(s_0)$, the magnetic current in the slot can be described by a symmetric function $J_0(s_0) = J_0^s(s_0)$ relative to the slot center. Then, the Eqs. (8.5) under condition $\varepsilon_{1(2)} = \mu_{1(2)} = 1$ can be transformed to

$$
\left(\frac{d^2}{ds_n^2} + k^2 \right) \left\{ \int_{-L_1}^{L_1} J_1(s_1') G_{s_1}^{HsE}(s_n, s_1') ds_1' + \cdots + \int_{-L_N}^{L_N} J_N(s_N') G_{s_N}^{HsE}(s_n, s_N') ds_N' \right\}
$$

$$
+ i k \int_{-L_0}^{L_0} J_0^s(s_0') \tilde{G}_{s_0}^{HsM}(s_n, s_0') ds_0' = i\omega z_{in}(s_n) J_n(s_n), \quad n = 1, 2, \cdots N;
$$

$$\left(\frac{d^2}{ds_0^2} + k_1^2\right) \int_{-L_0}^{L_0} J_0^s(s_0') [G_{s_0}^{WgM}(s_0, s_0') + G_{s_0}^{HsM}(s_0, s_0')] ds_0'$$

$$- ik \sum_{n=1}^{N} \int_{-L_n}^{L_n} J_n(s_n') \tilde{G}_{s_n}^{HsE}(s_0, s_n') ds_n' = -i\omega H_0^s(s_0). \tag{8.6}$$

where $\tilde{G}_{s_n}^{HsE}(s_0, s_0') = -\frac{\partial}{\partial z} G_{s_n}^{HsE}(s_0, 0, z; 0, s_n', z_n)|_{z=0}$, $\tilde{G}_{s_0}^{HsM}(s_n, s_0') = \frac{\partial}{\partial z} G_{s_0}^{HsM}(0, s_n, z; s_0', 0, 0)|_{z=z_n}$.

In the above expressions, the fixed values of the variable z are substituted after differentiation.

The equations system (8.6) can be solved by the generalized method of induced EMMF, using the functions $J_n(s_n) = J_{0n} f_n(s_n)$ and $J_0^s(s_0) = J_{00}^s f_0^s(s_0)$ as approximating expressions for the currents. In this expression, J_{0n} and J_{00}^s are unknown complex current amplitudes, $f_n(s_n)$ and $f_0^s(s_0)$ are predefined basic functions of current distributions, which can be obtained as solution of the equation defining the currents on the standalone vibrator and slot by the averaging method [Chap. 3]. The expressions for the YUCVSRS can be written as

$$f_n(s_n) = \cos \tilde{k}_n s_n - \cos \tilde{k}_n L_n, \quad f_0^s(s_0) = \cos k s_0 - \cos k L_0, \tag{8.7}$$

where $\tilde{k}_n = k + \frac{i\alpha_n 2z_{in}^{av}}{60\,\text{Ohm}}$, $\alpha_n = \frac{1}{2\ln[r_n/(2L_n)]}$ are natural small parameters, and $z_{in}^{av} = \frac{1}{2L_n}\int_{-L_n}^{L_n} z_{in}(s_n) ds_n$ are internal impedances averaged over the vibrators.

Let us apply the generalized method of the induced EMMF to the system of Eqs. (8.6). If the waveguide is excited by the H_{10}-wave with amplitude H_0, the system of linear algebraic equations (SLAE) relative to unknown current amplitudes J_{0n} and J_{00}^s can be presented as

$$\begin{cases} J_{00}^s(Z_{00}^{sWg} + Z_{00}^{sHs}) + J_{01}Z_{01} + J_{02}Z_{02} + \cdots + J_{0N}Z_{0N} = -\frac{i\omega}{2k}H_0^s, \\ J_{00}^s Z_{10} + J_{01}(Z_{11} + F_1^{\overline{Z}}) + J_{02}Z_{12} + \cdots + J_{0N}Z_{1N} = 0, \\ J_{00}^s Z_{20} + J_{01}Z_{21} + J_{02}(Z_{22} + F_2^{\overline{Z}}) + \cdots + J_{0N}Z_{2N} = 0, \\ \cdots \\ J_{00}^s Z_{N0} + J_{01}Z_{N1} + \cdots + J_{0(N-1)}Z_{N(N-1)} + J_{0N}(Z_{NN} + F_N^{\overline{Z}}) = 0, \end{cases} \tag{8.8}$$

where $H_0^s = H_0 \int_{-L_0}^{L_0} \cos \frac{\pi s_0}{a} f_0^s(s_0) ds_0$, Z_{mn} and $F_n^{\overline{Z}}$ are dimensionless coefficients. For example, the expression for the Z_{mn} coefficients can be written as

$$Z_{mn} = \int_{-L_m}^{L_m} f_m(s_m) \left(\frac{d^2}{ds_m^2} + k^2\right) \int_{-L_n}^{L_n} f_n(s_n') G_{s_n}^{HsE}(s_m, s_n') ds_n' ds_m,$$

$$n = 1, 2, \cdots N, \quad m = 1, 2, \cdots N, \tag{8.9}$$

which can be interpreted as the vibrator own or mutual effective resistances if $n = m$ or $n \neq m$.

The electrodynamic characteristics of the vibrator-slot antenna can be found by solving SLAE (8.8) relative to the currents amplitudes J_{0n} and J_{00}^s. For example, the field reflection coefficient S_{11} defined by the slot inhomogeneity in the semi-infinite rectangular waveguide operating in the single-mode regime can be presented as

$$S_{11} = \left\{ 1 - \frac{16\pi\gamma J_{00}^s}{abk\omega H_0} \cdot \frac{\sin kL_0 \cos \frac{\pi L_0}{a} - \frac{ka}{\pi} \cos kL_0 \sin \frac{\pi L_0}{a}}{1 - [\pi/(ka)]^2} \right\} e^{2i\gamma z}, \qquad (8.10)$$

where $\gamma = \sqrt{k^2 - (\pi/a)^2}$ is the propagation constant of H_{10}-wave. The power reflection coefficient $|S_\Sigma|^2$ of the slot into the half-space can be obtained by using the energy balance equation

$$|S_{11}|^2 + |S_\Sigma|^2 = 1. \qquad (8.11)$$

Let us introduce a system of spherical coordinates shown in Fig. 8.7. The total wave zone field ($R \gg \lambda$) reradiated by the combined vibrator director array equal to the sum of the secondary radiation fields of each vibrator taking into account the phases of fields arriving to the observation point $C(R, \theta, \varphi)$ can be written as

$$\vec{E}(R, \theta, \varphi) = \frac{2k^2}{\omega} \frac{e^{-ikR}}{R} \left[i \left(\vec{\theta}^0 \sin \varphi + \vec{\varphi}^0 \cos \theta \cos \varphi \right) J_{00}^s F_{C0} \right.$$

$$\left. + \left(\vec{\theta}^0 \cos \theta \sin \varphi + \vec{\varphi}^0 \cos \varphi \right) \sum_{n=1}^{N} J_{0n} F_{Cn} \sin(kz_n \cos \theta) \right], \qquad (8.12)$$

where and are the unit vectors, $F_{C0} = \int_{-L_0}^{L_0} f_0^s(x) e^{iks_0 \sin \theta \cos \varphi} dS_0$, and $F_{Cn} = \int_{-L_n}^{L_n} f_n(s_n) e^{ils_n \sin \theta \sin \varphi} ds_n$. If the relation (8.7) is taken into account, the function F_{C0} can be written as

$$F_{C0} = \frac{2}{k^2 - (k \sin \theta \cos \varphi)^2} [k \cos(kL_0 \sin \theta \cos \varphi) \sin(kL_0) -$$

$$k \sin(kL_0 \sin \theta \cos \varphi) \cos(kL_0) \sin \theta \cos \varphi]$$

$$- 2L_0 \cos(kL_0) \frac{\sin(kL_0 \sin \theta \cos \varphi)}{kL_0 \sin \theta \cos \varphi}, \qquad (8.13a)$$

$$F_{Cn} = \frac{2}{\tilde{k}_n^2 - (k \sin \theta \sin \varphi)^2} \left[\tilde{k}_n \cos(kL_n \sin \theta \sin \varphi) \sin\left(\tilde{k}_n L_n \right) \right.$$

$$\left. - k \sin(kL_n \sin \theta \sin \varphi) \cos\left(\tilde{k}_n L_n \right) \sin \theta \sin \varphi \right]$$

$$- 2L_n \cos\left(\tilde{k}_n L_n \right) \frac{\sin(kL_n \sin \theta \sin \varphi)}{kL_n \sin \theta \sin \varphi}. \qquad (8.13b)$$

As can be seen from the expressions (8.12) and (8.13a, b), the $E-$ plane ($\varphi = 90°$) and H—plane ($\varphi = 0°$) components of the electric field are $E_\varphi = 0$ and $E_\theta = 0$. The longitudinal component of the field in the far zone is $E_r = 0$.

8.2.2 Numerical Results

The electrodynamic characteristics were simulated for the YUCVSRS consisting of the rectangular waveguide and N vibrator directors. The waveguide was excited by the H_{10}—wave with amplitude $H_0 = 1$ at the frequency. $f = 9.2$ GHz ($\lambda = 32.6$ mm). The waveguide cross-section, wall thickness, slot width, and the radii of the perfectly conducting vibrators ($z_{in}^{av} = 0$) where $\{a \times b\} = 23.0 \times 10.0$ mm^2, $h = 1.0$ mm, $d = 1.5$ mm, and $r_n = 0.4$ mm. The remaining parameters of the YUCVSRS: the slot length, vibrator length, and distances between structural elements were selected by the numerical optimization to ensure the maximum directivity D in the direction of $\{0z\}$ axis under condition of satisfactory waveguide matching (the voltage standing wave ratio VSWR < 1.1). The directivity D and VSWR were calculated by the following relationships:

$$D = \frac{4\pi}{\int_0^{2\pi} \int_0^{\pi/2} [F(\theta, \varphi)]^2 \sin\theta d\theta d\varphi}, \tag{8.14}$$

$$\text{VSWR} = \frac{1 + |S_{11}|}{1 - |S_{11}|}, \tag{8.15}$$

where $F(\theta, \varphi) = \dfrac{|\vec{E}(R,\theta,\varphi)|}{|\vec{E}(R,\theta_m,\varphi_m)|}$ is the radiation pattern (RP) normalized at the electric field modulus in the radiation maximum direction.

The YUCVSRS optimization was carried out under condition that all its vibrator directors are positions at equal distances $\Delta z = z_1 = z_{n+1} - z_n$ from each other. The optimal parameters of the YUCVSRS: $2L_0$, $2L_n$ and Δz for different number of vibrator directors are shown in Table 8.1.

The calculated RP in the main polarization planes for the optimized YUCVSRS, the angular plots $|F| = |E_\theta(\theta, \varphi = 0°)|$ and $|F| = |E_\varphi(\theta, \varphi = 90°)|$, are presented

Table 8.1 Parameters of the YUCVSRS

N	D	$2L_0$	$2L_n$	Δz
1	6.89	0.521λ	0.398λ	0.269λ
2	11.90	0.536λ	0.405λ	0.294λ
3	16.13	0.54λ	0.406λ	0.321λ
4	19.73	0.528λ	0.402λ	0.349λ
5	22.97	0.518λ	0.399λ	0.367λ

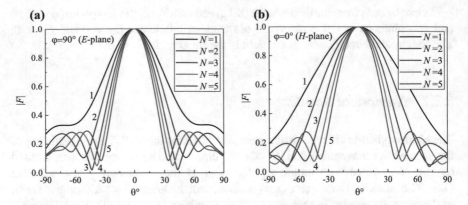

Fig. 8.8 The RPs of the YUCVSRS with various vibrator numbers

in Fig. 8.8. All plots are normalized to the maximum of the electric field modulus for a particular antenna structure. As can be seen from Fig. 8.8 the main lobe of the RP becomes narrower if the number of vibrator elements is increased. At the same time, the widths of the RP main lobes (at the level $|F| = 0.5$) in the orthogonal polarization planes tend to approach to one other. As was shown in [24], the two-element vibrator-slot system with the vibrator director, whose length was $2L_1 = 0.407\lambda$, and distance between the resonant slot and the vibrators was $\Delta z = 0.105\lambda$ is characterized by almost identical RPs in $E-$ and $H-$ planes. This effect was experimentally proved in [24]. The simulation results obtained for the two-element YUCVSRS have confirmed the similarity of the RP in the two polarization planes. Thus, this prove out the mathematical model validity and its physical adequacy.

Properties of the side radiation for the optimized YUCVSRS can be analyzed by comparing the RPs for $N = 2$, $N = 3$, $N = 4$, and $N = 5$ presented in Fig. 8.9. As can be see, if the number of radiating elements in the YUCVSRS is increased, the number of the RP side lobes also increases in accordance with electrodynamics principles. It was also established that the power levels for the side lobes do not exceed 10% of that for the main lobe.

The internal and external energy characteristics of the YUCVSRS with the parameters listed in Table 8.1 are presented in Fig. 8.10. The frequency plots of the coefficients $|S_{11}|$, $|S_\Sigma|^2$, and VSWR defined by the formulas (8.10), (8.11), and (8.15) for different N are shown in Fig. 8.10a. The frequency plots of the directivity $D[\text{dB}]$ (8.14) and the antenna gain G defined by the formula

$$G = D \cdot |S_\Sigma|^2 \tag{8.16}$$

are presented in Fig. 8.10b. The plots shown in Fig. 8.10 were calculated in the frequency range $f \in [7.5; 10.5]$ GHz belonging to the single-mode waveguide regime. As can be seen, the curves $|S_{11}|$, VSWR have maxima, and the curves $|S_\Sigma|^2$, D and G have minima in the vicinity of the frequency $f = 9.2$ GHz. It can be

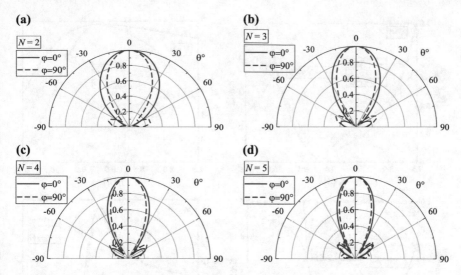

Fig. 8.9 The RPs of the YUCVSRS with optimized parameters

also stated that if the number of vibrators is increased, the operating frequency band becomes narrower.

Thus, the numerical results have confirmed the possibility of reducing the main lobe width and increasing their directivity and gain by using the system of several passive vibrator directors arranged in the Yagi-Uda configuration. It is shown that the YUCVSRS with optimized parameters can provide the maximum directivity and good matching of the waveguide transmission line with VSWR < 1.1.

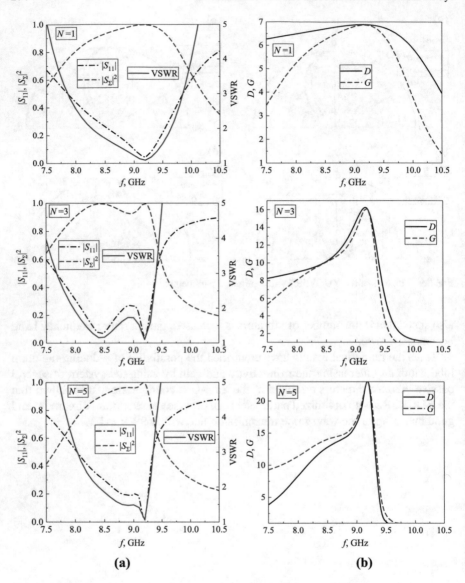

Fig. 8.10 The frequency plots of the energy characteristics for the YUCVSRS with various N

References

1. Amitay N, Galindo V, Wu CP (1972) Theory and analysis of phased array antennas. Wiley, New York
2. Ponomarev LI, Stepanenko VI (2009) Scanning multi-frequency combined antenna arrays. Radiotekhnika, Moscow ((in Russian))
3. Galejs J (1960) Multidiode switch. IRE Trans Antennas Propag 8:566–569
4. Mortenson KE (1964) Microwave semiconductor control devices. Microwave J. 7:49–57

5. Papierz AB, Sanzgiri SM, Laxpati SR (1977) Analysis of antenna structure with equal E- and H-plane patterns. Proc IEE 124(1):25–30
6. Elliott RS (1980) On the mutual admittance between Clavin elements. IEEE Trans Antennas Propag. 28:864–870
7. Clavin A, Huebner DA, Kilburg FJ (1974) An improved element for use in array antennas. IEEE Trans Antennas Propag 22:521–526
8. Clavin A (1975) A multimode antenna having equal E- and H-planes. IEEE Trans Antennas Propag 23:735–737
9. Berdnik SL, Katrich VA, Nesterenko MV, Penkin YM (2016) Excitation of electromagnetic waves by a longitudinal slot in a broad wall of a rectangular waveguide in the presence of passive impedance vibrators outside the waveguide. Radiophys Radioastronomy 21(3):198–215 (in Russian)
10. Berdnik SL, Katrich VA, Nesterenko MV, Penkin YM, Pshenichnaya SV (2015) Clavin element with impedance monopoles. In: Proceedings of the XX-th international seminar/workshop on direct and inverse problems of electromagnetic and acoustic wave theory. Lviv, Ukraine, pp 61–65
11. Yagi H, Uda S (1926) Projector of the sharpest beam of electric waves. Proc Imperial Acad Jpn 2:49–52
12. Balanis CA (2005) Antenna theory: analysis and design. Wiley, New York
13. King RWP, Mack RB, Sandler SS (1968) Arrays of cylindrical dipoles. Cambridge University Press, New York
14. Nesterenko MV, Katrich VA, Penkin YuM, Dakhov VM, Berdnik SL (2011) Thin impedance vibrators. Theory and applications. Springer Science+Business Media, New York
15. Campbell RW (1970) Endfire antenna array having loop directors. U.S. Patent 3,491,361, issued January 20
16. Petlya II, Panchenko VA (2001) Loop antenna. RF Patent 2174272, issued September 27
17. Liu H, Gao S (2013) Small director array for low-profile smart antennas achieving higher gain. IEEE Trans Antennas Propag 61:162–168
18. Wang Z, Liu XL, Yin Y-Z, JH, Wang JH, Li Z (2014) A novel design of folded dipole for broadband printed Yagi-Uda antenna. Progress Electromagnet Res C 46:23–30
19. Altshuler EE (1962) A monopole array driven from a rectangular waveguide. IRE Trans Antennas Propag 10:558–560
20. Zhang Z, Cao X-Y, Gao J, Li S-J, Liu X (2015) Compact microstrip magnetic Yagi antenna and array with vertical polarization based on substrate integrated waveguide. Progress Electromagnet Res C 59:135–141
21. Berdnik SL, Katrich VA, Nesterenko MV, Penkin YM, Penkin DY (2015) Radiation and scat tering of electromagnetic waves by a multi-element vibrator-slot structure in a rectangular waveguide. IEEE Trans Antennas Propag AP-63(9):4256–4259
22. Penkin DYu, Berdnik SL, Katrich VA, Nesterenko MV, Kijko VI (2013) Electromagnetic fields excitation by a multielement vibrator-slot structures in coupled electrodynamics volumes. Prog Electromagn Res B 49:235–252
23. Nesterenko MV, Katrich VA, Penkin YuM, Berdnik SL (2008) Analytical and hybrid methods in theory of slot-hole coupling of electrodynamic volumes. Springer Science + Business Media, New York
24. Lee Y-H, Hong D-H, Ra J-W (1983) Waveguide slot antenna with a coupled dipole above the slot. Electron Lett 19(8):280–282

Chapter 9
Ultrawideband Vibrator-Slot Structures

Transient signals continue to attract special interests of scientists and engineers because of its wide spreading in nature and modern devices. The popularity of powerful pulse-width modulated rectifiers and converters connected to solar cells and power grid causes necessity to investigate its influence on other radio electronic devices [1]. Transient interferences can produce errors in digital lines and violate the work of radio-frequency identification devices (RFID) [2]. One of the most powerful disturbances for the modern electronic devices including onboard ones is a lightning [3]. Quick switching on of large currents can seriously influence into shielded cables of airplanes and destroy onboard devices [4] as well as cause a damage to shielded cables buried in ground [5–7]. The simplest model of a transient radiator is a dipole that has possibility to approximate complicated current distributions of different sources of field [8]. The model of pulse Hertzian dipole generated static, induction and radiation parts is convenient to describe electrostatic discharge stress in manufacturing [9]. The transient radiation of a single electrical Hertzian dipole was considered in [10] in details.

Classical vibrators and slots are narrowband radiators due to resonant processes that caused by its excitation. While there are the wideband versions of the vibrator structures [11] as well as slots [12]. The interaction between them was used to improve the total radiation efficiency of the combined structure in a wide frequency range. The investigation of the mutual influence requires a knowledge about near field of the radiators for transient excitation. So, we begin to study the radiation process for arbitrary time dependence of exciting current of elementary radiator from investigation of electrical Hertzian dipole [13]. All expressions and figures of the chapter are presented in practical system of units SI (see Appendix E).

© The Author(s), under exclusive license to Springer Nature Switzerland AG 2020   277
M. V. Nesterenko et al., *Combined Vibrator-Slot Structures: Theory
and Applications*, Lecture Notes in Electrical Engineering 689,
https://doi.org/10.1007/978-3-030-60177-5_9

9.1 Transient Near Field of Herzian Dipole

Classical expressions for the dipole field are well-known and presented in a number of books [14]. Also, the classical expressions derived in time domain have a drawback. According to the expressions the wave propagated to infinity exists everywhere, in all space including vicinity of a current source that doesn't correspond to physics of process of the radiation described initially by Hertz [15]. This process consists in the formation of the electromagnetic wave in the vicinity of radiator by transformation of the energy of fast-decreasing components of the wave (quasistatic and inductive) [12]. The energy analysis of classical expressions which has been carried out in [16], allowed to allocate causal surfaces where the total stream of energy through the surfaces equals to zero, and it's the borders of volume which emits as much electromagnetic energy as receive back from dipolar and quasistatic components from external infinite space. However, it is impossible analytically from classical expressions to see that electromagnetic wave has its origin. The more exact expressions for near transient fields of the Hertzian dipole can be obtained taking into account the sizes of dipole. As result, we consider the segment of the thin wire radiator.

9.1.1 Statement and Initial Expressions of the Problem

The source of electromagnetic field is a conductor with a cross-section S and radius r_d that has current $I(t) = \frac{\partial Q(t)}{\partial t}$ with an arbitrary dependence of the time flowing in the direction of axis OZ. Let us consider the field radiated by a small segment of the conductor l, $l \ll c\tau$ and where τ is the minimum time for which the current $I(t)$ is changing noticeably, c is the speed of light. The current distribution along the length l is uniform. This condition will always be fulfilled if the length l is selected infinitely small.

Calculation of the field will be carried out using the vector potential method, using the known solution

$$\vec{A}(\vec{r}, t) = \frac{\mu_0}{4\pi} \int\limits_V \frac{\vec{j}\left(\vec{r}', t - R/c\right)}{R} dV'. \tag{9.1}$$

We use the expressions for the fields by the vector potentials

$$\begin{cases} \vec{H}(\vec{r}, t) = \frac{1}{\mu_o} \mathrm{rot} \vec{A}(\vec{r}, t), \\ \vec{E}(\vec{r}, t) = -\mathrm{grad}\varphi - \frac{\partial \vec{A}(\vec{r}, t)}{\partial t}. \end{cases} \tag{9.2}$$

We have receded from idea of the infinitesimal small conductor and we consider not only length of the conductor, but also its thickness as finite. Thus, the received expressions which describe the fields radiated from the thin conductor of small length with transient current excitation. Let's consider potential at the long distances from the source system, $r'/r \ll 1$, where $\vec{r}$ is the coordinate of the point of observation, $\vec{r}'$ is the coordinate of the point source. Considering that $\vec{R} = \vec{r} - \vec{r}'$, $R = \sqrt{r^2 - 2\vec{r}\vec{r}' + r'^2}$, and using well-known expression for Taylor series, it is possible to expand the integrand in (9.1) in point r accounting the first three components in the equality. It is easy to show that [17].

$$r\left(\underbrace{1 - \left(2\frac{(\vec{r},\vec{r}')}{r^2} - \frac{r'^2}{r^2}\right)}_{x}\right)^{1/2} \approx r\left(1 - \left(\frac{2\frac{(\vec{r},\vec{r}')}{r^2} - \frac{r'^2}{r^2}}{2}\right) - \left(\frac{\left(2\frac{(\vec{r},\vec{r}')}{r^2} - \frac{r'^2}{r^2}\right)^2}{8}\right)\right)$$

$$\approx r - \frac{(\vec{r},\vec{r}')}{r} + \frac{r'^2}{2r} - \frac{(\vec{r},\vec{r}')^2}{2r^3} + \frac{(\vec{r},\vec{r}')r'^2}{2r^3} - \frac{r'^4}{8r^3}.$$
(9.3)

Using properties of Taylor series we represent the integrand as $F(x_0 + x) \approx F(x_0) + \frac{x}{1!}F'(x_0) + \frac{x^2}{2!}F''(x_0)$ for small x, where $F(x_0 + x) = \frac{f(x_0+x)}{x_0+x}$. The integrand acquires the form

$$\frac{f(x_0 + x)}{x_0 + x} \approx \frac{f(x_0)}{x_0} + \frac{x}{1!}\frac{d}{dx_0}\left(\frac{f(x_0)}{x_0}\right) + \frac{x^2}{2!}\frac{d^2}{dx_0^2}\left(\frac{f(x_0)}{x_0}\right),$$
(9.4)

where

$$x = -\left(\frac{(\vec{r},\vec{r}')}{r} - \frac{r'^2}{2r} + \frac{(\vec{r},\vec{r}')^2}{2r^3} - \frac{(\vec{r},\vec{r}')r'^2}{2r^3} + \frac{r'^4}{8r^3}\right), x_0 = r.$$

In classical expression [12] for the field radiated by the Hertzian dipole the first component in this expression was used. It is easy to show that the components which contain scalar product $(\vec{r},\vec{r}')$ after integration on volume give a zero contribution to the expression. Therefore, we will consider the subsequent components with identical order of smallness in relation to the sizes of the dipole. Expression for the vector potential has the form [17]

$$\vec{A} = \frac{\mu_0}{4\pi}(\vec{m}_1 + \vec{m}_2),$$
(9.5)

where

$$\vec{m}_1 = \int\limits_V \frac{\vec{j}}{r} dV',$$

$$\vec{m}_2 = \int\limits_V \left(-\frac{r'^2}{2r^2 c}\dot{\vec{j}} + \frac{3\left(\vec{r},\vec{r'}\right)^2}{2r^4 c}\dot{\vec{j}} - \frac{r'^2}{2r^3}\vec{j} + \frac{3\left(\vec{r},\vec{r'}\right)^2}{2r^5}\vec{j} + \frac{\left(\vec{r},\vec{r'}\right)^2}{2c^2 r^3}\ddot{\vec{j}} \right) dV'$$

So,

$$\vec{A} = \frac{\mu_0}{4\pi} \int\limits_V \frac{dV'}{r} \left(\vec{j} - n_1 \left(\frac{1}{rc}\dot{\vec{j}} + \frac{\vec{j}}{r^2} \right) + (u_1 + u_2 + u_3) \left(\frac{1}{c^2 r}\ddot{\vec{j}} + \frac{2}{cr^2}\dot{\vec{j}} + 2\frac{\vec{j}}{r^3} \right) \right),$$

(9.6)

$$n_1 = -\left(\frac{\left(\vec{r},\vec{r'}\right)}{r} - \frac{r'^2}{2r} + \frac{\left(\vec{r},\vec{r'}\right)^2}{2r^3} - \frac{\left(\vec{r},\vec{r'}\right)r'^2}{2r^3} + \frac{r'^4}{8r^3} \right);$$

$$u_1 = \frac{\left(\vec{r},\vec{r'}\right)^2}{2r^2} + \frac{r'^4}{8r^2} + \frac{\left(\vec{r},\vec{r'}\right)^4}{8r^6} + \frac{\left(\vec{r},\vec{r'}\right)^2 r'^4}{8r^6} + \frac{r'^8}{128r^6} - \frac{\left(\vec{r},\vec{r'}\right)r'^2}{2r^2} + \frac{\left(\vec{r},\vec{r'}\right)^3}{2r^4};$$

$$u_2 = -\frac{\left(\vec{r},\vec{r'}\right)^2 r'^2}{2r^4} + \frac{\left(\vec{r},\vec{r'}\right)r'^4}{8r^4} - \frac{\left(\vec{r},\vec{r'}\right)^2 r'^2}{4r^4} + \frac{\left(\vec{r},\vec{r'}\right)r'^4}{4r^4};$$

$$u_3 = -\frac{r'^6}{16r^4} - \frac{\left(\vec{r},\vec{r'}\right)^3 r'^2}{4r^6} + \frac{\left(\vec{r},\vec{r'}\right)^2 r'^4}{16r^6} - \frac{\left(\vec{r},\vec{r'}\right)r'^6}{16r^6}.$$

$$\vec{A} = \frac{\mu_0}{4\pi} \int\limits_V \left(\frac{\vec{j}}{r} - \frac{r'^2}{2r^2 c}\dot{\vec{j}} + \frac{3\left(\vec{r},\vec{r'}\right)^2}{2r^4 c}\dot{\vec{j}} - \frac{r'^2}{2r^3}\vec{j} + \frac{3\left(\vec{r},\vec{r'}\right)^2}{2r^5}\vec{j} + \frac{\left(\vec{r},\vec{r'}\right)^2}{2c^2 r^3}\ddot{\vec{j}} \right) dV'.$$

(9.7)

In classical expression [10, 12] for the field radiated by the Hertzian dipole the first component in the expression (9.7) was considered. It is easy to show that the components that contain scalar product $\left(\vec{r},\vec{r'}\right)$ after integration on the volume surrounding the source of current, give a zero contribution to the expression. Therefore, we will consider the subsequent components with identical order of smallness in relation to the sizes of the dipole. Expression for vector potential becomes

$$\vec{A}(\vec{r},t) = \frac{\vec{z}_0 \mu_0 l}{4\pi} \left(a_0 \frac{I}{r} + a_1 \frac{\dot{I}}{4r^2 c} + a_2 \frac{\ddot{I}}{8rc^2} \right),$$

(9.8)

where $a_0 = 1 - \frac{l^2}{24r^2} + \frac{l^2}{8r^2}(\cos^2\theta) - \frac{r_d^2}{4r^2} + \frac{3r_d^2}{8r^2}(\sin^2\theta)$, θ is the polar angle,

$$a_1 = -\frac{l^2}{6} + \frac{l^2}{2}(\cos^2\theta) - r_d^2 + \frac{3r_d^2}{2}(\sin^2\theta), \ a_2 = \frac{l^2}{3}(\cos^2\theta) + r_d^2(\sin^2\theta).$$

It is easy to see that the received expression in case of the approaching of length of the dipole and its thickness to zero is transformed into classical form [10, 12]. It is obvious that influence of new components affects at the small distances from the radiator. It follows from the multipliers inversely proportional to the square of the distance in a_0. Only for reasons of real geometry of the statement of the problem it is possible to allocate components containing length of a dipole, as more important in comparison with components containing its radius. It is worth to notice that the distance between the origin and the point of observation along the longitudinal axis is more sensitive to the length of the radiator. The distance in the perpendicular plane passing through the center of the radiator is more sensitive to its radius.

9.1.2 Components of Radiated Field

The use of well-known formulas for vector operations in the spherical coordinate system for (9.8) leads to the following expression for the fields radiated from electrical Hertzian dipole:

$$H_\varphi(r,\theta,t) = \frac{l\sin\theta}{4\pi}\left(h_1\frac{\dot{Q}(t-r/c)}{r^2} + h_2\frac{\ddot{Q}}{cr} + h_3\frac{\ddot{Q}}{r^2c^2} + h_4\frac{\dddot{Q}}{rc^3}\right),$$

$$E_r = \frac{l\cos\theta}{2\pi\varepsilon_0}\left(e_{r0}\frac{Q(t-r/c)}{r^3} + e_{r1}\frac{\dot{Q}}{cr^2} + e_{r2}\frac{\ddot{Q}}{c^2r^3} + e_{r3}\frac{\ddot{Q}}{r^2c^3}\right),$$

$$E_\theta = \frac{l\sin\theta}{4\pi\varepsilon_0}\left(e_{\theta0}\frac{Q}{r^3} + e_{\theta1}\frac{\dot{Q}}{cr^2} + e_{\theta2}\frac{\ddot{Q}}{rc^2} + e_{\theta3}\frac{\ddot{Q}}{r^2c^3} + e_{\theta4}\frac{\dddot{Q}}{rc^4}\right), \quad (9.9)$$

where

$$h_1 = 1 - \frac{3l^2}{8r^2} + \frac{3r_d^2}{4r^2}(\frac{3}{2}\sin^2\theta - 1) + \left(\frac{5l^2}{8r^2} - \frac{3r_d^2}{4r^2}\right)\cos^2\theta,$$

$$h_2 = 1 - \frac{l^2}{8r^2} + \frac{3r_d^2}{4r^2}(3\sin^2\theta - 1) + \left(\frac{5l^2}{8r^2} - \frac{3r_d^2}{4r^2}\right)\cos^2\theta,$$

$$h_3 = -\frac{l^2}{24} + \left(\frac{l^2}{4} - \frac{r_d^2}{4}\right)\cos^2\theta + \frac{r_d^2}{2}(\sin^2\theta - \frac{1}{2}),$$

$$h_4 = \frac{l^2}{24}\cos^2\theta + \frac{r_d^2}{8}\sin^2\theta,$$

$$e_{r0} = 1 - \frac{3r_d^2}{2r^2} + \frac{l^2}{4r^2}(3\cos^2\theta - 1) + \left(\frac{15r_d^2}{4r^2} - \frac{l^2}{2r^2}\right)\sin^2\theta,$$

$$e_{r1} = 1 - \frac{3r_d^2}{4r^2} + \frac{15l^2}{16cr^2} \cos^2\theta + \left(\frac{r_d^2}{cr^2} - \frac{l^2}{4r^2c}\right)\sin^2\theta,$$

$$e_{r2} = -\frac{r_d^2}{2} - \frac{l^2}{12} + \frac{7l^2}{24}\cos^2\theta + \left(\frac{l^2}{12} + \frac{r_d^2}{8}\right)\sin^2\theta,$$

$$e_{r3} = \frac{r_d^2}{4} + \frac{l^2}{24}\cos^2\theta + \left(\frac{r_d^2}{4} - \frac{l^2}{24}\right)\sin^2\theta,$$

$$e_{\theta0} = 1 - \frac{6r_d^2 + l^2}{8r^2} + \left(\frac{13l^2}{8r^2} - \frac{15r_d^2}{4r^2}\right)\cos^2\theta + \left(\frac{15r_d^2}{8r^2} - \frac{l^2}{4r^2}\right)\sin^2\theta,$$

$$e_{\theta1} = 1 - \frac{3r_d^2}{4r^2} - \frac{l^2}{8r^2} + \left(\frac{13l^2}{8r^2} - \frac{7r_d^2}{4r^2}\right)\cos^2\theta + \left(\frac{7r_d^2}{8r^2} - \frac{l^2}{4r^2}\right)\sin^2\theta,$$

$$e_{\theta2} = 1 - \frac{r_d^2}{2r^2} - \frac{l^2}{12r^2} + \left(\frac{19l^2}{24r^2} - \frac{3r_d^2}{2r^2}\right)\cos^2\theta + \left(\frac{9r_d^2}{8r^2} - \frac{l^2}{12r^2}\right)\sin^2\theta,$$

$$e_{\theta3} = \frac{(l^2 - r_d^2)}{4}\cos^2\theta + \frac{5r_d^2\sin^2\theta}{8} - \frac{r_d^2}{4} - \frac{l^2}{24},$$

$$e_{\theta4} = \frac{l^2}{24}\cos^2\theta + \frac{r_d^2}{8}\sin^2\theta.$$

It is easily seen that new expressions coincides with classical ones in [10, 12] if the length of the dipole and its radius are small. Omitting the influence of current source radius ($r_d = 0$) for simplicity we can see that wave terms of transversal components $e_{\theta2}$ and h_2 become smaller near the source at transversal plane ($\theta = \pi/2$), i.e. $e_{\theta2} = 1 - l^2/6r^2$ and $h_2 = 1 - l^2/8r^2$ that explains smooth generation of the wave in the domain. Exactly at the plane, we have maximum values of the transversal components of electromagnetic field that bring in the main contribution to the total energy.

9.1.3 Numerical Simulations

Let us take a smooth function as a time dependence of the charge like it was proposed in [16]. For example, let [18]

$$Q(t) = Q_0(1 - \tanh(t/\tau))/2. \tag{9.10}$$

The time forms of normalized function $Q(t)$ and its first and second derivatives are shown in Fig. 9.1.

For the case when the transmitter of small size is surrounded by sphere of radius 1 m, we find the rate of change of signal defined by the parameter τ, at which a sphere with a radius of 1 m is the causal [12], i.e. the total flow of energy, including quasistatic field energy flow inside and flux radiated outward waves, becomes equal to zero. As the quasistatic field energy passing inside of the sphere does not depend

Fig. 9.1 The time
dependence of the dipole
charge and its derivatives

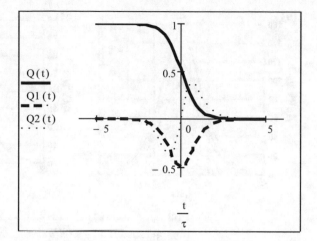

Fig. 9.2 The ratio of
radiated energy to energy of
quasistatic components of
the field outside the sphere of
the radius of 1 m

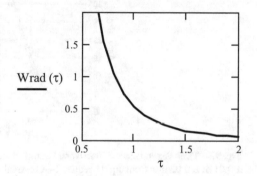

on τ, it is convenient to build the ratio W_{rad} of radiated energy to energy of quasistatic
components of the field outside the sphere, as depicted in Fig. 9.2. You can see that the
sphere of the selected size (radius 1 m) is causal at $\tau = 0.8$ ns. It should also be noted
that the energy of the emitted wave grows with decreasing inversely proportional to
the fourth degree τ for selected time dependence (9.10).

For a graphic illustration of the received expressions for vector potential (9.8)
the excitation of the small radiator of length $l = 0.01$ m and radius $r_d = 0.001$ m
by electrical current according expression (9.10) with amplitude $I_0 = 1$ A and time
scale $\tau = 10^{-9}$ s is considered. The time forms of the vector potential amplitudes
are presented for the observation angle $\theta = \pi/2$ of maximum radiation. The time
dependences of transversal component of vector potential value at various distances
from the dipole center for three cases, i.e. the classical solution (first term in (9.5)),
our improved solution (9.8), and exact solutions (9.1), are represented in Fig. 9.3. One
can see that our solution is closer to exact one than classical solution. The forming
of the wave in the nearest domain to the current source leads to the decreasing
of amplitude of precise solution. There is no noticeable difference between these
solutions at distance of two length of the dipole (see Fig. 9.3c).

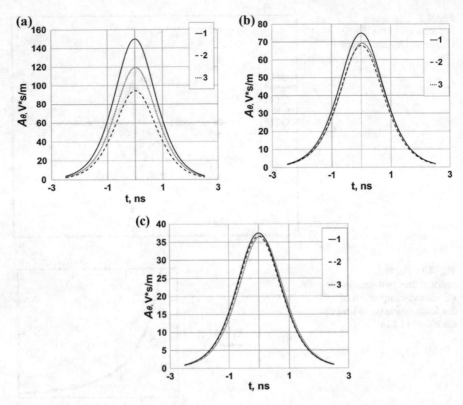

Fig. 9.3 Time dependences of transversal component of vector potential at the distances 0.005 m **a** 0.01 m **b** 0.02 m **c** from dipole center: 1—classical solution, 2—our solution, 3—strict solution

To trace the polar angle dependence of the solutions we plot the same function as at the figures above for the distance of observation 0.01 m and at angles $\theta = \pi/16, \pi/4, 7\pi/16$. The comparison of curves in Fig. 9.4 shows that our solution is closer to precise one in comparison with the classical solution. The increasing of the amplitude of precise solution at small polar angles (Fig. 9.4a) can be explained by the influence of finite sizes of the source of current. These figures illustrate the process of electromagnetic energy concentration at the direction of main maximum in near zone. One can conclude that the transversal energy transfer is ended at distance of wave zone.

It is interesting to check the influence of the vector potential solution precision on the time forms of field near the source [19]. It is shown in Fig. 9.5 where the time dependences of amplitude of magnetic field (9.2) are presented for the observation distance $r = 0.02$ m and for polar angles $\theta = \pi/16, \pi/4, \pi/2$. One can notice that our improved solution gives a different level of approximation quality for different angles, but it is always better than classical solution in whole. Obviously that the discrepancy between the curves is the smallest at $\theta = \pi/4$ (Fig. 9.5b) while

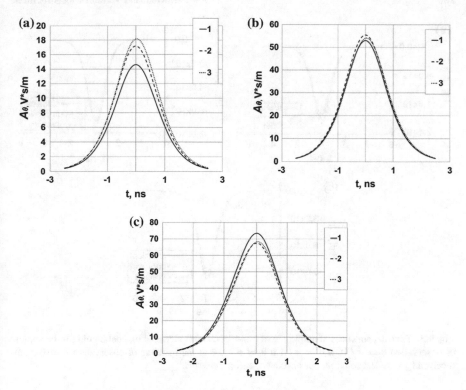

Fig. 9.4 Time dependences of transversal component of vector potential at the angles $\theta = \pi/16$ **a** $\theta = \pi/4$ **b** $\theta = 7\pi/16$ **c** and distance 0.01 m from dipole center: 1—classical solution, 2—our solution, 3—precise solution

the radiated wave is formed according the mentioned above explanation for vector potential.

As another example of a transient source let us consider the electromagnetic field generated by smooth switching off of the source of electrical current with length $l = 0.2$ m, transversal radius $r_d = 0.01$ m, and time dependence in the form of smooth decreasing from the maximum value $I_0 = 2000$ A to the zero:

$$I(t) = I_0(1 - \tanh(t/\tau))/2, \tag{9.11}$$

where $\tau = 2 \cdot 10^{-4}$ s [20].

Taking into account the small size of the radiator in comparison with the wavelengths of the highest frequencies in the spectrum of the given current that carry noticeable part of its energy one can consider the source as the Hertzian dipole. For the chosen parameters of the source of transient current the differences of classical (first term in (9.5)) and our improved solutions (9.8) should be noticeable even for distances of observation larger than size of the radiator. The difference becomes visually unnoticeable at the distance $r = 1.5$ m, i.e. 7.5 l. The angle dependence is

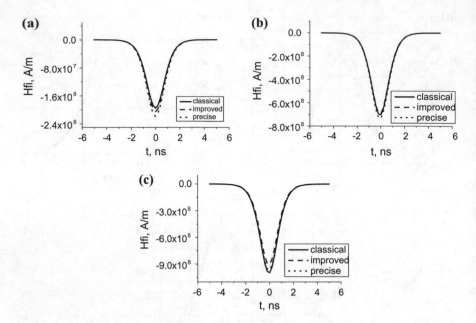

Fig. 9.5 Time dependences of amplitude of transversal component of magnetic field for the angles of observation $\theta = \pi/16$ **a** $\theta = \pi/4$ **b** $\theta = \pi/2$ **c** at the distance of observation $r = 0.02\,\text{m}$ received by classical solution, our improved, and precise one

depicted in Fig. 9.6, where transversal electrical component of the field is shown for the distance of observation $r = 0.5\,\text{m}$. It is easy to see that the time form of the field repeats the time form of exciting current $I(t)$. The figures illustrate the general difference between classical solution and improved one at different angles. The improved solution shows the decreasing of the energy flow at the angle $\theta = \pi/2$ and increasing at angles close to zero in comparison with the classical solutions.

9.1.4 Free Field Formation

As distinct from sinusoidal excitation case, transient excitation permits to observe the process of free field formation near radiator by tracing of time forms of electromagnetic field components. It is seen from our improved solution (9.9) or ordinary classical expressions for Herzian dipole field that the radiated wave in far zone has the time shape of first derivative of exciting current. It can be used for visual control of wave appearance from quasistatic components of field of radiator [12]. As an example, one can see the time form of amplitude of magnetic field (9.9) for polar angle $\theta = \pi/2$ and different distances of observation r presented in Fig. 9.7 for excitation (9.10). The amplitude is normalized by means of its multiplication by squared

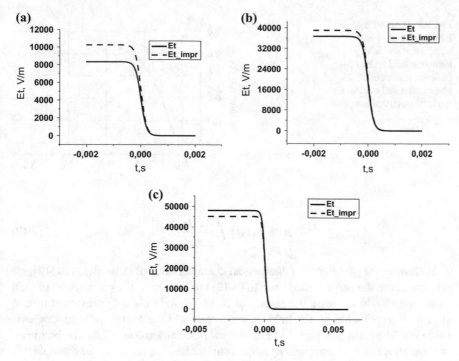

Fig. 9.6 Time dependences of amplitudes of transversal components of electrical field for the distance of observation $r = 0.5$ m and different polar angles of observation $\theta = \pi/16$ (a) $\theta = 5\pi/16$ (b) $\theta = \pi/2$ (c)

Fig. 9.7 Time dependence of normalized amplitude of transversal component of magnetic field (9.9) for different distances of observation r and for polar angle $\theta = \pi/2$

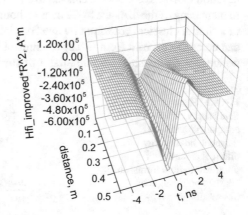

distance r. We can notice the appearing of wave of the form of first derivative of time shape of current from the distance $r = 0.2$ m.

So, the prevalence of this term in field expression can be used as a criterion of wave zone boundary. It was expressed by Harmuth [11] by the inequality for magnetic component of field

Fig. 9.8 Ratio of magnetic field energy of wave component and total magnetic field energy for different distances of observation and constant angle of observation $\theta = \pi/4$

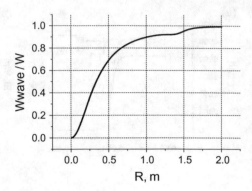

$$R \gg c \left| I(t) \middle/ \frac{d I(t)}{dt} \right|. \tag{9.12}$$

Following the preference of energetical characteristics of pulse fields in [12], we propose to use the same idea as used in (9.12) but instead of the comparison of field terms one should compare the energy of magnetic field (W_{wave}) presented by one term only responsible for the field in wave zone and the energy of total magnetic field (W). The ratio is depicted in Fig. 9.8 for polar angle $\theta = \pi/4$. The boundary of wave zone can be determined by some constant level of the ratio, for example 0.9. Such a criterion gives the approximate boundary of far zone in $r = 1$ m.

Let's study the change of the position of the boundary in dependence of time form of exciting current. Its simple change consists in the variation of the signal duration (9.10), i.e. τ. As we can see in Fig. 9.9, the increase of pulse duration leads to increasing of the distance to the nearest boundary of wave zone. It is a confirmation of a similar classical criterion of wave zone boundary $R \gg \lambda/2\pi$ from [21] which can be easily derived from (9.12) for sinusoidal time dependence case. One should notice that we consider here a radiator of small electrical size.

Fig. 9.9 Ratio of magnetic field energy of wave component and total magnetic field energy for different distances of observation, time forms, and constant polar angle of observation $\theta = \pi/4$

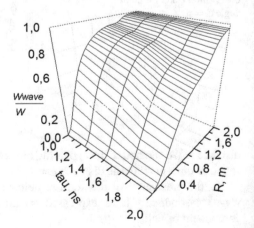

Fig. 9.10 Normalized amplitudes of longitudinal **a** and transversal **b** components of electrical field for different distances of observation versus retarded time for polar angle $\theta = \pi/2$

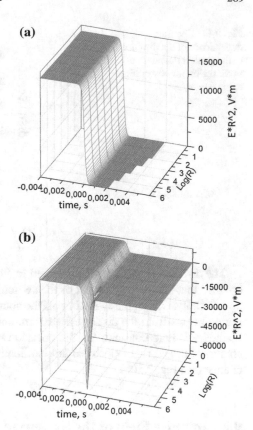

The similar behavior can be observed for the above mentioned case of smooth switching off (9.11). The normalized amplitudes of transversal and longitudinal components of electrical field at different distances from source ($1-10^6$ m) are shown in Fig. 9.10 in the approximation that dipolar component of field is absent, i.e. our source eliminates charges at the ends of our radiators. It is natural because usually the Herzian dipole is a part of a long conductor with current, so the charges at the ends excite the next piece of the conductor. Conversely, the current source (9.11) must generate infinite value of charges at the ends of the dipole before the moment $t = 0$. One can see that longitudinal component repeats the time form of exciting current (9.11) for all distances (see Fig. 9.10a). The exceeding of wave on reactive component of transversal electrical field can be observed at the distance $r = 10^6$ m in Fig. 9.10b, where the distance to the point of observation R is presented in logarithmic scale for convenience. It is in a good agreement with the above mentioned criterion of the nearest boundary of wave zone $\lambda/2\pi$ if we apply $c\tau = 6 \times 10^4$ m instead of λ.

Let us illustrate the criterion (9.12) of H. F. Harmuth by drawing the time dependent distance

Fig. 9.11 The time dependence of the distance to the nearest boundary of wave region for observation angle $\theta = \pi/2$

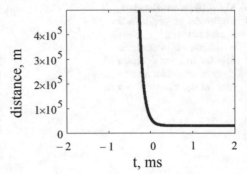

$$R = c \left| I(t) \middle/ \frac{dI(t)}{dt} \right|. \tag{9.13}$$

The distance versus time is presented in Fig. 9.11 for polar angle $\theta = \pi/2$. It is obvious that before the moment of the switching off the reactive field fills all space up to infinity. During the switching off the source of current (9.11) and later the wave region begins from the distance $3 \cdot 10^4$ m, and the magnitude of the distance to its boundary is hold constant. So, the criterion has a good agreement with the visual change of time form of electrical field indicating the distance of begin of time shape change (see Fig. 9.10b).

9.2   Impulse Field of the System of Short Radiators

Let's consider the example of radiation systems where powerful impulse electric current is used. Rail launcher is intended for the acceleration of objects by means of the Lorentz force caused by a large electrical current. Due to the fact that there are no theoretical limitations on the value of this force, the rail launcher speeds up metal shells to supersonic velocities [22, 23]. Numerical analysis makes it possible to accurately simulate the distribution of current and forces appeared [24, 25]. The approach presented here permits to use the distribution of impulse current for obtaining the characteristics of electromagnetic fields radiated by such rail launchers. Due to the fact that their transversal sizes are much smaller than the average wavelength of electromagnetic field that is radiated into outer space, the short electric Hertzian dipole with the excitation of a current with arbitrary time dependence is an acceptable model of the source of this radiating system. One can predict that the radiator in the first approximation is a transient dipole at transversal plane. But it is interesting to take into account the changes in the radiation field due to the flow of surface currents along the curvilinear surface of a metal shell. These changes should be registered at short distances from the radiator. Therefore, a field in its near-field zone will be investigated.

9.2.1 Field of Two Parallel Dipoles

The first approximation of the real current distribution is the small impulse source presented by two thin synchronously excited dipoles [26] shifted to each other on the small value d as shown in Fig. 9.12. The impulse excitation of each dipole is similar to (9.11) but the amplitude of current $I_0 = 1000$ A is a half of previous one for convenience of result comparison. Other parameters of dipoles are the same, i.e. the length $l = 0.2$ m, radius $r_d = 0.01$ m, and $\tau = 2 \cdot 10^{-4}$ s.

We will compare the radiation of system of two small synchronous impulse radiators with the radiation of single radiator (like presented in Figs. 9.6 and 9.9), classical solution for electrical dipole and improved one using expressions (9.9) and superposition principle. For the system of two dipoles the azimuth angle dependence is appeared that is illustrated in Fig. 9.13 where the time form of the transversal component amplitude is depicted for distance of observation $r = 0.5$ m and polar angle $\theta = \pi/2$.

The distance of observation $r = 0.5$ m was chosen to demonstrate the differences in curves because the time forms of electrical components at $r = 1$ m and more are almost similar [26]. One should note that we have here the prevailing of inductive component of field that is shown by the reproducing of the time dependence of exciting current (9.11). The excess of the fields of two dipoles at $\varphi = 0$ can be explained by influence of small shift at $d/2$ toward the point of observation. The surplus of the classical solutions on improved ones for all angles φ is in a good agreement with the previous results for polar angle $\theta = \pi/2$ (see Fig. 9.6). A bigger distance from dipoles to the point of observation in two dipole system at $\varphi = \pi/2$ leads to small decrease of its amplitude.

Fig. 9.12 System of two transient dipoles

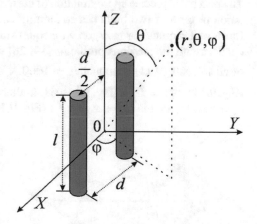

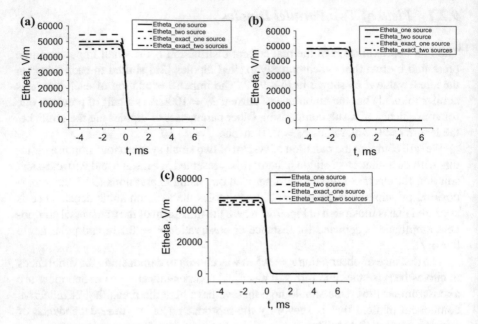

Fig. 9.13 Time dependences of amplitudes of transversal component of electrical field for the distance of observation $r = 0.5$ m and angle $\theta = \pi/2$ for cases of one and two dipoles obtained by classical and our improved (exact) solutions at $\varphi = 0$ (**a**) $\varphi = \pi/6$ (**b**) $\varphi = \pi/2$ (**c**)

9.2.2 Radiation of the System of Four and Six Dipoles

The next more precise approximation of current flow around metal shell is its presentation in form of two branches containing two or three dipoles. The systems have forms of a rhombus or hexagon with equal sides as figured in Fig. 9.14 [27, 28]. To connect with the previous problems [19, 26] we consider that the amplitude values of all currents (9.11) are equal $I_0 = 1000$ A, the length of each dipole $l = L\big/\sqrt{2}$ (Fig. 9.14a) or $l = L/2$ (Fig. 9.14b) is smaller than in previous cases, but its radiuses $r_d = 0.01$ m, are the same, $\alpha = \pi/4$ (Fig. 9.14a).

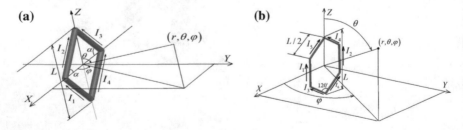

Fig. 9.14 Systems of four **a** and six **b** dipoles

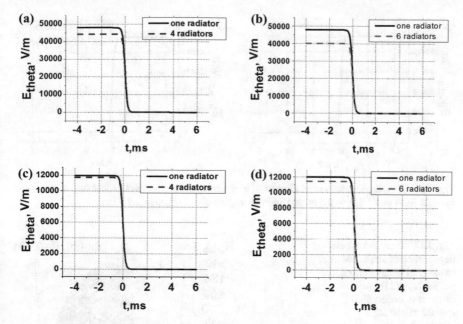

Fig. 9.15 Time dependences of amplitudes of transversal component of electrical field of four (**a, c**) and six (**b, d**) dipole systems for $\theta = \pi/2$, $\varphi = \pi/2$, and the distances of observation $r = 0.5$ m (**a, b**) and $r = 1$ m (**c, d**)

To solve the radiation problem we will use the improved solution (9.9) after its sequential transformation into Cartesian coordinate system, rotation, shift, superposition principle application, and transformation into spherical coordinate system. As for behavior of transversal electrical component at large distances [27], it is the same as shown in Fig. 9.10b. To trace the influence of observation distance on the transversal electrical field amplitude of these two systems the time dependences are figured in Fig. 9.15 at direction of radiation maximum in comparison with single dipole field. One can see on the base of single dipole analysis that six dipole system needs more distance to form stable angle distribution of electrical field in comparison with four dipole system. The pattern of the six dipole system became almost unchanged beginning from $r = 4$ m [28], while four dipole system requires smaller distance $r = 2$ m [27]. As the time forms coincide with exciting current (9.11) it means that the points of observation r in Fig. 9.15 are situated in the near zone of the radiation systems.

The transversal electrical field distribution on azimuthal angle is illustrated in Fig. 9.16 for six dipole system (Fig. 9.14b). The four dipole system has the similar behavior [27]. The amplitude of the field is noticeably smaller in comparison with case of single dipole in the direction of main maximum, $\varphi = \pi/2$ (Fig. 9.15b). The decrease of the azimuth angle leads to equalization of the amplitudes (Fig. 9.16b) at $\varphi = \pi/9$ and the dominance of six dipole field at $\varphi = 0$ (Fig. 9.16a). The latter can be explained by the proximity to the nearest dipoles (see Fig. 9.14). The strong

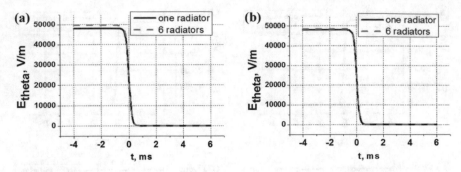

Fig. 9.16 Time dependences of amplitudes of transversal component of electrical field of six dipole systems for $\theta = \pi / 2$, $r = 0.5$ m, and azimuth angles $\varphi = 0$ (**a**), $\varphi = \pi / 9$ (**b**)

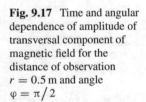

Fig. 9.17 Time and angular dependence of amplitude of transversal component of magnetic field for the distance of observation $r = 0.5$ m and angle $\varphi = \pi / 2$

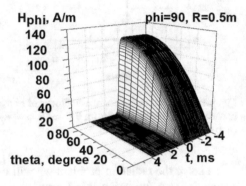

influence takes place because of inductive field prevalence at the distances illustrated in Fig. 9.10b.

Let's consider the time forms of the magnetic field at small distance depicted in Fig. 9.17 for six dipole system (Fig. 9.14b). As the angle θ increases, the intensity increases uniformly and reaches a maximum value of 120 A/m at $\theta = \pi / 2$ [28]. The behavior is typical for the single dipole, two- or four dipole system [27] (Fig. 9.17).

On the other hand, the dependence on polar angle θ for two cases presented in Fig. 9.14 has unusual behavior that is depicted in Fig. 9.18 in the direction of radiation maximum, $\varphi = \pi / 2$. The appearance of the unusual maximum at angle $\theta = 50°$ can be explained by small distance of observation $r = 1$ m and big influence of the nearest dipoles. The effect quickly decreases at longer distances.

9.3 Ultrawideband Combined Vibrator-Slot Radiator

A short electrical dipole can provide a smooth enough frequency characteristic of radiation as described above. Additionally it is supported by well-known fact that the pattern of short dipole has similar form to the pattern of half-wave dipole. There

Fig. 9.18 Time and angular dependence of amplitude of transversal component of electrical field for the distance of observation $r = 1$ m and angle $\varphi = \pi/2$ for four dipole system

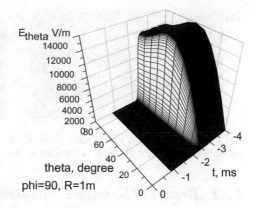

are different constructions of wideband slots. The purpose of the chapter is to build ultrawideband counterpart of Clavin radiator (see, Chap. 6). It is possible to use the strong interaction between slot and vibrators like presented here in Chap. 6 for improving of characteristics of a combined radiator. It is possible to provide the interaction in wide frequency range by using optimal placement of the wideband dipoles in near zone of a slot. The conical form of dipoles is utilized for the widening of its working frequency range. The existing analytical solutions of Pocklington and Hallen equations [29] cannot be directly applied for the case of smooth changing of dipole cross-section to improve its frequency parameters. So, a numerical solving is used for convenience. As an example of the magnetic radiator, we consider the ultrawideband slot of special form cut in infinite perfectly conducting screen optimized experimentally [30]. To diminish the shadowing of the slot by dipoles the short circuit conical monopole is turned upside-down. So, the three-dimensional analogue of diamond dipole [30] is utilized for the combined vibrator-slot radiator.

9.3.1 Statement of the Problem for Combined Radiator

Let the wideband slot [30] is cut in infinite perfectly conducting screen of finite thickness as presented in Fig. 9.19a. The slot is excited by electrical voltage source and resistance connected in series and applied to the both sides of the slot at its center. The value of the resistance is chosen to minimize the impulse signal reflected from the radiator. To improve the characteristics of the slot the two perfectly conducting cones as ultrawideband short-circuited dipoles are added as pictured in Fig. 9.19b. The problem is to find the sizes of the radiating structure for optimal characteristics.

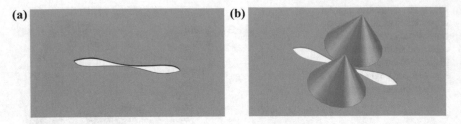

Fig. 9.19 Ultrawideband slot of sizes given in [30] in perfectly conducting screen (**a**) and the ultrawideband analogue of Clavin radiator on the base of the slot (**b**)

9.3.2   Results of Numerical Simulation

It is convenient to use direct numerical calculation in time domain for the curvilinear structures. The infinite screen of 1 mm thickness is simulated by the screen of length and width of five lengths of the slot (Fig. 9.19a). All boundaries of calculation space are absorbing. Numerical simulation performed by the method of finite differences in time domain shows that the increasing of the sizes of the screen does not change the results of calculation noticeably. The voltage standing wave ratio as function of frequency for the slot is represented in Fig. 9.20. The absence of strict periodicity of minima can be explained by the curvilinear form of the slot and complicated ways of surface currents around the slot [30]. To improve its radiation efficiency at low frequencies, the two conical dipoles, adjoined to the curvilinear slot boundary, were situated symmetrically with respect to slot center along the line that is perpendicular to the slot (see Fig. 9.19b), i.e. the ultrawideband counterpart of Clavin radiator was constructed. It was found out that the radius of the cone basis has influence on the

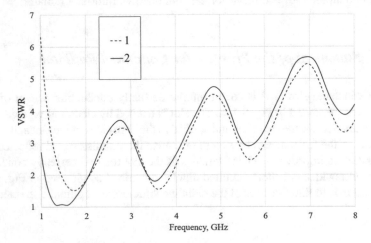

Fig. 9.20   VSWR versus frequency of ultrawideband slot (curve 1) and Clavin radiator on its basis (curve 2)

quality of its resonances. As for effect of the cones positions and its heights, it is similar to the described in Chap. 6. In particular, minimum at $f = 1.3$ GHz (see curve 2 in Fig. 9.20) is provided by the classical relation $L/\lambda = 0.25$ and recommended value of $x_d/\lambda = 0.16$, but for better radiation the minimum at $f = 1.55$ GHz is provided by the $L/\lambda = 0.3$ and increased value of $x_d/\lambda = 0.2$ for better radiation. These improvements noticeably make wider the frequency range of the radiator.

9.3.3 Characteristics of Radiation of the Ultrawideband Analogue of Clavin Radiator

Let's compare directivity of these radiators (Fig. 9.19) for different frequencies. It is known that the RP of rectangular slot at the E-plane is close to isotropic one. The directive gains of the ultrawideband slot and Clavin radiator at the E-plane are depicted at different frequencies in Fig. 9.21. One can see that patterns of the slot is isotropic with smooth increasing of its directivity for bigger frequencies (Fig. 9.21a, c). Clavin radiator demonstrates more directed radiation (Fig. 9.21d) but with undesirable maxima at $\theta = 90°$ for the frequencies lower than 1.8 GHz (Fig. 9.21b). It can be explained by effective excitation of cones at the low frequencies due to its heights and distance between them.

The RP of rectangular slot at the H-plane always has maximum along to normal to its plane and zeroes at its plane. The directive gains of presented ultrawideband slot and Clavin radiator have the similar form shown in Fig. 9.22. Also, the complicated amplitude distribution generates more flat directivity distributions at bigger frequencies as figured in Fig. 9.22c. The weak influence of cones (see Fig. 9.22b, d) can be explained here by the mutually opposite sign of each cone excitation by the slot center. It causes the reciprocal cancelling of total radiating field at H-plane.

Besides parameters of the ultrawideband Clavin radiator at fixed frequencies, it is more interesting to study the time forms of amplitudes of radiated impulse electromagnetic wave. The electrical component of the wave at distance 0.3 m from the center of the radiators is presented in Fig. 9.23 for polar angles $\theta = 0°, 30°, 60°$ at H- and E-planes when the radiators are excited by Gaussian pulse of 0.08 ns duration. As seen in Fig. 9.23a, passive cones increased the amplitude of radiated waves along the normal to the plane of the slot but extended the transient process in the structure. It is important to explain the unusual time delay caused by the presence of cones. These cones serve as inhomogeneous transmission line that generates reflected impulse wave at the starting moment of its excitation. The wave interferes with the exciting wave cancelling its action in the center of the slot. Further, the amplitude of the reflected wave of the inhomogeneous line decreases, changes its sign, so, the total electrical field distributed in the slot increases, and the slot begins to radiate obtaining additional energy from the reflected wave of cones. The angle dependence at H-plane (Fig. 9.23b, c) demonstrates the decreasing of amplitudes for the slot (curve 1) as well as for the Clavin radiator (curve 3) according to all curves in

Fig. 9.23. One can notice that first maximum of curve 1 in Fig. 9.23b being bigger than corresponding maximum of curve 1 in Fig. 9.23a is not exclusion because of high value of the next minimum of curve 1 in Fig. 9.23b. The similar behavior we can observe at E-plane for Clavin radiator (curve 4) because, as seen from Fig. 9.23, the main part of the pulse energy is concentrated at 2.5–3 GHz frequency range, where Clavin radiator demonstrates the decreasing of radiated field for bigger polar angles (see Fig. 9.21d). At the same time, the radiation of the slot at E-plane (curve 2) is isotropic in accordance with directive gains in Fig. 9.21a, c.

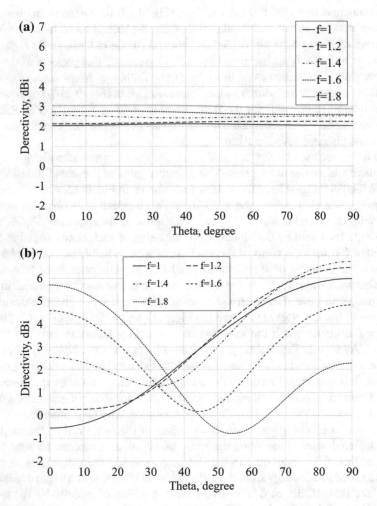

Fig. 9.21 Directive gains at E-plane of ultrawideband slot (**a, c**) and Clavin radiator (**b, d**) at different frequencies

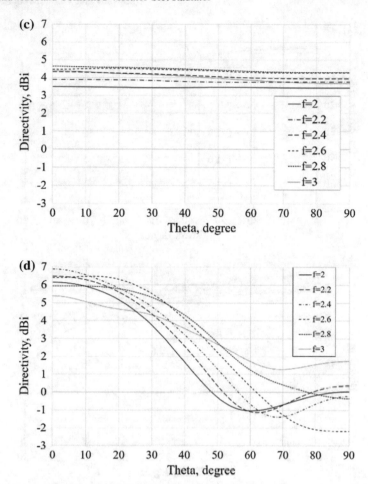

Fig. 9.21 (continued)

So, the ultrawideband combined vibrator-slot structure has a number of directions for the further optimizations in accordance with applications needed. The counterpart of Clavin radiator can concentrate the radiated energy in the given direction and provide wide working frequency range. It is possible to optimize its simple construction for the radiation of pulses with given time parameter.

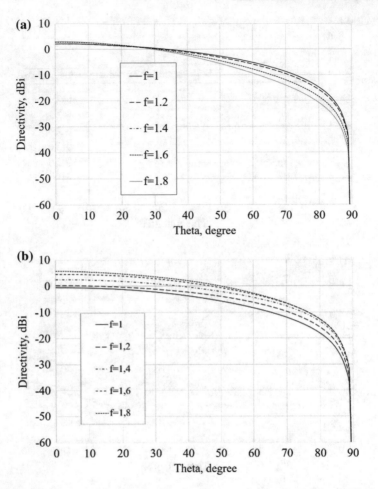

Fig. 9.22 Directive gains at *H*-plane of ultrawideband slot (**a, c**) and Clavin radiator (**b, d**) for different frequencies

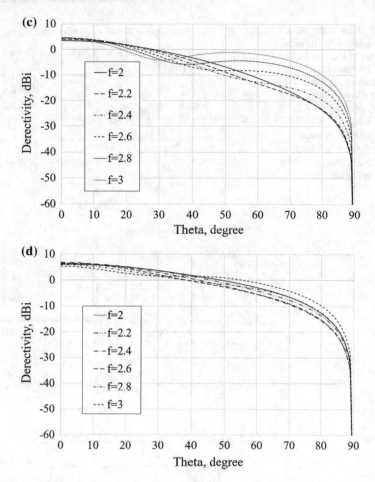

Fig. 9.22 (continued)

Fig. 9.23 Time dependences
of the amplitude of electrical
component of the radiated
field for the distance of
observation $r = 0.3$ m for
$\theta = 0°$ **a** where 1-slot,
2-Clavin radiator; for
$\theta = 30°$ **b** and $\theta = 60°$
c where 1-slot at H-plane,
2-slot at E-plane, 3-Clavin
radiator at H-plane, 4-Clavin
radiator at E-plane

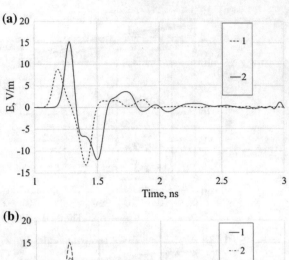

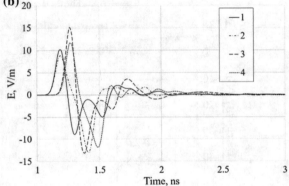

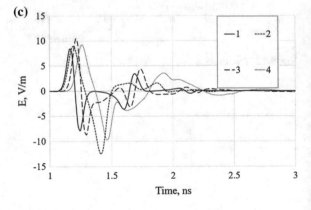

References

1. Luszcz J, Smolenski R (2015) Low frequency conducted emissions of grid connected static converters. IEEE Electromagn Compat Mag 4(1):86–100
2. Pous M, Silva F (2014) Prediction of the impact of transient disturbances in real-time digital wireless communication system. IEEE Electromagn Compat Mag. 3(3):76–82
3. Kebel R, Stadtler T, Rouquette J-A, Flourens F, Avenet A, Rouvrais N (2016) Numeric lightning protection prediction for wires in an aircraft wing raceway. IEEE Electromagn Compat Mag 5(4):71–79
4. Vogel MH (2014) Impact of lightning and high-intensity radiated fields on cables in aircraft. IEEE Electromagn Compat Mag 3(2):56–61
5. Grcev LD, Menter FE (1996) Transient electromagnetic fields near large earthing systems. IEEE Trans Magn 32(3):1525–1528
6. Tanaka H, Baba Y, Barbosa CF (2016) Effect of shield wires on the lightning-induced currents on buried cables. IEEE Trans Electromagn Compat 58(3):738–746
7. Rachidi F, Janischewskyj W, Hussein AM, Nucci CA, Guerrieri S, Kordi B, Chang J-S (2001) Current and electromagnetic field associated with lightning-return strokes to tall towers. IEEE Trans Electromagn Compat 43(3):356–367
8. Martins V (2016) A method for sketching the transient electromagnetic field response radiating from a dipole/monopole antenna excited by a general source impedance and driving function. In: Proceedings of the antennas and propagation society international symposium, Kharkiv, Ukraine pp 171–175
9. Maloney T (2013) Pulsed Hertzian dipole radiation and electrostatic discharge events in manufacturing. IEEE Electromagn Compat Mag 2(3):37–46
10. Fraceschetti G, Papas CH (1974) Pulsed antennas. IEEE Trans Antennas Propag 22(5):651–661
11. Harmuth H (1981) Nonsinusoidal waves for radar and radio communications. Academic Press, New York
12. Schantz HG (2005) The art and science of ultrawideband antennas. Artech House, London
13. Dumin O, Volvach IS, Dumina O (2012) Transient Near field of Hertzian dipole. In: Proceedings of the 6th international conference on ultrawideband and ultrashort impulse signals, Sevastopol, Ukraine, pp 69–71
14. Allen B, Dohler M, Okon EE, Matik WK, Brown AK, Edwards DJ (2007) Ultra-wideband Antennas and propagation for communications radar and imaging. Wiley, Chichester
15. Hertz H (1893) Electric waves being researches on the propagation of electric action with finite velocity through space. The forces of electric oscillations, treated according to Maxwell's theory. Macmillan, New York, pp 137–159
16. Schantz HG (2001) Electromagnetic energy around hertzian dipoles. IEEE Antennas Propag Mag 43(2):50–62
17. Dumin OM, Plakhtii VA, Volvach IS, Pshenichnaya SV, Dumina OO (2015) Near field of Hertzian dipole excited by impulse current. In: Proceedings of the 10th international conference on antenna theory and techniques, Kharkiv, Ukraine, pp 90–92
18. Volvach YS, Dumin OM, Dumina OO (2011) The energy of the field radiated by Hertz dipole. In: Proceedings of the VII international conferences on antenna theory and techniques, Kyiv, Ukraine, pp 86–88
19. Plakhtii VA, Dumin OM, Katrich VA, Dumina OO (2016) Field regions of impulse current radiator of small size. In: Proceedings of the 9th international Kharkiv symposium on physics and engineering of microwaves, millimeter and submillimeter waves, Kharkiv, Ukraine, D-27
20. Plakhtii VA, Dumin OM, Katrich VA, Dumina OO, Volvach IS (2016) Energy transformation of transient field of herzian dipole. In: Proceedings of the 16th IEEE international conference on mathematical methods in electromagnetic theory, Lviv, Ukraine, pp 314–317
21. Balanis CA (1997) Antenna theory. Wiley, New York
22. Rodger D, Leonard PJ, Eastham JF (1991) Modelling electromagnetic rail launchers at speed using 3D finite elements. IEEE Trans Magnetics 27(1):314–317

23. Engel TG, Neri JM, Veracka MJ (2008) Characterization of the velocity skin effect in the surface layer of a railgun sliding contact. IEEE Trans Magnetics 44(7):1837–1844
24. Tan S, Lu J, Zhang X, Li B, Zhang Y, Jiang Y (2016) The numerical analysis methods of electromagnetic rail launcher with motion. IEEE Trans Plasma Sci 44(12):3417–3423
25. Hsieh KT, Kim BK (1997) 3D modeling of sliding electrical contact. IEEE Trans Magnetics 33(1):237–239
26. Dumin OM, Plakhtii VA, Katrich VA, Dumina OO, Volvach IS (2016) Radiation of two small impulse current radiators. In: Proceedings of the 8th international conference on ultrawideband and ultrashort impulse signals, Odessa, Ukraine, pp 81–84
27. Plakhtii VA, Dumin OM, Prishchenko OA (2017) Transient radiation of system of four noncollinear dipoles. In: Proceedings of the 2017 IEEE first Ukraine conference on electrical and computer engineering. Kyiv, Ukraine, pp 225–228
28. Plakhtii VA, Dumin OM, Prishchenko OA (2017) Near radiation zone of six short impulse radiators. In: Proceedings of the 2017 IEEE international young scientists forum on applied physics and engineering. Lviv, Ukraine, pp 251–254
29. Tijhuis AG, Zhongqiu P (1992) Transient excitation of a strait thin-wire segment: a new look at an old problem. IEEE Trans Antennas Propag 40(10):1132–1146
30. Barnes MA (2000) Ultra-wideband magnetic antenna. US patent 6,091,374, July 18

Appendix A
Green's Functions of the Considered Electrodynamic Volumes

1. Electrical Dyadic Green's Functions

1. The unbounded space with the permittivity and the permeability of the medium ε_1 and μ_1 ($k_1 = k\sqrt{\varepsilon_1\mu_1}$).

$$\hat{G}^e(\vec{r}, \vec{r}') = \hat{I}\frac{e^{-ik_1|\vec{r}-\vec{r}'|}}{|\vec{r} - \vec{r}'|}. \tag{A.1}$$

2. The half-space over the perfectly conducting plane with the permittivity and the permeability of the medium ε_1 and μ_1 (Fig. A.1).

$$\hat{G}^e(\vec{r}, \vec{r}') = \hat{I}\frac{e^{-ik_1 R}}{R} - (\vec{e}_x \otimes \vec{e}_{x'})\frac{e^{-ik_1 R_1}}{R_1} - (\vec{e}_y \otimes \vec{e}_{y'})\frac{e^{-ik_1 R_1}}{R_1}$$
$$+ (\vec{e}_z \otimes \vec{e}_{z'})\frac{e^{-ik_1 R_1}}{R_1}, \tag{A.2}$$

$$R = \sqrt{(x - x')^2 + (y - y')^2 + (z - z')^2},$$
$$R_1 = \sqrt{(x - x')^2 + (y - y')^2 + (z + z')^2}.$$

3. The space with the permittivity and the permeability of the medium ε_1 and μ_1 of the medium and inside the corner region, concluded between two planes and perpendicularly intersecting along the abscissa axis $X0Y$ and $X0Z$ (Fig. A.2).

Fig. A.1 The half-space over a perfectly plane

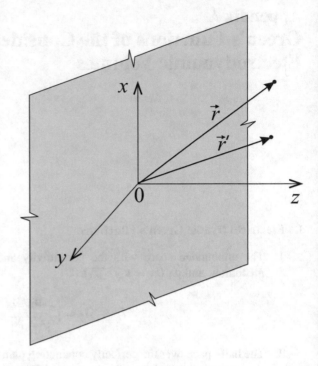

Fig. A.2 The space inside a corner region

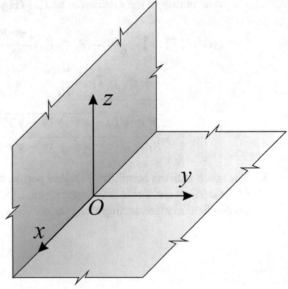

Fig. A.3 The hollow infinite rectangular waveguide

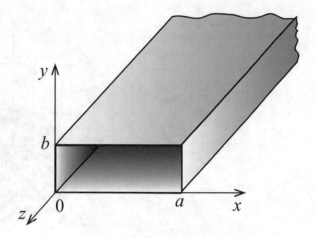

$$G_{zz}^e(x, y, z; x', y', z') = \frac{e^{-ik_1 R}}{R} - \frac{e^{-ik_1\sqrt{(x-x')^2+(y+y')^2+(z-z')^2}}}{\sqrt{(x-x')^2+(y+y')^2+(z-z')^2}}$$

$$+ \frac{e^{-ik_1 R_1}}{R_1} - \frac{e^{-ik_1\sqrt{(x-x')^2+(y+y')^2+(z+z')^2}}}{\sqrt{(x-x')^2+(y+y')^2+(z+z')^2}}.$$

$$(A.3)$$

4. The hollow infinite rectangular waveguide of the cross-section $\{a \times b\}$ with the perfectly conducting walls (Fig. A.3).

$$\hat{G}^e(\vec{r}, \vec{r}') = \frac{2\pi}{ab} \sum_{m=0}^{\infty} \sum_{n=0}^{\infty} \frac{\varepsilon_m \varepsilon_n}{k_z} e^{-k_z|z-z'|} \big[(\vec{e}_x \otimes \vec{e}_{x'}) \Phi_x(x, y; x', y')$$
$$+ (\vec{e}_y \otimes \vec{e}_{y'}) \Phi_y(x, y; x', y')$$
$$+ (\vec{e}_z \otimes \vec{e}_{z'}) \Phi_z(x, y; x', y') \big].$$

$$(A.4)$$

5. The hollow half-infinite rectangular waveguide of the cross-section $\{a \times b\}$ with the perfectly conducting walls.

$$\hat{G}^e(\vec{r}, \vec{r}') = \frac{2\pi}{ab} \sum_{m=0}^{\infty} \sum_{n=0}^{\infty} \frac{\varepsilon_m \varepsilon_n}{k_z} \Big\{ (\vec{e}_x \otimes \vec{e}_{x'}) \Phi_x^e(x, y; x', y') \big[e^{-k_z|z-z'|} - e^{-k_z(z+z')} \big]$$
$$+ (\vec{e}_y \otimes \vec{e}_{y'}) \Phi_y^e(x, y; x', y') \big[e^{-k_z|z-z'|} - e^{-k_z(z+z')} \big]$$
$$+ (\vec{e}_z \otimes \vec{e}_{z'}) \Phi_z^e(x, y; x', y') \big[e^{-k_z|z-z'|} + e^{-k_z(z+z')} \big] \Big\}.$$

$$(A.5)$$

6. The hollow rectangular resonator with dimensions $\{a_R \times b_R \times H\}$ and perfectly conducting walls.

Fig. A.4 The space outside
a perfectly conducting sphere

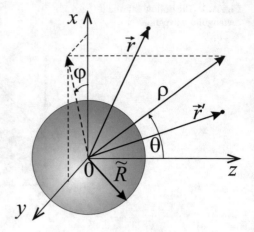

$$\hat{G}^e(\vec{r}, \vec{r}')$$

$$= \frac{2\pi}{a_R b_R} \sum_{m=0}^{\infty} \sum_{n=0}^{\infty} \frac{\varepsilon_m \varepsilon_n}{k_z} \times \left\{ (\vec{e}_x \otimes \vec{e}_{x'}) \Phi_x^e(x, y; x', y') \left[\frac{\operatorname{ch} k_z(H - |z - z'|) - \operatorname{ch} k_z(H - |z + z'|)}{\operatorname{sh} k_z H} \right] \right.$$

$$+ (\vec{e}_y \otimes \vec{e}_{y'}) \Phi_y^e(x, y; x', y') \left[\frac{\operatorname{ch} k_z(H - |z - z'|) - \operatorname{ch} k_z(H - |z + z'|)}{\operatorname{sh} k_z H} \right]$$

$$\left. + (\vec{e}_z \otimes \vec{e}_{z'}) \Phi_z^e(x, y; x', y') \left[\frac{\operatorname{ch} k_z(H - |z - z'|) + \operatorname{ch} k_z(H - |z + z'|)}{\operatorname{sh} k_z H} \right] \right\}. \tag{A.6}$$

In the expressions (A.1)–(A.6) the following notations is used:

$$\Phi_x^e(x, y; x', y') = \cos k_x x \cos k_x x' \sin k_y y \sin k_y y',$$
$$\Phi_y^e(x, y; x', y') = \sin k_x x \sin k_x x' \cos k_y y \cos k_y y',$$
$$\Phi_z^e(x, y; x', y') = \sin k_x x \sin k_x x' \sin k_y y \sin k_y y',$$

$$\varepsilon_{m,n} = \begin{cases} 1, & m, n = 0 \\ 2, & m, n \neq 0 \end{cases}, \quad k_x = \frac{m\pi}{a}, \quad k_y = \frac{n\pi}{b}, \quad k_z = \sqrt{k_x^2 + k_y^2 - k^2}, \quad m \text{ and } n \text{ are}$$

the integral numbers, $\vec{e}_x$, $\vec{e}_y$ and $\vec{e}_z$ are the unit vectors of the rectangular coordinate
system, $\hat{I} = (\vec{e}_x \otimes \vec{e}_{x'}) + (\vec{e}_y \otimes \vec{e}_{y'}) + (\vec{e}_z \otimes \vec{e}_{z'})$ is the unit dyadic, and "$\otimes$" stands
for dyadic product.

7. Space outside the perfectly conducting sphere of the radius $\tilde{R}$ with the permittivity
    and the permeability of the medium ε_1 and μ_1 (Fig. A.4).

$$\hat{G}^e(\rho, \theta, \varphi; \rho', \theta', \varphi') = \begin{vmatrix} G_{\rho\rho'}^e & 0 & 0 \\ 0 & G_{\theta\theta'}^e & G_{\theta\varphi'}^e \\ 0 & G_{\varphi\theta'}^e & G_{\varphi\varphi'}^e \end{vmatrix}, \tag{A.7}$$

$$G^e_{\rho\rho'}(\rho, \theta, \varphi; \rho', \theta', \varphi')$$

$$= -\sum_{n=0}^{\infty} \sum_{m=0}^{n} \frac{\varepsilon_m \, h_n(\rho, \rho')}{2C_{nm}} P_n^m(\cos\theta) \, P_n^m(\cos\theta') \, \cos m(\varphi - \varphi'),$$

$$G^e_{\theta\theta'}(\rho, \theta, \varphi; \rho', \theta', \varphi') = -\sum_{n=0}^{\infty} \sum_{m=0}^{n} \frac{\varepsilon_m \, u_n(\rho,\rho') \cos m(\varphi-\varphi')}{2n(n+1)\, C_{nm} \sin\theta \sin\theta'}$$

$$\times \left[m^2 P_n^m(\cos\theta) P_n^m(\cos\theta') + \sin\theta \sin\theta' \frac{d P_n^m(\cos\theta)}{d\theta} \frac{d P_n^m(\cos\theta')}{d\theta'} \right],$$

$$G^e_{\theta\varphi'}(\rho, \theta, \varphi; \rho', \theta', \varphi') = \sum_{n=0}^{\infty} \sum_{m=0}^{n} \frac{m \, u_n(\rho,\rho') \sin m(\varphi-\varphi')}{n(n+1)\, C_{nm}}$$

$$\times \left[\frac{d P_n^m(\cos\theta)}{d\theta} \frac{P_n^m(\cos\theta')}{\sin\theta'} + \frac{P_n^m(\cos\theta)}{\sin\theta} \frac{d P_n^m(\cos\theta')}{d\theta'} \right],$$

$$G^e_{\varphi\theta'}(\rho, \theta, \varphi; \rho', \theta', \varphi') = -G^e_{\theta\varphi'}(\rho, \theta, \varphi; \rho', \theta', \varphi'),$$

$$G^e_{\varphi\varphi'}(\rho, \theta, \varphi; \rho', \theta', \varphi') = G^e_{\theta\theta'}(\rho, \theta, \varphi; \rho', \theta', \varphi').$$

Here $P_n^m(\cos\theta)$ is the associated Legendre functions of the first sort,

$$C_{nm} = \frac{2\pi \, (n+m)!}{(2n+1)\,(n-m)!},$$

$$h_n(\rho, \rho') = \begin{cases} 4\pi k_1 h_n^{(2)}(k_1\rho') \begin{bmatrix} j_n(k_1\rho) \, Q_n\left(y_n\left(k_1\tilde{R}\right)\right) \\ -y_n(k_1\rho)\, Q_n\left(j_n\left(k_1\tilde{R}\right)\right) \end{bmatrix}, & \tilde{R} \leq \rho < \rho', \\[4mm] 4\pi k_1 h_n^{(2)}(k_1\rho) \begin{bmatrix} j_n(k_1\rho') \, Q_n\left(y_n\left(k_1\tilde{R}\right)\right) \\ -y_n(k_1\rho')\, Q_n\left(j_n\left(k_1\tilde{R}\right)\right) \end{bmatrix}, & \rho > \rho', \end{cases}$$

$$Q_n(f_n(k_1 R)) = \frac{n\, f_n\left(k_1\tilde{R}\right) - k_1 R\, f_{n+1}\left(k_1\tilde{R}\right)}{n\, h_n^{(2)}\left(k_1\tilde{R}\right) - k_1 R h_{n+1}^{(2)}\left(k_1\tilde{R}\right)},$$

$$u_n(\rho, \rho') = \begin{cases} 4\pi k_1 \dfrac{h_n^{(2)}(k_1\rho')}{h_n^{(2)}\left(k_1\tilde{R}\right)} \left[j_n(k_1\rho)\, y_n\left(k_1\tilde{R}\right) - y_n(k_1\rho)\, j_n\left(k_1\tilde{R}\right) \right], & \tilde{R} \leq \rho < \rho', \\[4mm] 4\pi k_1 \dfrac{h_n^{(2)}(k_1\rho)}{h_n^{(2)}\left(k_1\tilde{R}\right)} \left[j_n(k_1\rho')\, y_n\left(k_1\tilde{R}\right) - y_n(k_1\rho')\, j_n\left(k_1\tilde{R}\right) \right], & \rho > \rho', \end{cases}$$

$h_n^{(2)}(k_1\rho) = j_n(k_1\rho) - i y_n(k_1\rho) = \sqrt{\frac{\pi}{2k_1\rho}} H_{n+1/2}^{(2)}(k_1\rho)$ is the Hankel spherical function

of the second sort, $j_n(k_1\rho) = \sqrt{\frac{\pi}{2k_1\rho}} J_{n+1/2}(k_1\rho)$ and $y_n(k_1\rho) = \sqrt{\frac{\pi}{2k_1\rho}} N_{n+1/2}(k_1\rho)$
are the Bessel spherical function and the Neumann one, correspondingly, $J_{n+1/2}(k_1\rho)$
is the Bessel function, $N_{n+1/2}(k_1\rho)$ is the Neumann function and $H_{n+1/2}^{(2)}(k_1\rho)$ is the
Hankel function of the second sort with the half-integral index [1].

Let us note, that it turns out to be possible to represent the expression for the component of the Green's function $G^e_{\rho\rho'}(\rho, \theta, \varphi; \rho', \theta', \varphi')$ in a more suitable form for numerical realization, having made the transition from the double series to the single one with the help of the summation theorem for the Legendre polynomials:

$$G^e_{\rho\rho'}(\rho, \theta, \varphi; \rho', \theta', \varphi')$$

$$= -\sum_{n=0}^{\infty} \frac{n+1/2}{2\pi} \, h_n(\rho, \rho') \, P_n\big(\cos\theta\cos\theta' + \sin\theta\sin\theta'\cos(\varphi - \varphi')\big).$$

II. Magnetic Dyadic Green's Functions

8. The unbounded space with the permittivity and the permeability of the medium ε_1 and μ_1 $(k_1 = k\sqrt{\varepsilon_1\mu_1})$.

$$\hat{G}^m(\vec{r}, \vec{r}') = \hat{I}\frac{e^{-ik_1|\vec{r}-\vec{r}'|}}{|\vec{r} - \vec{r}'|}. \tag{A.8}$$

9. The half-space over the perfectly conducting plane with the permittivity and the permeability of the medium ε_1 and μ_1 (Fig. A.1).

$$\hat{G}^m(\vec{r}, \vec{r}') = \hat{I}\frac{e^{-ikR}}{R} + (\vec{e}_x \otimes \vec{e}_{x'})\frac{e^{-ikR_1}}{R_1} + (\vec{e}_y \otimes \vec{e}_{y'})\frac{e^{-ikR_1}}{R_1} - (\vec{e}_z \otimes \vec{e}_{z'})\frac{e^{-ikR_1}}{R_1},$$
$$\tag{A.9}$$
$$R = \sqrt{(x - x')^2 + (y - y')^2 + (z - z')^2}, \quad R_1 = \sqrt{(x - x')^2 + (y - y')^2 + (z + z')^2}.$$

10. The space with the permittivity and the permeability of the medium ε_1 and μ_1 of the medium and inside the corner region, concluded between two planes and perpendicularly intersecting along the abscissa axis $X0Y$ and $X0Z$ (Fig. A.2).

$$G^m_{zz}(x, y, z; x', y', z') = \frac{e^{-ikR}}{R} + \frac{e^{-ik\sqrt{(x-x')^2+(y+y')^2+(z-z')^2}}}{\sqrt{(x - x')^2 + (y + y')^2 + (z - z')^2}}$$

$$- \frac{e^{-ikR_1}}{R_1} - \frac{e^{-ik\sqrt{(x-x')^2+(y+y')^2+(z+z')^2}}}{\sqrt{(x - x')^2 + (y + y')^2 + (z + z')^2}}.$$
$$\tag{A.10}$$

11. The hollow infinite rectangular waveguide of the cross-section $\{a \times b\}$ with the perfectly conducting walls (Fig. A.3).

$$
\hat{G}^m(\vec{r}, \vec{r}') = \frac{2\pi}{ab} \sum_{m=0}^{\infty} \sum_{n=0}^{\infty} \frac{\varepsilon_m \varepsilon_n}{k_z} \{ (\vec{e}_x \otimes \vec{e}_{x'}) \Phi_x^m(x, y; x', y') e^{-k_z|z-z'|}
$$
$$
+ (\vec{e}_y \otimes \vec{e}_{y'}) \Phi_y^m(x, y; x', y') e^{-k_z|z-z'|} \tag{A.11}
$$
$$
+ (\vec{e}_z \otimes \vec{e}_{z'}) \Phi_z^m(x, y; x', y') e^{-k_z|z-z'|} \}.
$$

12. The hollow half-infinite rectangular waveguide of the cross-section $\{a \times b\}$ with the perfectly conducting walls.

$$
\hat{G}^m(\vec{r}, \vec{r}') = \frac{2\pi}{ab} \sum_{m=0}^{\infty} \sum_{n=0}^{\infty} \frac{\varepsilon_m \varepsilon_n}{k_z} \{ (\vec{e}_x \otimes \vec{e}_{x'}) \Phi_x^m(x, y; x', y') [e^{-k_z|z-z'|} + e^{-k_z(z+z')}]
$$
$$
+ (\vec{e}_y \otimes \vec{e}_{y'}) \Phi_y^m(x, y; x', y') [e^{-k_z|z-z'|} + e^{-k_z(z+z')}]
$$
$$
+ (\vec{e}_z \otimes \vec{e}_{z'}) \Phi_z^m(x, y; x', y') [e^{-k_z|z-z'|} - e^{-k_z(z|z')}] \}.
$$
$$
\tag{A.12}
$$

13. The hollow rectangular resonator with dimensions $\{a_R \times b_R \times H\}$ and perfectly conducting walls.

$$
\hat{G}^m(\vec{r}, \vec{r}') = \frac{2\pi}{a_R b_R} \sum_{m=0}^{\infty} \sum_{n=0}^{\infty} \frac{\varepsilon_m \varepsilon_n}{k_z}
$$
$$
\times \left\{ (\vec{e}_x \otimes \vec{e}_{x'}) \Phi_x^m(x, y; x', y') \left[\frac{\mathrm{ch} k_z(H-|z-z'|) + \mathrm{ch} k_z(H-|z+z'|)}{\mathrm{sh} k_z H} \right] \right.
$$
$$
+ (\vec{e}_y \otimes \vec{e}_{y'}) \Phi_y^m(x, y; x', y') \left[\frac{\mathrm{ch} k_z(H-|z-z'|) + \mathrm{ch} k_z(H-|z+z'|)}{\mathrm{sh} k_z H} \right] \tag{A.13}
$$
$$
\left. + (\vec{e}_z \otimes \vec{e}_{z'}) \Phi_z^m(x, y; x', y') \left[\frac{\mathrm{ch} k_z(H-|z-z'|) - \mathrm{ch} k_z(H-|z+z'|)}{\mathrm{sh} k_z H} \right] \right\}.
$$

14. The half-infinite rectangular waveguide with an impedance $(\overline{Z}_S)$ end in the case when the sources are on the end surface.

$$
\hat{G}^m(\vec{r}, \vec{r}') = \frac{2\pi}{ab} \sum_{m=0}^{\infty} \sum_{n=0}^{\infty} \frac{\varepsilon_m \varepsilon_n}{k_z} \{ (\vec{e}_x \otimes \vec{e}_{x'}) \Phi_x^m(x, y; x', y') f_{\mathrm{II}}(k_z, \overline{Z}_S) 2 e^{-k_z z}
$$
$$
+ (\vec{e}_y \otimes \vec{e}_{y'}) \Phi_y^m(x, y; x', y') f_{\mathrm{II}}(k_z, \overline{Z}_S) 2 e^{-k_z z} \}, \tag{A.14}
$$
$$
f_{\mathrm{II}}(k_z, \overline{Z}_S) = \frac{k k_z (1 + \overline{Z}_S^2)}{(ik + k_z \overline{Z}_S)(k \overline{Z}_S - ik)}.
$$

15. The half-infinite rectangular waveguide with an impedance $(\overline{Z}_S)$ end in the case longitudinal currents excitation.

$$\hat{G}^m(\vec{r}, \vec{r}')$$

$$= \frac{2\pi}{ab} \sum_{m=0}^{\infty} \sum_{n=0}^{\infty} \frac{\varepsilon_m \varepsilon_n}{k_z} \left\{ (\vec{e}_z \otimes \vec{e}_{z'}) \Phi_z^m(x, y; x', y') \left[e^{-k_z|z-z'|} - f_\perp(k_z, \overline{Z}_S) e^{-k_z(z+z')} \right] \right\}$$

$$f_\perp(k_z, \overline{Z}_S) = \frac{ik - k_z \overline{Z}_S}{ik + k_z \overline{Z}_S}. \tag{A.15}$$

In the expressions (A.8)–(A.15) the following notations is used:

$$\Phi_x^m(x, y; x', y') = \sin k_x x \sin k_x x' \cos k_y y \cos k_y y',$$
$$\Phi_y^m(x, y; x', y') = \cos k_x x \cos k_x x' \sin k_y y \sin k_y y',$$
$$\Phi_z^m(x, y; x', y') = \cos k_x x \cos k_x x' \cos k_y y \cos k_y y'.$$

The remaining notations is the same as in (A.1)–(A.6).

16. Space outside the perfectly conducting sphere of the radius $\tilde{R}$ with the permittivity and the permeability of the medium ε_1 and μ_1 (Fig. A.4).

$$\hat{G}^m(\rho, \theta, \varphi; \rho', \theta', \varphi') = \begin{vmatrix} G_{\rho\rho'}^m & 0 & 0 \\ 0 & G_{\theta\theta'}^m & G_{\theta\varphi'}^m \\ 0 & G_{\varphi\theta'}^m & G_{\varphi\varphi'}^m \end{vmatrix}, \tag{A.16}$$

$$G_{\rho\rho'}^m(\rho, \theta, \varphi; \rho', \theta', \varphi')$$

$$= -\sum_{n=0}^{\infty} \sum_{m=0}^{n} \frac{\varepsilon_m h_n^m(\rho, \rho')}{2C_{nm}} P_n^m(\cos\theta) P_n^m(\cos\theta') \cos m(\varphi - \varphi'),$$

$$G_{\theta\theta'}^m(\rho, \theta, \varphi; \rho', \theta', \varphi') = -\sum_{n=0}^{\infty} \sum_{m=0}^{n} \frac{\varepsilon_m u_n^m(\rho, \rho') \cos m(\varphi - \varphi')}{2n(n+1) C_{nm} \sin\theta \sin\theta'}$$

$$\times \left[m^2 P_n^m(\cos\theta) P_n^m(\cos\theta') + \sin\theta \sin\theta' \frac{dP_n^m(\cos\theta)}{d\theta} \frac{dP_n^m(\cos\theta')}{d\theta'} \right],$$

$$G_{\theta\varphi'}^m(\rho, \theta, \varphi; \rho', \theta', \varphi') = \sum_{n=0}^{\infty} \sum_{m=0}^{n} \frac{m u_n^m(\rho, \rho') \sin m(\varphi - \varphi')}{n(n+1) C_{nm}}$$

$$\times \left[\frac{dP_n^m(\cos\theta)}{d\theta} \frac{P_n^m(\cos\theta')}{\sin\theta'} + \frac{P_n^m(\cos\theta)}{\sin\theta} \frac{dP_n^m(\cos\theta')}{d\theta'} \right],$$

$$G_{\varphi\theta'}^m(\rho, \theta, \varphi; \rho', \theta', \varphi') = -G_{\theta\varphi'}^m(\rho, \theta, \varphi; \rho', \theta', \varphi'),$$

$$G_{\varphi\varphi'}^m(\rho, \theta, \varphi; \rho', \theta', \varphi') = G_{\theta\theta'}^m(\rho, \theta, \varphi; \rho', \theta', \varphi').$$

Here $h_n^m(\rho, \rho') = u_n^e(\rho, \rho'), u_n^m(\rho, \rho') = h_n^e(\rho, \rho')$, as well as other notations from (A.7).

III. Green's Functions of the Free Half-Space over the Impedance Plane

For a free half-space over an infinite plane (Fig. A.1), $\sqrt{\mu_0/\varepsilon_0} = Z_0 = 120\pi$ ($\omega\varepsilon_0 = k/Z_0$, $\omega\mu_0 = Z_0 k$), characterized by distributed constant impedance $\bar{Z}_S = Z_S/Z_0$, provided that the half-space is excited by coupled surface currents $j_x^m = \bar{Z}_S j_y^e$, $j_y^m = -\bar{Z}_S j_x^e$ (based on the requirement of the boundary condition $[\vec{E}, \vec{z}_0] = -\bar{Z}_S[[\vec{H}, \vec{z}_0], \vec{z}_0]$)

$$G_{xx}^m(x, y, z; x', y', 0) = \frac{1}{4\pi^2} \int\limits_{-\infty}^{\infty} \int\limits_{-\infty}^{\infty} C_\chi e^{-ik_x(x-x')-ik_y(y-y')-i\chi z} dk_x dk_y,$$

$$G_{yy}^e(x, y, z; x', y', 0) = -\frac{\bar{Z}_S^2}{4\pi^2} \int\limits_{-\infty}^{\infty} \int\limits_{-\infty}^{\infty} C_\chi e^{-ik_x(x-x')-ik_y(y-y')-i\chi z} dk_x dk_y, \quad (A.17)$$

where $C_\chi = \frac{4\pi}{i\chi} \frac{\chi k(1+\bar{Z}_S^2)}{\chi k(1+\bar{Z}_S^2)+\bar{Z}_S(k^2+\chi^2)}, \chi^2 = k^2 - k_x^2 - k_y^2$.

Reference

1. Abramowits MIA, Stegun IA (1964) Handbook of mathematical functions with formulas, graphs and mathematical tables. Applied mathematics series, vol 55. National Bureau of Standards

Appendix B

Representations of the Green's Function of Unlimited Space in Orthogonal Coordinate Systems

The scalar Green's function of an unbounded uniformly filled space $e^{-ik|\vec{r}-\vec{r}'|}/|\vec{r}-\vec{r}'|$ (hereinafter in the appendix formulas when considering a medium with material parameters (ε_1, μ_1), it is necessary to replace the wavenumber $k \to k_1 = k\sqrt{\varepsilon_1\mu_1}$) can be represented in the form of various expansions in terms of the eigenfunctions of the orthogonal coordinate systems [1].

1. Rectangular Coordinate System

The Green's function in the form of a triple Fourier integral in a rectangular coordinate system (x, y, z):

$$G_0^{e(m)}(x, y, z; x', y', z') = \frac{1}{(2\pi)^3} \int_{-\infty}^{\infty}\int_{-\infty}^{\infty}\int_{-\infty}^{\infty} \frac{e^{-ik_x(x-x')-ik_y(y-y')-ik_z(z-z')}}{k_x^2 + k_y^2 + k_z^2 - k^2} dk_x\, dk_y\, dk_z.$$

$$(B.1)$$

The Green's function in the form of a double Fourier integral, if the integral over any of the variables involved is preliminarily taken, for example, over k_z:

$$G_0^{e(m)}(x, y, z; x', y', z') = \frac{1}{8\pi^2} \int_{-\infty}^{\infty}\int_{-\infty}^{\infty} \frac{e^{-ik_x(x-x')-ik_y(y-y')-\sqrt{k_x^2+k_y^2-k^2}|z-z'|}}{\sqrt{k_x^2 + k_y^2 - k^2}} dk_x\, dk_y.$$

$$(B.2)$$

In the two-dimensional case, when there is no dependence on one of their coordinates (for example, on y), the Green's function has the form:

© The Editor(s) (if applicable) and The Author(s), under exclusive license to Springer Nature Switzerland AG 2020
M. V. Nesterenko et al., *Combined Vibrator-Slot Structures: Theory and Applications*, Lecture Notes in Electrical Engineering 689,
https://doi.org/10.1007/978-3-030-60177-5

$$G_0^{e(m)}(x, z; x', z') = \frac{1}{4\pi} \int\limits_{-\infty}^{\infty} \frac{e^{-ik_x(x-x') - \sqrt{k_x^2 - k^2}\,|z-z'|}}{\sqrt{k_x^2 - k^2}}\, dk_x$$

$$= -\frac{i}{4} H_0^{(2)}\left[k\sqrt{(x - x')^2 + (z - z')^2}\right], \quad (B.3)$$

where $H_0^{(2)}(x)$ is the Hankel function of the second kind.

Representations of the Green's function for open regions can be supplemented by expanding the function in terms of the eigenfunctions of a rectangular waveguide with cross-sectional dimensions $\{a \times b\}$ [2]:

$$G_0^{e(m)}(x, y, z; x', y', z') = \frac{2\pi}{ab} \sum_{m=0}^{\infty} \sum_{n=0}^{\infty} \frac{\chi_{mn}}{k_z} e^{-i[k_x(x-x') + k_y(y-y') + k_z|z-z'|]}, \quad (B.4)$$

where $k_x = \frac{m\pi}{a}$, $k_y = \frac{n\pi}{b}$, $k_z = \sqrt{k^2 - k_x^2 - k_y^2}$, $\chi_{mn} = \begin{cases} 2, & m, n = 0 \\ 1, & m, n \neq 0 \end{cases}$.

2. Cylindrical Coordinate System

The Green's function in a cylindrical coordinate system (r, φ, z) can be represented in two forms:

$$G_0^{e(m)}\left(r, \varphi, z; r', \varphi', z'\right) = \frac{1}{8\pi} \sum_{n=-\infty}^{\infty} e^{-in(\varphi-\varphi')} \int\limits_{\kappa=-\infty}^{\infty} e^{-\sqrt{\kappa^2 - k^2}\,|z-z'|}$$

$$\times \frac{\kappa}{\sqrt{\kappa^2 - k^2}} \begin{cases} H_n^{(2)}(\kappa r') J_n(\kappa r)\big|_{r<r'} \\ H_n^{(2)}(\kappa r) J_n(\kappa r')\big|_{r>r'} \end{cases} d\kappa, \quad (B.5)$$

$$G_0^{e(m)}\left(r, \varphi, z; r', \varphi', z'\right)$$

$$= \frac{1}{8\pi i} \sum_{n=-\infty}^{\infty} e^{-in(\varphi-\varphi')} \int\limits_{\kappa=-\infty}^{\infty} e^{-ik_z(z-z')} \begin{cases} H_n^{(2)}(vr') J_n(vr)\big|_{r<r'} \\ H_n^{(2)}(vr) J_n(vr')\big|_{r>r'} \end{cases} dk_z. \quad (B.6)$$

Here $H_n^{(2)}(x)$ is the Hankel function of the second kind, $J_n(x)$ is the Bessel function, $v = -i\sqrt{k_z^2 - k^2}$.

3. Spherical Coordinate System

Green's function in a spherical coordinate system (r, θ, φ):

$$G_0^{e(m)}\left(r, \theta, \varphi; r', \theta', \varphi'\right) = \frac{k}{4\pi i} \sum_{n=0}^{\infty} (2n+1) P_n(\cos\beta) \begin{cases} h_n^{(2)}(kr') j_n(kr)\big|_{r<r'}, \\ h_n^{(2)}(kr) j_n(kr')\big|_{r>r'}, \end{cases}$$

$$\text{(B.7)}$$

where

$$P_n(\cos\beta) = \sum_{m=-n}^{n} \frac{(n-m)!}{(n+m)!} P_n^m(\cos\theta) P_n^m(\cos\theta') e^{-im(\varphi-\varphi')},$$

$j_n(kr) = \sqrt{\frac{\pi}{2kr}} J_{n+1/2}(kr),\ h_n^{(2)}(kr) = \sqrt{\frac{\pi}{2kr}} H_{n+1/2}^{(2)}(kr)$ are the Bessel and Hankel functions of a half-integer index, $P_n^m(x)$ are the associated functions of Legendre.

4. Elliptical Coordinate System

Green's function in the coordinate system of an elliptical cylinder (u, v, z):

$$G_0^{e(m)}\left(u, v, z; u', v', z'\right) = -2i \int_{-\infty}^{\infty} e^{-ik_z(z-z')}$$

$$\times \left[\begin{array}{l} \displaystyle\sum_{m=0}^{\infty} \frac{Se_m(\kappa d, \cos v')}{Me_m(\kappa d)} Se_m(\kappa d, \cos v) \begin{cases} Je_m(\kappa d, \mathrm{ch}u') He_m(\kappa d, \mathrm{ch}u) \\ Je_m(\kappa d, \mathrm{ch}u) He_m(\kappa d, \mathrm{ch}u') \end{cases} \\ +\displaystyle\sum_{m=0}^{\infty} \frac{So_m(\kappa d, \cos v')}{Mo_m(\kappa d)} So_m(\kappa d, \cos v) \begin{cases} Jo_m(\kappa d, \mathrm{ch}u') Ho_m(\kappa d, \mathrm{ch}u) \\ Jo_m(\kappa d, \mathrm{ch}u) Ho_m(\kappa d, \mathrm{ch}u') \end{cases} \end{array} \right] dk_z,$$

$$\text{(B.8)}$$

where $Se_m(\kappa d, \cos v)$ are the even Mathieu angular function, $So_m(\kappa d, \cos v)$ are the odd Mathieu angular function, $Je_m(\kappa d, \mathrm{ch}u)$, $He_m(\kappa d, \mathrm{ch}u)$ are the even radial functions Mathieu, $Jo_m(\kappa d, \mathrm{ch}u)$, $Ho_m(\kappa d, \mathrm{ch}u)$ are the odd radial functions Mathieu, $\kappa = -i\sqrt{k_z^2 - k^2}$, $Me_m(o_m) = \int_0^{2\pi} [Se_m(o_m)(\kappa d, \cos v)]^2 dv$. In curly brackets, the upper lines take at $u > u'$, and the lower ones at $u < u'$, $2d$ is the distance between the foci of the ellipse describing the contour of the cylinder cross-section.

References

1. Markov GG, Chaplin AF (1983) Excitation of electromagnetic waves. Radio i svyaz', Moscow (in Russian)
2. Petlenko VA, Khizhnyak NA (1978) Scattering of electromagnetic waves by perfectly conducting bodies in a rectangular waveguide. Izv Vuz Radiophys 21(9):1325–1331 (in Russian)

Appendix C
Strict Solution of Field Equations for a Magnetodielectric Layer on a Conducting Plane with a Linear Law of Change Permittivity

Equation (1.75) for the linear law of change $\varepsilon_1(z)$ have the form:

$$\frac{d^2 E_x(z)}{dz^2} + k_1^2 \left(1 + \varepsilon_r \frac{z}{h_d}\right) E_x(z) = 0, \tag{C.1a}$$

$$H_y(z) = \frac{i}{k\mu_1} \frac{dE_x(z)}{dz}, \tag{C.1b}$$

where $k_1^2 = k^2 \mu_1 \varepsilon_1(0)$, $\varepsilon_r = \frac{\varepsilon_1(0) - \varepsilon(-h_d)}{\varepsilon_1(0)}$. We transform equations (C.1) by performing the following change of variable

$$\xi = -\left(\frac{k_1 h_d}{\varepsilon_r}\right)^{2/3} \left(1 + \varepsilon_r \frac{z}{h_d}\right),$$
$$\frac{d}{d\xi} = -\left(\frac{h_d}{k_1^2 \varepsilon_r}\right)^{1/3} \frac{d}{dz}. \tag{C.2}$$

Then Eq. (C.1a) becomes the standard Airy equation

$$\frac{d^2 E_x(\xi)}{d\xi^2} - \xi E_x(\xi) = 0, \tag{C.3}$$

which has a strict solution

$$E_x(\xi) = C_1 Ai(\xi) + C_2 Bi(\xi), \tag{C.4}$$

in which $Ai(\xi)$ and $Bi(\xi)$ are the Airy functions [1], C_1 and C_2 are the arbitrary constants. Performing the necessary calculations in (C.1b) taking into account (C.2), we obtain

M. V. Nesterenko et al., *Combined Vibrator-Slot Structures: Theory and Applications*, Lecture Notes in Electrical Engineering 689, https://doi.org/10.1007/978-3-030-60177-5

$$H_y(\xi) = -\frac{i}{\bar{Z}_1}\left(\frac{\varepsilon_r}{k_1 h_d}\right)^{1/3}[C_1 Ai'(\xi) + C_2 Bi'(\xi)], \tag{C.5}$$

where $Ai'(\xi)$ and $Bi'(\xi)$ are the derivatives of Airy functions, $\bar{Z}_1 = \sqrt{\mu_1/\varepsilon_1(0)}$.

Defining further in (C.4) and (C.5) from the corresponding boundary conditions the unknown variables and making a change $2h_d \rightarrow h_d$, we obtain, according to (1.74), the desired expression for the surface impedance

$$\bar{Z}_S = i\bar{Z}_1\zeta^{1/3}\frac{Ai[-\zeta^{2/3}(1-\varepsilon_r)] - \frac{Ai[-\zeta^{2/3}(1+\varepsilon_r)]}{Bi[-\zeta^{2/3}(1+\varepsilon_r)]}Bi[-\zeta^{2/3}(1-\varepsilon_r)]}{Ai'[-\zeta^{2/3}(1-\varepsilon_r)] - \frac{Ai[-\zeta^{2/3}(1+\varepsilon_r)]}{Bi[-\zeta^{2/3}(1+\varepsilon_r)]}Bi'[-\zeta^{2/3}(1-\varepsilon_r)]}, \tag{C.6}$$

in which $\zeta = \frac{k_1 h_d}{2\varepsilon_r}$ is the dimensionless parameter.

From a comparison of expressions (1.84) and (C.6) among themselves, it follows that the approximate formula (1.84) for calculating the surface impedance of a magnetodielectric layer on a conducting plane with a linear law of change $\varepsilon_1(z)$ compares favorably with formula (C.6) obtained on the basis of a rigorous solution of field equations, its simplicity and convenience in the calculations.

Reference

1. Abramowits MIA, Stegun IA (1964) Handbook of mathematical functions with formulas, graphs and mathematical tables. Applied mathematics series, vol 55. National Bureau of Standards

Appendix D
Proofs of the Theorems for Thin Vibrator Radiators

1. Statement and Proof of the Lemma

To simplify the proof of the first theorem, we first consider an auxiliary lemma for the current of an impedance vibrator radiator located in an infinite homogeneous medium.

Lemma *Let a thin radiating impedance vibrator, excited by a point source is placed in an infinite homogeneous medium with material parameters* (ε_1, μ_1). *The vibrator is a segment of a circular cylinder, whose radius r and the length 2L are such that inequalities* $[r/(2L)] \ll 1$ *and* $r\sqrt{\varepsilon_1\mu_1}/\lambda \ll 1$ *(λ is the wavelength in free space) hold. Then the electric current on the vibrator can be represented by a power series* $J(s) = \alpha J_1(s) + \alpha^2 J_2(s) + \ldots$ *in the small parameter* $\alpha = \frac{1}{2\ln[r/(2L)]}$, $|\alpha| \ll 1$, *and* $J_n(s)$ *is the current approximation of the nth order* $(n = 1, 2 \ldots)$.

Proof of the Lemma. The basis of the proof of the lemma is the solution of the integral-differential equation for the vibrator current using the technique of isolating a natural small parameter. We turn to the boundary conditions for the electric field on the surface S of the vibrator and as an initial analysis we consider the following equation:

$$\frac{1}{i\omega\varepsilon_1}(\text{graddiv} + k_1^2) \int_S \hat{G}^e(\vec{r}, \vec{r}')\vec{J}(\vec{r}')\mathrm{d}\vec{r}' = -\vec{E}_0(\vec{r}) + z_i(\vec{r})\vec{J}(\vec{r}), \qquad (D.1)$$

where $z_i(\vec{r})$ is the linear intrinsic impedance (Ω/m) of the vibrator, $\vec{E}_0(\vec{r})$ is the field of extraneous sources, $\hat{G}'(\vec{r}, \vec{r})$ is the tensor Green's function of the spatial domain for the electric vector potential, $k_1 = k\sqrt{\varepsilon_1\mu_1}$, $k = \omega/c = 2\pi/\lambda$ is wavenumber, $c \approx 2.998 \times 10^{10}$ cm/s is the speed of light in vacuum.

© The Editor(s) (if applicable) and The Author(s), under exclusive license
to Springer Nature Switzerland AG 2020
M. V. Nesterenko et al., *Combined Vibrator-Slot Structures: Theory and Applications*, Lecture Notes in Electrical Engineering 689,
https://doi.org/10.1007/978-3-030-60177-5

Taking into account the thin-wire approximations, the electric current induced in the vibrator, in accordance with (2.16), can be represented as follows:

$$\vec{J}(\vec{r}) = \vec{e}_s J(s)\psi(\rho, \varphi), \tag{D.2}$$

where $\vec{e}_s$ is the unit vector directed along the vibrator axis; s is the local coordinate along the vibrator axis; $\psi(\rho, \varphi)$ is the function of transverse ($\perp$) polar coordinates, satisfying the normalization condition $\int_\perp \psi(\rho, \varphi)\rho d\rho d\varphi = 1$. If the relations

$$\int_S \hat{G}^e(\vec{r}, \vec{r}')\vec{J}(\vec{r}')d\vec{r}' = \int_{-L}^{L} J(s') \int_{-\pi}^{\pi} \frac{e^{-ik_1\sqrt{(s-s')^2+[2r\sin(\varphi/2)]^2}}}{\sqrt{(s-s')^2+[2r\sin(\varphi/2)]^2}} \psi(r, \varphi) r \, d\varphi ds'$$

$$\approx \int_{-L}^{L} J(s')\frac{e^{-ik_1\sqrt{(s-s')^2+r^2}}}{\sqrt{(s-s')^2+r^2}}ds' = \int_{-L}^{L} J(s')\frac{e^{-ik_1 R(s,s')}}{R(s,s')}ds', \tag{D.3}$$

are valid and $z_i(\vec{r}) = z_i(s) \equiv const$, the surface equation (D.1) can be converted to an integral equation with a "quasi-one-dimensional" kernel

$$\left(\frac{d^2}{ds^2} + k_1^2\right)\int_{-L}^{L} J(s')\frac{e^{-ik_1 R(s,s')}}{R(s,s')}ds' = -i\omega\varepsilon_1 E_{0s}(s) + i\omega\varepsilon_1 z_i J(s), \tag{D.4}$$

where $E_{0s}(s)$ is projection of the extraneous source field on the vibrator axis.

We single out the logarithmic singularity of the kernel of Eq. (D.4) using the artificial technique [1]:

$$\int_{-L}^{L} J(s')\frac{e^{-ik_1 R(s,s')}}{R(s,s')}ds' = \Omega(s)J(s) + \int_{-L}^{L} \frac{J(s')e^{-ik_1 R(s,s')} - J(s)}{R(s,s')}ds'. \tag{D.5}$$

Here

$$\Omega(s) = \int_{-L}^{L} \frac{ds'}{\sqrt{(s-s')^2+r^2}} = \Omega + \gamma(s), \tag{D.6}$$

$\gamma(s) = \ln\frac{[(L+s)+\sqrt{(L+s)^2+r^2}][(L-s)+\sqrt{(L-s)^2+r^2}]}{4L^2}$ is a function vanishing at the vibrator center and reaching maxima at the vibrator ends, where the current is zero as required by the boundary conditions $J(\pm L) = 0$, $\Omega = 2\ln\frac{2L}{r}$ is a large parameter. Then, taking into account (D.5), Eq. (D.4) can be converted to the following integral-differential equation with a small parameter

$$\frac{d^2 J(s)}{ds^2} + k_1^2 J(s) = \alpha\{i\omega\varepsilon_1 E_{0s}(s) + F[s, J(s)] - i\omega\varepsilon_1 z_i J(s)\}. \tag{D.7}$$

Here $\alpha = -\frac{1}{\Omega} = \frac{1}{2\ln[r/(2L)]}$ is the small natural parameter ($|\alpha| \ll 1$), and functional

$$
\begin{aligned}
F[s, J(s)] = &-\frac{dJ(s')}{ds'} \frac{e^{-ik_1 R(s,s')}}{R(s, s')} \Bigg|_{-L}^{L} + \left[\frac{d^2 J(s)}{ds^2} + k_1^2 J(s)\right] \gamma(s) \\
&+ \int_{-L}^{L} \frac{\left[\frac{d^2 J(s')}{ds'^2} + k_1^2 J(s')\right] e^{-ik_1 R(s,s')} - \left[\frac{d^2 J(s)}{ds^2} + k_1^2 J(s)\right]}{R(s, s')} ds'
\end{aligned} \tag{D.8}
$$

presents the vibrator eigenfield in the spatial domain.

If we denote $\tilde{k} = k_1\sqrt{1 + i\alpha\omega\varepsilon_1 z_i/k_1} = k_1\sqrt{1 + i2\alpha\bar{Z}_S/(\mu_1 kr)}$ ($\bar{Z}_S = Z_S/Z_0$ is distributed surface impedance normalized to the wave resistance $Z_0 = \sqrt{\mu_1/\varepsilon_1}$ [Ohm]), Eq. (D.7) can be written as

$$\frac{d^2 J(s)}{ds^2} + \tilde{k}^2 J(s) = \alpha\{i\omega\varepsilon_1 E_{0s}(s) + F[s, J(s)]\}. \tag{D.9}$$

Since Eq. (D.9) is proportional to the small parameter α, its solution can be obtained by a successive approximations technique using the following algorithm

$$
\begin{aligned}
&\frac{d^2 J_1(s)}{ds^2} + \tilde{k}^2 J_1(s) = i\omega\varepsilon_1 E_{0s}(s), \\
&\frac{d^2 J_2(s)}{ds^2} + \tilde{k}^2 J_2(s) = F[s, J_1(s)], \\
&\cdots\cdots\cdots\cdots\cdots\cdots\cdots\cdots\cdots\cdots \\
&\frac{d^2 J_n(s)}{ds^2} + \tilde{k}^2 J_n(s) = F[s, J_{n-1}(s)].
\end{aligned} \tag{D.10}
$$

The solution of each differential equation can be obtained using the boundary conditions for the current $J_1(\pm L) = J_2(\pm L) = \ldots = J_n(\pm L) = 0$. Thus, we obtain the current decomposition as power series in small parameter α, i.e., $J(s) = \alpha J_1(s) + \alpha^2 J_2(s) + \ldots$, which was to be proved.

It should be noted, that zero approximation for the current J_0 was not included into the equation system (D.10), since its solution is $J_0(s) = C_1 \cos\tilde{k}s + C_2 \sin\tilde{k}s$, independent of the exciting field $E_{0s}(s)$. Taking into account losses in the medium and/or on the vibrator surface, the trigonometric functions in the solution are complex and cannot be zero for any arguments. Therefore, to satisfy the boundary conditions $J_0(\pm L) = 0$, the constants C_1 and C_2 should be zero and identities $J_0 \equiv 0$, $F[s, J_0(s)] \equiv 0$ become valid for any vibrator length.

The first approximation of the vibrator current obtained from (D.10) as sum of the general and partial solutions is

$$J(s) \approx \alpha J_1(s) = -\alpha \frac{i\omega\varepsilon_1/\tilde{k}}{\sin 2\tilde{k}L} \times \begin{cases} \sin \tilde{k}(L - s) \int\limits_{-L}^{s} E_{0s}(s') \sin \tilde{k}(L + s')ds' \\ \\ + \sin \tilde{k}(L + s) \int\limits_{s}^{L} E_{0s}(s') \sin \tilde{k}(L - s')ds', \end{cases}$$

(D.11)

and it does not depend on the function $F[s, J(s)]$ (D.8).

2. Statement and Proof of the First Theorem

Theorem 1 *Let a thin radiating impedance vibrator, exited by a point source, be placed in an electrodynamic volume filled by homogeneous medium with material parameters* (ε_1, μ_1). *The volume boundary is an arbitrary Lyapunov surface S [2], not passing through sources of extraneous currents. The vibrator is a segment of a circular cylinder, whose radius r and the length 2L are such that inequalities* $[r/(2L)] \ll 1$ *and* $r\sqrt{\varepsilon_1\mu_1}/\lambda \ll 1$ *($\lambda$is the wavelength in free space) hold. Then the influence of the volume boundaries upon the current distribution on the vibrator surface does not exceed an amount proportional to the small natural parameter* $\alpha = \frac{1}{2\ln[r/(2L)]}$.

Proof of the Theorem. In contrast to the case of an infinite spatial region, here the kernel of the integral equation for the current in the vibrator (D.1) is defined as the electric Green's function $\hat{G}^e(\vec{r}, \vec{r}')$ of a volume closed inside the boundaries. It, according to the general properties (2.6), in the system of orthogonal curvilinear coordinates (q_1, q_2, q_3) satisfies the inhomogeneous Helmholtz equation:

$$\delta\hat{G}(\vec{q}, \vec{q}') + k_1^2 \hat{G}(\vec{q}, \vec{q}') = -4\pi\hat{I}\frac{\delta(q_1 - q_1')\delta(q_2 - q_2')\delta(q_3 - q_3')}{h_1 h_2 h_3}, \quad (D.12)$$

where $\hat{I}$ is unit tensor, (q_1', q_2', q_3') are source coordinates, $\delta(q - q')$ is Dirac delta function, h_n are Lame coefficients. Laplacian δ applies to all tensor components. Then, the solution for the vector Hertz potential $\Pi^e(\vec{q})$ in the integral form can be written as

$$\Pi^e(\vec{q}) = \frac{1}{i\omega\varepsilon_1} \int\limits_{V} \vec{J}(\vec{q}') \; \hat{G}^e(\vec{q}, \vec{q}') \; dv$$

$$+ \oint\limits_{S} \left\{ \mathrm{div}\Pi^e(\vec{q}') \; \hat{G}^e(\vec{q}, \vec{q}') \; \vec{n} - \mathrm{div}\hat{G}^e(\vec{q}, \vec{q}')\Pi^e(\vec{q}') \; \vec{n} \right. \quad (D.13)$$

$$+ \left[\vec{n}, \; \hat{G}^e(\vec{q}, \vec{q}')\right] \mathrm{rot} \, \Pi^e(\vec{q}') - \left[\vec{n}, \; \Pi^e(\vec{q}')\right] \mathrm{rot}\hat{G}^e(\vec{q}, \vec{q}') \left. \right\} \; ds',$$

where $\vec{n}$ is the unit vector of the external normal to the surface S. The volume integral in (D.13) is taken over the entire volume V (dv is the volume element), and the surface integral is taken over the entire surface S (ds' is the area element in the primed coordinates). The expression in the curly brackets is a vector, and differentiation is performed over the primed coordinates.

Thus, the solution of the inhomogeneous Helmholtz equation is the sum of the volume and surface integrals. The surface integrals in (D.13) can be eliminated by building the Green's function in a special way. If the components of the Green's function $\hat{G}^e(\vec{q}, \vec{q}')$ and components of the vector potentials $\Pi^e(\vec{q})$ satisfy the boundary conditions on the surface S, the surface integrals vanish, since integrand of surface integrals in (D.13) vanish. Otherwise, the solution will have the general form (D.13). Thus, the expression (D.13) allows us to use alternative forms of the Green's function.

In our case, when using the Green's tensor of an unbounded region in the form

$$\hat{G}^e(\vec{r}, \vec{r}') = \hat{I} \frac{e^{-ik_1|\vec{r}-\vec{r}'|}}{|\vec{r} - \vec{r}'|} = \hat{I}G_0^e(\vec{r}, \vec{r}'), \tag{D.14}$$

which is a solution of Eq. (D.12) with boundary conditions at infinity, the Hertz vector potential (D.13) is represented by the following expression:

$$\Pi^e(\vec{r}) = \frac{1}{i\omega\varepsilon_1} \int_V \vec{J}(\vec{r}') \, \hat{I} \frac{e^{-ik_1|\vec{r}-\vec{r}'|}}{|\vec{r}-\vec{r}'|} \, dv$$
$$+ \oint_S \left\{ \mathrm{div}\,\Pi^e(\vec{r}')\left(\hat{I}G_0^e(\vec{r}, \vec{r}')\right) \, \vec{n} - \mathrm{div}\left(\hat{I}G_0^e(\vec{r}, \vec{r}')\right)\Pi^e(\vec{r}') \, \vec{n} \right. \tag{D.15}$$
$$\left. + \left[\vec{n}, \, \left(\hat{I}G_0^e(\vec{r}, \vec{r}')\right)\right] \mathrm{rot}\,\Pi^e(\vec{r}') - \left[\vec{n}, \, \Pi^e(\vec{r}')\right] \mathrm{rot}\left(\hat{I}G_0^e(\vec{r}, \vec{r}')\right) \right\} \, ds'.$$

If we substitute formula (D.15) into Eq. (D.1), where the Hertz vector was determined only by the volume integral, we obtain:

$$\frac{1}{i\omega\varepsilon_1}(\mathrm{grad\,div} + k_1^2) \int_S \left(\hat{I}\frac{e^{-ik_1|\vec{r}-\vec{r}'|}}{|\vec{r}-\vec{r}'|}\right)\vec{J}(\vec{r}')d\vec{r}' = -\vec{E}_0(\vec{r}) + z_i(\vec{r})\vec{J}(\vec{r})$$
$$-(\mathrm{grad\,div} + k_1^2) \oint_S \left\{ \mathrm{div}\,\Pi^e(r')\left(\hat{I}G_0^e(r, r')\right) \, \vec{n} - \mathrm{div}\left(\hat{I}G_0^e(\vec{r}, \vec{r}')\right)\Pi^e(\vec{r}') \, \vec{n} \right.$$
$$\left. + \left[\vec{n}, \, \left(\hat{I}G_0^e(\vec{r}, \vec{r}')\right)\right] \mathrm{rot}\,\Pi^e(\vec{r}') - \left[\vec{n}, \, \Pi^e(\vec{r}')\right] \mathrm{rot}\left(\hat{I}G_0^e(\vec{r}, \vec{r}')\right) \right\} \, ds'. \tag{D.16}$$

The reaction field of arbitrarily given boundaries at the observation point from relation (D.16) is denoted by the functional

$$F_S\left(\vec{r}, \vec{J}(\vec{r})\right) = (\mathrm{grad\,div} + k_1^2) \oint_S \left\{ \mathrm{div}\,\Pi^e(\vec{r}')\left(\hat{I}G_0^e(\vec{r}, \vec{r}')\right) \, \vec{n} \right.$$

$$-\mathrm{div}\left(\hat{I}G_0^e(\vec{r}, \vec{r}')\right)\Pi^e(\vec{r}')\vec{n}$$

$$+ \left[\vec{n}, \, \left(\hat{I}G_0^e(\vec{r}, \vec{r}')\right)\right] \mathrm{rot}\,\Pi^e(\vec{r}')$$

$$\left. -\left[\vec{n}, \, \Pi^e(\vec{r}')\right] \mathrm{rot}\left(\hat{I}G_0^e(\vec{r}, \vec{r}')\right) \right\} \, ds', \tag{D.17}$$

Then the equation for the current in the vibrator (D.4) can be written as follows:

$$\left(\frac{d^2}{ds^2} + k_1^2\right) \int\limits_{-L}^{L} J(s') \frac{e^{-ik_1 R(s,s')}}{R(s,s')} ds' = -i\omega\varepsilon_1 E_{0s}(s) - i\omega\varepsilon_1 F_S(s, J(s))$$

$$+ i\omega\varepsilon_1 z_i J(s), \qquad (D.18)$$

where $F_S(s, J(s))$ is the projection of the vector functional $F_S\left(\vec{r}, \vec{J}(\vec{r})\right)$ on the vibrator axis. Then, combining field functionals $f_\Sigma\left(\vec{r}, \vec{J}(\vec{r})\right) = F\left(\vec{r}, \vec{J}(\vec{r})\right) + i\omega\varepsilon_1 F_S\left(\vec{r}, \vec{J}(\vec{r})\right)$, in equation similar to (D.7), we can use the Lemma to proof the Theorem 1. Indeed, according to Lemma, the electric current on a thin impedance vibrator can be represented by a series in the small parameter α, therefore, its first approximation, determined by the expression (D.11), does not depend on the functional of the boundary influence. The influence is taken into account by successive approximations. Thus, the influence of the volume boundaries on the current distribution on the vibrator does not exceed an amount proportional to the small parameter α. Theorem 1 is proved.

The approach used to the proof of Theorem 1 allows us to formulate the second theorem. It concerns the fundamental possibility to compensate the influence of the spatial boundaries upon the current distribution on a perfectly conducting vibrator using "application" of the distributed impedance on the surface.

3. Statement and Proof of the Second Theorem

In the proof of the Lemma and Theorem 1, linear impedance of the vibrator was assumed to be constant $z_i(\vec{r}) = z_i(s) \equiv const$. Now, let us assume that the impedance can be distributed along the vibrator axis $z_i(\vec{r}) = z_i(s)$, or be concentrated at some points on the vibrator axis, or be superposition of these two options.

Theorem 2 *Let a thin radiating impedance vibrator, exited by a point source, be placed in an electrodynamic volume filled by homogeneous medium with material parameters* (ε_1, μ_1). *The volume boundary is an arbitrary Lyapunov surface S, not passing through sources of extraneous currents. The vibrator is a segment of a circular cylinder, whose radius r and the length 2L are such that inequalities* $[r/(2L)] \ll 1$ *and* $r\sqrt{\varepsilon_1\mu_1}/\lambda \ll 1$ (λ *is the wavelength in free space) hold. Then, the influence of the volume boundaries upon the current distribution on the vibrator surface can be compensated by coating its surface with the complex impedance, varying along the vibrator axis,* $z_i(s) = \frac{F_S(s, J(s))}{J(s)}$. $J(s)$ *is the current distribution on the perfectly conducting vibrator, and* $F_S(s, J(s))$ *is the functional* (D.17) *defining the boundary influence.*

Proof of the Theorem. As a source, consider an equation similar to (D.18):

$$\left(\frac{d^2}{ds^2} + k_1^2\right) \int\limits_{-L}^{L} J(s') \frac{e^{-ik_1 R(s,s')}}{R(s,s')} ds' = -i\omega\varepsilon_1 (E_{0s}(s) + F_S(s, J(s)) - z_i(s)J(s)),$$

(D.19)

in which the impedance $z_i(s)$ is assumed to be variable (along the axis of the vibrator) parameter described by the complex function. If we assume that a situation arises when equality $F_S(s, J(s)) = z_i(s)J(s)$ will be fulfilled on the right side of the equation, then the current in the vibrator will be determined only by the field of the external source. Then Eq. (D.19) can be formally represented as a system of two equations in the following form:

$$\begin{cases} \left(\frac{d^2}{ds^2} + k_1^2\right) \int\limits_{-L}^{L} J(s') \frac{e^{-ik_1 R(s,s')}}{R(s,s')} ds' = -i\omega\varepsilon_1 E_{0s}(s), \\ F_S(s, J(s)) - z_i(s)J(s) = 0. \end{cases}$$

(D.20)

The first equation in this system is Eq. (D.4) for a perfectly conducting vibrator, $z_i = 0$, located in an infinite homogeneous medium. The second equation is the functional equation, which can be solved using the current found from the first equation. These equations can be used to obtain the distribution of variable complex impedance along the vibrator axis

$$z_i(s) = \frac{F_S(s, J(s))}{J(s)}.$$

(D.21)

where $z_i(s)$ can be a generalized function.

Thus, the influence of the boundaries can be fully compensated by "applying" the impedance (D.21) to the vibrator surface. The current distribution of such impedance vibrator corresponds now to that of perfectly conducting vibrator, located in an infinite homogeneous medium with material parameters (ε_1, μ_1). Thus, Theorem 2 is proved.

References

1. King RWP (1956) The theory of linear antennas. Harvard University Press, Cambridge, MA
2. Tretyakov S (2003) Analytical modeling in applied electromagnetics. Artech House

Appendix E
Electromagnetic Values in CGS and SI Systems of Units

Because we use the CGS system in the book, it is expedient to make brief comparison of main ratios of the electromagnetic theory, represented in the SGS and SI systems.

CGS system of units SI system of units

Maxwell's equations:

$$\mathrm{rot}\vec{E} = -\frac{1}{c}\frac{\partial \vec{B}}{\partial t},$$

$$\mathrm{rot}\vec{H} = \frac{1}{c}\frac{\partial \vec{D}}{\partial t} + \frac{4\pi}{c}\vec{j},$$

$$\mathrm{div}\vec{D} = 4\pi\rho,$$

$$\mathrm{div}\vec{B} = 0,$$

where the field sources are:

$\vec{j} = ne\vec{v}$, n is the electron density,
$\vec{v}$ is their velocity, e is the charge in CGS units;
e is the charge in CGS units;
$\rho = ne$ is the electric charge density,

$$\mathrm{rot}\,\vec{E} = -\frac{\partial \vec{B}}{\partial t},$$

$$\mathrm{rot}\vec{H} = \frac{\partial \vec{D}}{\partial t} + \vec{j},$$

© The Editor(s) (if applicable) and The Author(s), under exclusive license
to Springer Nature Switzerland AG 2020
M. V. Nesterenko et al., *Combined Vibrator-Slot Structures: Theory
and Applications*, Lecture Notes in Electrical Engineering 689,
https://doi.org/10.1007/978-3-030-60177-5

$$\mathrm{div}\,\vec{D} = \rho,$$

$$\mathrm{div}\,\vec{B} = 0,$$

where the field sources are:

$\vec{j}$ is the electric current density in *ampere per square meter* (A/m^2);

ρ is the electric charge density in *coulombs per cubic meter* (C/m^3)

1 unit of electric charge in CGS system $= \frac{1}{3\times 10^9}$ *coulombs* (C),

1 unit of electric current density in CGS system $= \frac{1}{3\times 10^5}$ A/m^2.

Constitutive equations: $\vec{D} = \varepsilon\vec{E}$,

$$\vec{B} = \mu\vec{H},$$

$$\vec{D} = \varepsilon_0\varepsilon\vec{E},$$

$$\vec{B} = \mu_0\mu\vec{H},$$

where the permittivity and permeability in CGS system do not have dimensions and equal to the relative permittivity and permeability ε and μ in the SI system. ε_0 and μ_0 are the corresponding permittivity and permeability of vacuum here.

The electric and magnetic field intensities in the CGS system have the same dimension $\mathrm{g}^{1/2}\,\mathrm{cm}^{-1/2}\,\mathrm{s}^{-1}$, but intensity of the magnetic field is called *oersted* (Oe).

1 unit of electric field intensity in CGS system $= 3 \times 10^4$ V/m,

$$1\ \mathrm{Oe} = \frac{1}{4\pi} \times 10^3 \mathrm{A/m},$$

The electric and magnetic inductions ($\vec{D}$ and $\vec{B}$) in CGS system have dimensions of the intensities of corresponding fields. They have different dimensions and different names in SI system.

In SI system the electric induction is measured in *coulombs per square meter* (C/m^2), at this

1 unit of the electric induction in CGS system $= \frac{1}{12\pi} \times 10^{-5}$ C/m^2,

and the magnetic induction amount is measured in *Webers per square meter* $= 1$ T (T),

1 *gauss* (Gs) in CGS system $= 10^{-4}$ T.

The potentials are defined nearly alike in both systems (the potentials of only electric type are represented here)

$$\vec{H} = \frac{1}{\mu}\mathrm{rot}\vec{A},$$

$$\vec{E} = -\text{grad}\varphi - \frac{1}{c}\frac{\partial \vec{A}}{\partial t},$$

$$\vec{B} = \text{rot}\vec{A},$$

$$\vec{E} = -\text{grad}\varphi - \frac{\partial \vec{A}}{\partial t},$$

with the Lorentz's condition

$$\frac{1}{c}\frac{\partial \varphi}{\partial t} + \text{div }\vec{A} = 0, \qquad \frac{1}{v^2}\frac{\partial \varphi}{\partial t} + \text{div }\vec{A} = 0,$$

where v is the phase velocity of the wave in the corresponding medium.

The wave equations also differ slightly:

$$\Delta\varphi - \frac{1}{v^2}\frac{\partial^2 \varphi}{\partial t^2} = -\frac{4\pi\rho}{\varepsilon}, \qquad \Delta\varphi - \frac{1}{v^2}\frac{\partial^2 \varphi}{\partial t^2} = -\frac{\rho}{\varepsilon_0\varepsilon},$$

$$\delta\vec{A} - \frac{1}{v^2}\frac{\partial^2 \vec{A}}{\partial t^2} = -\frac{4\pi\mu}{c}\vec{j}, \qquad \delta\vec{A} - \frac{1}{v^2}\frac{\partial^2 \vec{A}}{\partial t^2} = -\mu_0\mu\,\vec{j}$$

and the phase velocity of the plane wave equals

$$v = \frac{c}{\sqrt{\varepsilon\mu}}$$

in the first case and

$$v = \frac{1}{\sqrt{\varepsilon_0\mu_0}}\frac{1}{\sqrt{\varepsilon\mu}} = \frac{3\times 10^8\ [\text{m/s}]}{\sqrt{\varepsilon\mu}}$$

in the second case because

$$\mu_0 = 4\pi\cdot 10^{-7}\left[\frac{H}{m}\right], \quad \varepsilon_0 = \frac{1}{36\pi}10^{-9}\left[\frac{F}{m}\right],$$

and *henry · farad* $= sec^2$. Thus the finite result is the same.

The wave equations solutions can be proposed in the form of general potentials in both cases, and they are written with the help of the Fourier components of these potentials in the form of:

$$\vec{E}(\vec{r}) = \left(\text{graddiv} + k^2\varepsilon\mu\right)\Pi(\vec{r}),$$

$$\vec{H}(\vec{r}) = \frac{ik}{w} \mathrm{rot}\Pi(\vec{r}),$$

where in CGS system

$$k^2 = \frac{\omega^2}{c^2}, \; w = 1, \; \Pi(\vec{r}) = \frac{1}{i\omega} \int_V \frac{\vec{j}(\vec{r}')}{|\vec{r} - \vec{r}'|} e^{-ik|\vec{r}-\vec{r}'|} d\vec{r}',$$

in SI system

$k = \omega\sqrt{\varepsilon_0\mu_0}, \; w = \sqrt{\frac{\mu_0}{\varepsilon_0}}$ is the wave impedance of free space,

$$\Pi(\vec{r}) = \frac{1}{4\pi i\omega\varepsilon_0} \int_V \frac{\vec{j}(\vec{r}')}{|\vec{r} - \vec{r}'|} e^{-ik|\vec{r}-\vec{r}'|} d\vec{r}'$$

Thus the electromagnetic wave in free space in SGS system is designated as mutually orthogonal triplet of the vectors $\vec{E}$, $\vec{H}$ and $\vec{k}$ where $\vec{E}$ and $\vec{H}$ have equal dimension and value, and this very triplet consists already of the values, which have different names and dimensions—*volt/meter*, *ampere/meter* and the wave vector with dimension 1/*meter* in SI system. But the wave has simple physical meaning—a long line in this case.

Bibliography

1. Gorobets NN, Nesterenko MV, Petlenko VA, Khizhnyak NA (1984) Thin impedance vibrator in a rectangular waveguide. Radio Eng 39(1):65–68 (in Russian)
2. Yatsuk LP, Penkin YuM (1987) Effect of the scattering vibrator on the energy characteristics of the slot in the waveguide. Izv Vuz Radioelectron 30(1):42–46 (in Russian)
3. Nesterenko MV, Petlenko VA (1988) Current distribution and resonant frequencies of thin impedance vibrators in a rectangular waveguide. Izv Vuz Radioelectron 31(2):80–82 (in Russian)
4. Yatsuk LP, Penkin YuM (1988) H_{10}-wave scattering by a narrow slot in a rectangular waveguide in the presence of an L-shaped passive vibrator. Radio Eng 84:35–42 (in Russian)
5. Penkin YuM, Yatsuk LP (1989) Investigation of the possibility of increasing the isolation between the waveguide-slot radiators using scattering vibrators. Vestnik Kharkov Univ 336:49–52 (in Russian)
6. Gorobets NN, Nesterenko MV, Petlenko VA (1990) Resonance characteristics of thin impedance vibrators in a cut-off rectangular waveguide. Telecommun Radio Eng 45(4):110–112
7. Yatsuk LP, Penkin YuM (1991) Interaction of a waveguide-slot radiator and a vibrator with switched on concentrated load. Izv Vuz Radioelectron 34(3):71–74 (in Russian)
8. Nesterenko MV (2002) Surface impedance of vibrators in the thin-wire approximation. Vestnik Kharkov Univ 544:47–49 (in Russian)
9. Katrich VA, Nesterenko MV, Khizhnyak NA (2002) Comparative analysis of analytical methods for solving integral equations for magnetic current in slot radiators. Foreign Electron 12:15–25 (in Russian)
10. Nesterenko MV (2002) Control characteristics of mobile communication vibrator antennas through distributed impedance. Vestnik Kharkov Univ 570:22–25 (in Russian)
11. Katrich VA, Nesterenko MV, Yatsuk LP, Berdnik SL (2002) The method of induced magnetomotive forces for electrically long slots in waveguide walls. Radioelectron Commun Syst 45(12):9–14
12. Nesterenko MV (2004) The electromagnetic wave radiation from a thin impedance dipole in a lossy homogeneous isotropic medium. Telecommun Radio Eng 61(10):110–112
13. Nesterenko MV (2004) Impedance model of a magnetodielectric layer with an inhomogeneous permittivity on a perfectly conducting plane. Radiophys Radioastron 9(1):73–80 (in Russian)
14. Nesterenko MV (2004) Scattering of electromagnetic waves by a resonant diaphragm with an arbitrarily oriented slot in a rectangular waveguide. Radiophys Radioastron 9(3):274–285 (in Russian)

M. V. Nesterenko et al., *Combined Vibrator-Slot Structures: Theory and Applications*, Lecture Notes in Electrical Engineering 689, https://doi.org/10.1007/978-3-030-60177-5

15. Nesterenko MV, Penkin YuM (2004) Diffraction radiation from a slot in the impedance end of a semi-infinite rectangular waveguide. Radiophys Quantum Electron 47(7):489–499

16. Nesterenko MV (2004) Resonance properties of thin impedance vibrators in free space. Vestnik Kharkov Univ 646:135–138 (in Russian)

17. Katrich VA, Lyashchenko VA, Nesterenko MV, Berdnik SL (2005) Slot antennas for various space and avionic applications. Proc Eur Microwave Assoc 1(3):239–244

18. Nesterenko MV, Katrich VA, Dakhov VM (2005) Radiation field of a thin horizontal impedance vibrator in a semi-infinite medium with losses over an ideally conducting plane. Radiophys Radioastron 10(3):314–324 (in Russian)

19. Nesterenko MV, Belogurov EYu (2006) H_{10}-wave scattering by a thin vibrator with variable impedance in a rectangular waveguide. Electron Inform 1(32):8–12 (in Russian)

20. Nesterenko MV, Katrich VA (2006) The asymptotic solution of an integral equation for magnetic current in a problem of waveguides coupling through narrow slots. Prog Electromagn Res 57:101–129

21. Nesterenko MV (2006) Scattering of electromagnetic waves by thin impedance vibrators of variable radius. Radiophys Radioastron 11(2):169–175 (in Russian)

22. Nesterenko MV, Katrich VA, Dakhov VM (2006) Near field and polarization characteristics of a system of crossed impedance vibrators in a semi-infinite medium with losses. Radiophys Radioastron 11(3):264–275 (in Russian)

23. Nesterenko MV (2006) Scattering of electromagnetic waves at the hole coupling of electrodynamic volumes in the presence of an impedance body of finite dimensions. Vestnik Kharkov Univ 712:47–51 (in Russian)

24. Nesterenko MV, Katrich VA, Penkin YuM, Berdnik SL (2007) Analytical methods in theory of slot–hole coupling of electrodynamics volumes. Prog Electromagn Res 70:79–174

25. Nesterenko MV, Katrich VA, Dakhov VM, Berdnik SL (2008) Impedance vibrator with arbitrary point of excitation. Prog Electromagn Res B 5:275–290

26. Nesterenko MV (2010) Analytical methods in the theory of thin impedance vibrators. Prog Electromagn Res B 21:299–328

27. Nesterenko MV, Katrich VA, Dakhov VM (2010) Formation of the radiation field with the set spatial-polarization characteristics by the crossed impedance vibrators system. Radiophys Quantum Electron 53(5–6):371–378

28. Nesterenko MV, Katrich VA, Berdnik SL, Penkin YuM, Dakhov VM (2010) Application of the generalized method of induced EMF for investigation of characteristics of thin impedance vibrators. Prog Electromagn Res B 26:149–178

29. Nesterenko MV, Katrich VA, Penkin YuM, Berdnik SL, Kijko VI (2012) Combined vibrator-slot structures in electrodynamic volumes. Prog Electromagn Res B 37:237–256

30. Berdnik SL, Penkin YuM, Katrich VA, Nesterenko MV, Kijko VI (2013) Electromagnetic waves radiation into the space over a sphere by a slot in the end-wall of a semi-infinite rectangular waveguide. Prog Electromagn Res B 46:139–158

31. Berdnik SL, Vasilkovsky VS, Katrich VA, Nesterenko MV, Penkin YuM (2014) Slot spherical antenna with a multi-element resonant diaphragm in a waveguide. Vestnik Kharkov Univ 1115:35–40 (in Russian)

32. Berdnik SL, Katrich VA, Nesterenko MV, Penkin YuM (2015) E-plane T-junction of rectangular waveguides with vibrator-slot coupling between arms. Telecommun Radio Eng 74(14):1225–1240

33. Berdnik SL, Katrich VA, Kijko VI, Nesterenko MV, Penkin YuM (2016) Power characteristics of a T-junction of rectangular waveguides with a multi-element monopole-slotted coupling structure. Telecommun Radio Eng 75(6):489–506

34. Penkin YuM, Katrich VA, Nesterenko MV (2016) Development of fundamental theory of thin impedance vibrators. Prog Electromagn Res M 45:185–193

35. Penkin YuM, Katrich VA Nesterenko MV (2016) Alternative representation of Green's function for electric field on surfaces of thin vibrators. Prog Electromagn Res M 52:169–179

36. Berdnik SL, Katrich VA, Nesterenko MV, Penkin YuM (2017) Waveguide T-junctions with resonant coupling between sections of different dimensions. Int J Microwave Wirel Technol 9(5):1059–1065

37. Penkin YuM, Katrich VA, Nesterenko MV (2019) Two-frequency operating mode of antenna arrays with radiators of Clavin type and switching vibrator and slot elements. Prog Electromagn Res M 87:171–178

38. Katrich VA, Kiyko VI, Nesterenko MV, Yatsuk LP (2002) Basis functions in the analysis of electrically long slots in rectangular waveguide with the induced magnetomotive forces method. In: Proceedings of the international conference on mathematical methods in electromagnetic theory, Kiev, Ukraine, vol 1, pp 328–330

39. Nesterenko MV, Katrich VA (2003) Thin vibrators with arbitrary surface impedance as a handset antennas. In: Proceedings of the 5th European personal mobile communications conference, Glasgow, Scotland, pp 16–20

40. Katrich VA, Nesterenko MV, Yatsuk LP (2003) Induced magnetomotive forces method for the analysis of radiating and coupling slots in waveguides. In: Proceedings of the IEEE AP-S/URSI international symposium, Columbus, USA, pp 111–114

41. Nesterenko MV, Katrich VA (2003) Asymptotic solution of integral equation into a problem of coupling between waveguides through narrow slots. In: Proceedings of the IVth international conference on antenna theory and techniques, Sevastopol, Ukraine, pp 280–284

42. Nesterenko MV, Penkin YuM, Katrich VA (2003) Radiating slot in impedance end face of rectangular waveguide. In: Proceedings of the VIIIth international seminar/workshop on direct and inverse problems of electromagnetic and acoustic wave theory, Lviv, Ukraine, pp 90–93

43. Nesterenko MV, Katrich VA, Penkin YuM (2004) Scattering of electromagnetic waves on stepped junction of two rectangular waveguides with slot impedance iris. In: Proceedings of the 5th international symposium on physics and engineering of microwaves, millimeter, and submillimeter waves, Kharkov, Ukraine, vol 2, pp 713–715

44. Nesterenko MV, Dakhov VM (2005) The near zone field of thin impedance vibrators over perfect conducting plane. In: Proceedings of the 5th international conference on antenna theory and techniques, Kyiv, Ukraine, pp 145–148

45. Nesterenko MV, Katrich VA, Dakhov VM (2005) The near zone fields and polarization characteristics of crossed impedance vibrators over perfect conducting plane. In: Proceedings of the Xth international seminar/workshop on direct and inverse problems of electromagnetic and acoustic wave theory, Lviv, Ukraine, pp 97–101

46. Nesterenko MV, Katrich VA, Belogurov EYu (2006) Thin vibrator with variable impedance in a rectangular waveguide. In: Proceedings of the 16th international crimean conference microwave & telecommunication technology, Sevastopol, Ukraine, vol 2, pp 548–549

47. Belogurov EYu, Katrich VA, Kiyko VI, Nesterenko MV (2007) Thin impedance vibrator of variable radius in a rectangular waveguide. In: Proceedings of the 17th international crimean conference microwave & telecommunication technology, Sevastopol, Ukraine, vol 2, pp 516–517

48. Katrich VA, Nesterenko MV, Dakhov VM (2007) The near zone field of crossed impedance vibrators in absorbing medium over the perfect conducting plane. In: Proceedings of the 6th international conference on antenna theory and techniques, Sevastopol, Ukraine, pp 184–187

49. Nesterenko MV, Katrich VA (2008) Electromagnetic waves scattering by thin impedance vibrators and narrow slots in waveguides. In: Proceedings of the XIIIth international seminar/workshop on direct and inverse problems of electromagnetic and acoustic wave theory, Tbilisi, Georgia, pp 97–100

50. Nesterenko MV, Katrich VA, Dakhov VM, Berdnik SL (2009) Thin vibrators with arbitrary excitation and surface impedance. In: Proceedings of the 7th international conference on antenna theory and techniques, Lviv, Ukraine, pp 105–109

51. Nesterenko MV, Dakhov VM, Katrich VA, Berdnik SL, Pshenichnaya SV (2010) Electromagnetic waves radiation by thin vibrators with asymmetrical surface impedance. In: Proceedings of the 5th international conference ultrawideband and ultrashort impulse signals, Sevastopol, Ukraine, pp 212–214

52. Nesterenko MV, Katrich VA, Dakhov VM, Kiyko VI (2010) Electrodynamic characteristics of the system impedance vibrator-slot radiator in a rectangular waveguide. In: Proceedings of the XVth international seminar/workshop on direct and inverse problems of electromagnetic and acoustic wave theory. Tbilisi, Georgia, pp 81–84

53. Dakhov VM, Katrich VA, Nesterenko MV, Berdnik SL (2011) Radiation polarization control of the turnstile antenna with impedance vibrators. In: Proceedings of the VIIIth international conference on antenna theory and techniques, Kyiv, Ukraine, pp 199–201
54. Berdnik SL, Penkin YuM, Katrich VA, Nesterenko MV, Pshenichnaya SV (2013) Resonant slot spherical antenna. In: Proceedings of the 23rd international crimean conference microwave & telecommunication technology, Sevastopol, Ukraine, vol 2, pp 610–611
55. Penkin DYu, Berdnik SL, Katrich VA, Nesterenko MV, Kiyko VI (2013) Excitation of electromagnetic fields by a multielement vibrator-slot structure in a rectangular waveguide. In: Proceedings of the 23rd international crimean conference microwave & telecommunication technology, Sevastopol, Ukraine, vol 2, pp 708–709
56. Berdnik SL, Katrich VA, Nesterenko MV, Pshenichnaya SV (2013) Electrodynamic characteristics of a three-element vibrator-slot structure in a rectangular waveguide. In: Proceedings of the XVIIIth international seminar/workshop on direct and inverse problems of electromagnetic and acoustic wave theory, Lviv, Ukraine, pp 45–48
57. Berdnik SL, Katrich VA, Nesterenko MV, Penkin YuM (2013) Spherical antenna excited by a slot in an impedance end-wall of a rectangular waveguide. In: Proceedings of the XVIIIth international seminar/workshop on direct and inverse problems of electromagnetic and acoustic wave theory, Lviv, Ukraine, pp 111–114
58. Berdnik SL, Penkin YuM, Katrich VA, Nesterenko MV, Blinova NK (2014) Spherical antenna excited by a slot in the impedance end-wall with the losses of a rectangular waveguide. In: Proceedings of the 23rd international crimean conference microwave & telecommunication technology, Sevastopol, Ukraine, vol 2, pp 491–492
59. Penkin DYu, Katrich VA, Dakhov VM, Nesterenko MV (2014) Input impedance of radial monopole mounted on a perfectly conducting sphere. In: Proceedings of the 7th international conference on ultrawideband and ultrashort impulse signals, Kharkiv, Ukraine, pp 146–148
60. Berdnik SL, Vasylkovskyi VS, Nesterenko MV, Penkin YuM (2015) Radiation fields of the spherical slot antenna in a material medium. In: Proceedings of the Xth anniversary international conference on antenna theory and techniques, Kharkiv, Ukraine, pp 282–284
61. Berdnik SL, Katrich VA, Nesterenko MV, Penkin YuM (2015) Waveguide E-plane T-junction with resonance coupling between shoulders. In: Proceedings of the X anniversary international conference on antenna theory and techniques, Kharkiv, Ukraine, pp 306–308
62. Berdnik SL, Blinova NK, Katrich VA, Nesterenko MV, Penkin YuM (2015) Spherical antenna with a Clavin radiator. In: Proceedings of the XXth international seminar/workshop on direct and inverse problems of electromagnetic and acoustic wave theory, Lviv, Ukraine, pp 75–77
63. Berdnik SL, Katrich VA, Nesterenko MV, Penkin YuM (2016) E-plane T-junctions of rectangular waveguides with vibrator-slot coupling between arms of different dimensions. In: Proceedings of the 8th international conference on ultrawideband and ultrashort impulse signals, Odessa, Ukraine, pp 68–72
64. Berdnik SL, Katrich VA, Kiyko VI, Nesterenko MV, Penkin YuM (2017) T-junction of rectangular waveguides with monopole-slot coupling structure and elements coated by a metamaterial. In: Proceedings of the XXIIth international seminar/workshop on direct and inverse problems of electromagnetic and acoustic wave theory, Dnipro, Ukraine, pp 123–127
65. Penkin YuM, Katrich VA, Nesterenko MV, Berdnik SL (2018) Dual-symmetric of integral equations for antenna currents. In: Proceedings of the XXIIIth international seminar/workshop on direct and inverse problems of electromagnetic and acoustic wave theory, Tbilisi, Georgia, pp 55–59
66. Penkin YuM, Katrich VA, Nesterenko MV, Dumin OM, Pshenichnaya SV (2019) Impedance synthesis for the ring slotted radiators on hemispherical ledges above screen. In: Proceedings of the IIth Ukraine conference on electrical and computer engineering, Lviv, Ukraine, pp 68–72
67. Penkin YuM, Dumin OM, Katrich VA, Nesterenko MV (2020) Waveguide technique for thin metallic film surface impedance determination. In: Proceedings of the 15th international conference on advanced trends in radioelectronics, telecommunications and computer engineering, Lviv-Slavske, Ukraine, pp 1–5

68. Berdnik SL, Katrich VA, Nesterenko MV, Penkin YuM, Dumin OM (2020) Yagi-Uda combined radiating structures of centimeter and millimeter wave bands. Prog Electromagn Res M 93:89–97
69. Berdnik SL, Katrich VA, Nesterenko MV, Penkin YuM (2020) Waveguide radiation of the combined vibrator-slot structures. Prog Electromagn Res B 87:151–170
70. Dumin O, Fomin P, Plakhtii V, Nesterenko M (2020) Ultrawideband combined monopole-slot radiator of Clavin type. In: Proceedings of the XXVth international seminar/workshop on direct and inverse problems of electromagnetic and acoustic wave theory. Tbilisi, Georgia, pp 32–36
71. Penkin Yu, Katrich V, Nesterenko M, Berdnik S, Dumin O (2020) Dual-band combined vibrator-slot radiating structures. In: Proceedings of the XII international conference on antenna theory and techniques. Kharkiv, Ukraine, pp 149–153

Printed in the United States
by Baker & Taylor Publisher Services